Springer Series in
SOLID-STATE SCIENCES 132

Springer
*Berlin
Heidelberg
New York
Barcelona
Hong Kong
London
Milan
Paris
Tokyo*

Physics and Astronomy   ONLINE LIBRARY

http://www.springer.de/phys/

Springer Series in
SOLID-STATE SCIENCES

Series Editors:
M. Cardona   P. Fulde   K. von Klitzing   R. Merlin   H.-J. Queisser   H. Störmer

The Springer Series in Solid-State Sciences consists of fundamental scientific books prepared by leading researchers in the field. They strive to communicate, in a systematic and comprehensive way, the basic principles as well as new developments in theoretical and experimental solid-state physics.

126 **Physical Properties of Quasicrystals**
Editor: Z.M. Stadnik

127 **Positron Annihilation in Semiconductors**
Defect Studies
By R. Krause-Rehberg and H.S. Leipner

128 **Magneto-Optics**
Editors: S. Sugano and N. Kojima

129 **Computational Materials Science**
From Ab Initio to Monte Carlo Methods
By K. Ohno, K. Esfarjani, and Y. Kawazoe

130 **Contact, Adhesion and Rupture of Elastic Solids**
By D. Maugis

131 **Field Theories for Low-Dimensional Condensed Matter Systems**
Spin Systems and Strongly Correlated Electrons
By G. Morandi, P. Sodano, A. Tagliacozzo, and V. Tognetti

132 **Vortices in Unconventional Superconductors and Superfluids**
Editors: R.P. Huebener, N. Schopohl, and G.E. Volovik

Series homepage – http://www.springer.de/phys/books/sss/

Volumes 1–125 are listed at the end of the book.

R.P. Huebener, N. Schopohl,
G.E. Volovik (Eds.)

Vortices in Unconventional Superconductors and Superfluids

With 106 Figures

Springer

Professor Dr. R.P. Huebener
Eberhard-Karls-Universität Tübingen
Physikalisches Institut
Auf der Morgenstelle 14
72076 Tübingen, Germany

Professor Dr. N. Schopohl
Eberhard-Karls-Universität Tübingen
Lehrstuhl für Theoretische Festkörperphysik
Auf der Morgenstelle 14
72076 Tübingen, Germany

Professor Dr. G.E. Volovik
Landau Institute for Theoretical Physics
Kosygin Str. 2
117334 Moscow, Russia
and
Low Temperature Laboratory
Helsinki University of Technology
P.O. Box 2200
02015, Espoo, Finland

Series Editors:

Professor Dr., Dres. h. c. Manuel Cardona
Professor Dr., Dres. h. c. Peter Fulde*
Professor Dr., Dres. h. c. Klaus von Klitzing
Professor Dr., Dres. h. c. Hans-Joachim Queisser

Max-Planck-Institut für Festkörperforschung, Heisenbergstrasse 1, 70569 Stuttgart, Germany
* Max-Planck-Institut für Physik komplexer Systeme, Nöthnitzer Strasse 38
  01187 Dresden, Germany

Professor Dr. Roberto Merlin
Department of Physics, 5000 East University, University of Michigan
Ann Arbor, MI 48109-1120, USA

Professor Dr. Horst Störmer
Dept. Phys. and Dept. Appl. Physics, Columbia University, New York, NY 10023 and
Bell Labs., Lucent Technologies, Murray Hill, NJ 07974, USA

Library of Congress Cataloging-in-Publication Data applied for.
Die Deutsche Bibliothek - CIP-Einheitsaufnahme

Vortices in unconventional superconductors and superfluids/R.P. Huebener ... (ed.). – Berlin; Heidelberg;
New York; Barcelona; Hong Kong; London; Milan; Paris; Singapore; Tokyo: Springer, 2002 (Springer series in
solid-state sciences; 132) (Phsyics and astronomy online library) ISBN 3-540-42336-2

ISSN 0171-1873
ISBN 3-540-42336-2 Springer-Verlag Berlin Heidelberg New York

This work is subject to copyright. All rights are reserved, whether the whole or part of the material is
concerned, specifically the rights of translation, reprinting, reuse of illustrations, recitation, broadcasting,
reproduction on microfilm or in any other way, and storage in data banks. Duplication of this publication or
parts thereof is permitted only under the provisions of the German Copyright Law of September 9, 1965, in its
current version, and permission for use must always be obtained from Springer-Verlag. Violations are liable
for prosecution under the German Copyright Law.

Springer-Verlag Berlin Heidelberg New York
a member of BertelsmannSpringer Science+Business Media GmbH

http://www.springer.de

© Springer-Verlag Berlin Heidelberg 2002
Printed in Germany

The use of general descriptive names, registered names, trademarks, etc. in this publication does not imply,
even in the absence of a specific statement, that such names are exempt from the relevant protective laws and
regulations and therefore free for general use.

Typesetting: camera-ready copies by the authors
Cover concept: eStudio Calamar Steinen
Cover production: *design & production* GmbH, Heidelberg

Printed on acid-free paper   SPIN: 10838764   57/3141tr - 5 4 3 2 1 0

Preface

The physics of vortices in classical fluids has been a highly important subject for many years, both in fundamental science and in engineering applications. About 50 years ago, vortices started to become prominent as quantum mechanical objects constructed from a macroscopic wavefunction. Here the key developments are associated with the names R. Feynman, L. Onsager, L. D. Landau, F. London, V.L. Ginzburg and A.A. Abrikosov. Recently, the physics of vortices has undergone a further important step of diversification, namely in unconventional superconductors and superfluids, which are characterized by an anisotropic and/or spatially complex order parameter. It is this latest evolutionary step of vortex physics that is addressed in this book. The individual chapters are concerned with the microscopic structure and dynamics of vortices in diverse systems ranging from superfluids and superconductors to neutron stars.

Each of the 20 chapters is written by one or more experts on the particular subject. Each chapter provides an introduction and overview, emphasizing theoretical as well as experimental work, and includes references to both recent and pioneering earlier developments. In this way non-expert readers will also benefit from these lecture notes. Hence, the book will be useful for all researchers and graduate students interested in the physics of vortices in unconventional superconductors and superfluids. It may also serve as supplementary material for a graduate course on low-temperature solid-state physics.

The idea for this book originated from a workshop held in Dresden, Germany from February 28 to March 3, 2000, at the Max-Planck-Institut Für Physik Komplexer Systeme. The editors express their special thanks to Prof. Dr. P. Fulde and his staff from this Institute for their support and hospitality during the workshop.

It is our privilege to thank the participants of the workshop for their contributions during the discussions. The quality of all the lectures and the enthusiasm shown by all the participants made the workshop a great success. We trust that this book will be similarly well received.

Tübingen, *N. Schopohl, R. Huebener*
Moscow *G.E. Volovik*
October 2001

Contents

1 The Beautiful World of the Vortex
G.E. Volovik 1

Part I Vortices in Superconductors, Superfluids, Neutron Stars, and QFT

2 Type II Superconductors and Vortices from the 1950s to the 1990s
A.A. Abrikosov 7

2.1 Preamble 7
2.2 History 7
2.3 Imaging 12
2.4 Pinning and Melting of the Vortex Lattice 15
2.5 Other Kinds of Vortices 17
References 19

3 What Can Superconductivity Learn from Quantized Vorticity in ^{3}He Superfluids?
G.E. Volovik, V.B. Eltsov, and M. Krusius 21

3.1 Unconventional Quantized Vorticity 21
3.2 Special Features of ^{3}He Superfluids 24
3.3 Continuous Vortices, Skyrmions and Merons 25
3.4 Transformation from Singular to Continuous Vortex 27
3.5 Vortex with Composite Core 28
3.6 Vortex Sheet 28
    3.6.1 Vortex-Sheet Structure in ^{3}He-A 28
    3.6.2 Vortex Sheet in Rotating Superfluid 28
    3.6.3 Vortex Sheet in Superconductor 30
3.7 Fractional Vorticity and Fractional Flux 30
3.8 Broken Symmetry in the Vortex Core 32
    3.8.1 Vortex Core Transition 32
    3.8.2 Ferromagnetic Core 34

	3.8.3	Asymmetric Double Core	34
3.9		Vortex Formation by Intrinsic Mechanisms	34
	3.9.1	Nucleation Barrier	35
	3.9.2	Vortex Formation in a Hydrodynamic Instability	36
	3.9.3	Formation of Continuous Vortex Lines: Dependence of Critical Velocity on Core Size	38
	3.9.4	Formation of Vortex Sheet	41
	3.9.5	Vortex Formation in Ionizing Radiation	42
3.10		Vortex Dynamics Without Pinning	43
3.11		Conclusion	44
References			45

4 Nucleation of Vortices in Superfluid ^{3}He-B by Rapid Thermal Quench
Igor S. Aranson, Nikolai B. Kopnin, and Valerii M. Vinokur 49

4.1	Introduction		49
4.2	Model		50
4.3	Results of Simulations		51
4.4	Dynamics of Vortex/Antivortex Annihilation		54
4.5	Instability of Normal–Superfluid Interface		55
	4.5.1	Stationary Solutions	56
	4.5.2	Linearized Equations	56
	4.5.3	Long-Wavelength Limit	57
	4.5.4	Large u Limit	58
	4.5.5	Estimate for the Number of Vortices	60
4.6	Generalization		61
4.7	Conclusion		62
References			63

5 Superfluidity in Relativistic Neutron Stars
David Langlois 65

5.1	Introduction		65
5.2	Superfluidity and Superconductivity		66
	5.2.1	Composition of the Interior of a Neutron Star	66
	5.2.2	Energy Gaps and Critical Temperature	67
	5.2.3	Various Equations of State	68
5.3	Cooling Processes in Neutron Stars		68
5.4	Rotational Dynamics of Neutron Stars: Glitches		69
	5.4.1	The Two–Fluid Model	69
	5.4.2	Role of the Vortices	70
	5.4.3	Origin of the Glitches	73
5.5	Relativistic Description		73
	5.5.1	Perfect Fluid in General Relativity	74
	5.5.2	Relativistic Superfluid	75

		5.5.3 Superfluid-Superconducting Mixtures	76
5.6		Relativistic Neutron Stars	78
	5.6.1	Static Neutron Star	78
	5.6.2	Oscillations of Superfluid Neutron Stars	79
References			80

6 Superconducting Superfluids in Neutron Stars
Brandon Carter 83

6.1	Introduction	83
6.2	Generic Category of 3-Constituent Superconducting Superfluid Models	84
6.3	The Semi-Macroscopic Application	88
6.4	Phenomenological Interpretation	92
References		96

Part II Vortex Dynamics, Spectral Flow and Aharonov–Bohm Effect

7 Vortex Dynamics and the Problem of the Transverse Force
N.B. Kopnin 99

7.1	Introduction	99
7.2	Boltzmann Kinetic Equation Approach	101
	7.2.1 Localized Excitations	101
	7.2.2 Delocalized Excitations	104
7.3	Forces	105
	7.3.1 Flux-Flow Conductivity	107
7.4	Transverse Force	108
	7.4.1 Low-Field Limit and Superfluid ^{3}He	111
7.5	Vortex Momentum	113
	7.5.1 Equation of Vortex Dynamics	114
	7.5.2 Vortex Mass	115
7.6	Conclusions	117
References		117

8 Magnus Force and Aharonov–Bohm Effect in Superfluids
Edouard Sonin 119

8.1	Introduction	119
8.2	The Magnus Force in Classical Hydrodynamics	121
8.3	The Magnus Force in a Superfluid	124
8.4	Nonlinear Schrödinger Equation and Two–Fluid Hydrodynamics	125

Contents

8.5 Scattering of Phonons by the Vortex
in Hydrodynamics 129
8.6 The Iordanskii Force
and the Aharonov–Bohm Effect 133
8.7 Partial-Wave Analysis
and the Aharonov–Bohm Effect 136
8.8 Momentum Balance in Two-Fluid Hydrodynamics 139
8.9 Magnus Force and the Berry Phase 142
8.10 Discussion and Conclusions 143
References 144

9 Lorentz Force Exerted by the Aharonov–Bohm Flux Line
Andrei Shelankov and A.F. Ioffe 147

9.1 The Magnetic Scattering 150
    9.1.1 Paraxial Solution 150
    9.1.2 Deflection of the Beam 152
    9.1.3 Exact Solution 154
    9.1.4 Scattering Amplitude 156
9.2 The Momentum Balance 157
    9.2.1 The Force 159
    9.2.2 The AB–Line 160
9.3 Conclusions 161
9.4 Appendix: The Momentum–Flow Tensor
for the Schrödinger Equation 163
9.5 Appendix: The Force: Arbitrary Wave 164
References 165

10 Relativistic Solution of Lordanskii Problem in Multi-Constituent Superfluid Mechanics
B. Carter, D. Langlois, and R. Prix 167

10.1 Introduction 167
10.2 Perfect Multiconstituent Fluid Dynamics 168
10.3 Specification of Lift Force on Vortex 169
10.4 Generalised Joukowski Theorem 170
10.5 Application to the Landau Model 171
References 173

11 Vortex Core Structure and Dynamics in Layered Superconductors
M. Eschrig, D. Rainer, and J. A. Sauls 175

11.1 Introduction 175
11.2 Nonequilibrium Transport Equations 177
    11.2.1 Constitutive Equations 179
    11.2.2 Linear Response 181

11.3	Electronic Structure of Vortices	184
	11.3.1 Singly Quantized Vortices for S-Wave Pairing	185
	11.3.2 Singly Quantized Vortices for D-Wave Pairing	188
	11.3.3 Vortices Pinned to Mesoscopic Metallic Inclusions	191
	11.3.4 Doubly Quantized Vortices	193
11.4	Nonequilibrium Response	195
	11.4.1 Dynamical Charge Response	197
	11.4.2 Local Dynamical Conductivity	199
	11.4.3 Induced Current Density	201
	11.4.4 Summary	202
References		202

Part III Fermion Zero Modes on Vortices

12 Band Theory of Quasiparticle Excitations in the Mixed State of d-Wave Superconductors
Alexander S. Mel'nikov 207

12.1	Introduction	207
12.2	Basic Equations	210
	12.2.1 BdG Equations for a Spin Singlet Unconventional Superconductor in a Magnetic Field	210
	12.2.2 Quasiparticles Confined Near Gap Nodes	212
12.3	A Single Isolated Vortex Line: Aharonov–Bohm Effect for Quasiparticles	213
12.4	Quasiparticle States in Vortex Lattices	214
	12.4.1 Cyclotron Orbits in the Mixed State	214
	12.4.2 Quasiparticle Band Spectrum in Vortex Lattices	215
	12.4.3 Modified Semiclassical Approach for QP States in a Vortex Lattice	219
12.5	Conclusions	222
References		222

13 Magnetic Field Dependence of the Vortex Structure Based on the Microscopic Theory
Masanori Ichioka, Mitsuaki Takigawa and Kazushige Machida 225

13.1	Introduction	225
13.2	Magnetic Field Dependence of the Vortex Structure	227
	13.2.1 Quasiclassical Eilenberger Theory	227
	13.2.2 Local Density of States	228
	13.2.3 Pair Potential and Internal Field Distribution	232
13.3	Site-Selective Nuclear Spin Relaxation Time	233

        13.3.1 Bogoliubov–de Gennes Theory
               for Extended Hubbard Model 234
        13.3.2 Field Distribution and Site-Selective NMR 236
        13.3.3 Temperature Dependence of T_1 237
13.4 Concluding Remarks 240
References 241

14 Quasiparticle Spectrum in the Vortex State in Nodal Superconductors
H. Won and K. Maki 243

14.1 Introduction 243
14.2 The Volovik Effect 245
14.3 Thermal Conductivity for $T \ll \epsilon$ 246
14.4 Thermal Conductivity Tensor for $T \gg \epsilon$ 249
14.5 Concluding Remarks 250
References 251

15 Random-Matrix Ensembles in p-Wave Vortices
Dmitri A. Ivanov 253

15.1 Single–Quantum Vortex 255
15.2 Half–Quantum Vortex 258
15.3 Spin Vortex 260
15.4 Level Mixing by Disorder 261
15.5 Summary and Discussion 262
References 264

Part IV Selected Experiments on Vortices in Superconductors

16 Scanning Tunneling Spectroscopy on Vortex Cores in High-T_c Superconductors
B.W. Hoogenboom, C. Renner, I. Maggio-Aprile, and Ø. Fischer 269

16.1 Introduction 269
16.2 Studying Vortex Cores by STM 270
16.3 Quasiparticle Excitations in Vortex Cores 271
16.4 $NbSe_2$ 272
16.5 $YBa_2Cu_3O_{7-\delta}$ 273
16.6 $Bi_2Sr_2CaCu_2O_{8+\delta}$ 275
        16.6.1 Vortex Core Spectra 275
        16.6.2 Vortex Core Images 276
16.7 Conclusions and Outlook 279
References 280

17 Charged Vortices in High-T_c Superconductors
Yuji Matsuda and Ken-ichi Kumagai 283

17.1 Introduction... 283
17.2 Charged Vortices 285
    17.2.1 Electrostatics of a Vortex 285
    17.2.2 Vortex Charge Probed by NMR......................... 287
17.3 Dynamical Properties of Vortices............................ 291
    17.3.1 Vortex Hall Anomaly 291
    17.3.2 Time Dependent Ginzburg–Landau Equation 292
17.4 Enhanced Vortex Charge
     of High Temperature Superconductors 295
17.5 Summary 297
References 298

18 Neutron Scattering from Vortex Lattices in Superconductors
Andrew Huxley 301

18.1 Introduction.. 301
18.2 The Measurement Technique 301
18.3 Field History 303
18.4 The Form Factor.. 304
18.5 Flux Line Lattice Geometries............................... 305
18.6 An Anisotropic Effective Mass 306
18.7 Non-local Corrections
     and Fermi Surface Anisotropy 308
18.8 Non-conventional Superconductors 310
18.9 Two Component Superconductors.............................. 316
18.10 Conclusions 317
References 318

19 Vortex Dynamics in a Temperature Gradient
Axel Freimuth 321

19.1 Introduction.. 321
19.2 Definition of Transport Coefficients........................ 322
19.3 Summary of Experimental Results 323
    19.3.1 Transverse Effects 323
    19.3.2 Longitudinal Effects 325
    19.3.3 Thermal and Electrical Hall Angle 326
19.4 Implications for the Dynamics of Vortices 326
19.5 Origin of the Heat Current 327
    19.5.1 Excitation Spectrum in the Presence of Vortices...... 328
    19.5.2 Heat Current 330
19.6 Analysis of Vortex Motion 332
    19.6.1 Forces on Vortices 332

19.6.2 Current Driven Vortex Motion 333
19.6.3 Vortex Motion in a Temperature Gradient 334
19.6.4 Pinning Effects 337
19.7 Summary 337
References 337

20 Electric Field Dependent Flux–Flow Resistance in the Cuprate Superconductor $Nd_{2-x}Ce_xCuO_y$
R.P. Huebener 341

20.1 Introduction 341
20.2 Damping of the Vortex Motion 343
20.3 Flux–Flow Resistance Measurements
    in $Nd_{2-x}Ce_xCuO_y$ 345
20.4 Nucleation and Growth
    of High-Electric-Field Domains 346
20.5 Electric Field Dependent Flux–Flow Resistance 349
20.6 Subbands, Bloch Oscillations,
    and Zener Breakdown 351
20.7 Step Structure in the Quasiparticle DOS 355
20.8 Summary and Conclusions 357
References 357

List of Contributors

A.A. Abrikosov
Materials Science Division
Argonne National Laboratory
9700 South Cass Ave.
Argonne, IL 60439, USA

I.S. Aranson
Argonne National Laboratory
9700 South Cass Avenue
Argonne, IL 60439, USA

B. Carter
Département d'Astrophysique
Relativiste et de Cosmologie
Centre National de la
Recherche Scientifique
Observatoire de Paris
92195 Meudon, France

V.B. Eltsov
Low Temperature Laboratory
Helsinki University of Technology
P.O. Box 2200
02015 HUT, Espoo, Finland

Kapitza Institute for
Physical Problems
Kosygina 2
117334 Moscow, Russia

M. Eschrig
Department of Physics
Northwestern University
Evanston, IL 60208, USA

Materials Science Division
Argonne National Laboratory
Argonne, IL 60439, USA

Ø. Fischer
DPMC, Université de Genève
24 Quai Ernest-Ansermet
1211 Genève 4, Switzerland

A. Freimuth
II. Physikalisches Institut
Universität zu Köln
50937 Köln, Germany

B.W. Hoogenboom
DPMC, Université de Genève
24 Quai Ernest-Ansermet
1211 Genève 4, Switzerland

R.P. Huebener
Physikalisches Institut
Lehrstuhl Experimentalphysik II
Universität Tübingen
72076 Tübingen, Germany

A. Huxley
CEA, Départment de Recherche
Fondamentale sur la Matière
Condensée, SPSMS
38054 Grenoble Cedex 9, France

M. Ichioka
Department of Physics
Okayama University
Okayama 700-8530, Japan

XVI   List of Contributors

A.F. Ioffe
Academy of Sciences of Russia,
194021 St.Petersburg, Russia

D.A. Ivanov
Institut für Theoretische Physik
ETH-Hönggerberg
8093 Zürich, Switzerland

N.B. Kopnin
L.D. Landau Institute for
Theoretical Physics
117334 Moscow, Russia

Low Temperature Laboratory
Helsinki University of Technology
P.O.Box 2200
02015 HUT, Espoo, Finland

M. Krusius
Low Temperature Laboratory
Helsinki University of Technology
P.O. Box 2200
02015 HUT, Espoo, Finland

K. Kumagai
Division of Physics,
Graduate School of Science
Hokkaido University
Kita-ku, Sapporo 060-0810
Japan

D. Langlois
Département d'Astrophysique
Relativiste et de Cosmologie
Centre National de la
Recherche Scientifique
Observatoire de Paris
92195 Meudon, France

Institut d'Astrophysique de Paris
98bis Boulevard Arago
75014 Paris, France

K. Machida
Department of Physics
Okayama University
Okayama 700-8530, Japan

K. Maki
Department of Physics
and Astronomy
University of Southern California
Los Angeles, CA 90089-0484, USA

I. Maggio-Aprile
DPMC, Université de Genève
24 Quai Ernest-Ansermet
1211 Genève 4, Switzerland

Y. Matsuda
Institute for Solid State Physics
University of Tokyo
Kashiwanoha 5-1-5
Kashiwa, Chiba 277-8581
Japan

A.S. Mel'nikov
Institute for Physics of
Microstructures
Russian Academy of Sciences
603600, Nizhny Novgorod
GSP-105 Russia

R. Prix
Département d'Astrophysique
Relativiste et de Cosmologie
Centre National de la
Recherche Scientifique
Observatoire de Paris
92195 Meudon, France

D. Rainer
Physikalisches Institut
Universität Bayreuth
95440 Bayreuth, Germany

C. Renner
NEC Research Institute
4 Independence Way
Princeton, New Jersey 08540
USA

J.A. Sauls
Department of Physics
Northwestern University
Evanston, IL 60208, USA

A. Shelankov
Theoretical Physics Department
University of Umeå
901 87 Umeå
Sweden

E. Sonin
Racah Institute of Physics
Hebrew University of Jerusalem
Jerusalem 91904, Israel

Ioffe Physical Technical Institute
St. Petersburg 194021, Russia

M. Takigawa
Department of Physics
Okayama University
Okayama 700-8530, Japan

V.M. Vinokur
Argonne National Laboratory
9700 South Cass Avenue
Argonne, IL 60439, USA

G.E. Volovik
Low Temperature Laboratory
Helsinki University of Technology
P.O. Box 2200
02015 HUT, Espoo, Finland

Landau Institute for
Theoretical Physics
Kosygina 2
117334 Moscow, Russia

Low Temperature Laboratory
Helsinki University of Technology
P.O. Box 2200
02015 HUT, Finland

H. Won
Department of Physics
Hallym University
Chunchon 200-702, South Korea

1 The Beautiful World of the Vortex

G.E. Volovik

The most exciting aspect of unconventional superfluidity and superconductivity is its rich diversity in extraordinary phenomena of different character, which are perhaps most vividly exemplified in the rich variety of the topological defects. These inhomogeneous structures differ by their topological charges, i.e., conserved topological numbers, and by their symmetry. Among the topological defects the quantized vortices are objects which are under heavy experimental investigation in a full scale, because they can be stabilized by rotation in neutral superfluids, and by applying an external magnetic field in superconductors.

In this book the focus is made on the properties of individual vortices. We are interested in their topology, symmetry, core structure, their elementary excitations, fermions living in cores of vortices, the dynamics of vortex lines, and the influence of the physics of the vortex on the low temperature properties of the superconducting/superfluid state. Wherever it is possible we tried to make connections to the relativistic quantum field theories (RQFT), where the similar topological objects with a very similar physics can exist. On the other hand, we avoided consideration of the collective phenomena incorporating an array of vortices: the latter represent the whole new branch of physics requiring a special book, which should cover the properties of vortex matter: symmetry of the vortex crystal; melting of the vortex crystal; different phases of the vortex matter (vortex glass and vortex liquid states); collective pinning; quantum turbulence, etc. We omitted this material, unless it can give us information on the structure and symmetry of the order parameter and its relation to the property of individual vortices.

In Chap. 2 the dramatic history of type II superconductivity and vortices is reviewed by the father of the Abrikosov vortex. In addition to the standard Abrikosov vortex, the other vortices are discussed, including Pearl and Josephson vortices, and also the phase slip center – the vortex in x, t plane – which is an analog of the instanton in quantum field theory.

The new impulse to the physics of vortices came after the discovery of superfluid ^{3}He – the first example of unconventional Cooper pairing, in which several symmetries are simultaneously broken. Later evidences of unconventional superconductivity and superfluidity were presented in heavy fermionic compounds, then in cuprates exhibiting the high temperature superconductivity, and most recently in ruthenates, and in low density alkali-vapours in laser manipulated traps. One of the most striking results of the ^{3}He research discussed in Chap. 3 is the existence of vortices with continuous structure. Textures with continuous vorticity – skyrmions, merons, vortex sheets – have been intensively studied both theoretically and experimentally in ^{3}He-A. There is now evidence that the continuous vorticity exists also in Bose–Einstein condensates in a laser manipulated traps.

Among the other exotic properties of topological defects in unconventional superfluids and superconductors one can mention the following. (i) Existence of vortices with fractional topological charge. Though predicted in ^{3}He-A, the half quantum vortices were first experimentally realized in cuprate high-temperature superconductors. They are the counterpart of Alice strings in RQFT. (ii) Additional spontaneous symmetry breaking in the vortex core. This has also an analogy in RQFT where, in addition to the electroweak symmetry breaking in the vacuum, the electromagnetic $U(1)$ symmetry can be broken in the core of the cosmic string. As a result the core matter of the string becomes superconducting. (iii) The hierarchy of interactions leads to a rich variety of several core regimes with different length scales: hard core, soft core, ferromagnetic core, topological phase transition in the vortex core, ... (iv) Topological interaction of defects of different dimensionalities. Examples are: soliton terminating on a vortex; vortex terminating on a monopole-like object (this configuration resembles in many respects Dirac magnetic monopole with the Dirac string); boojums – point defects living on the domain wall, etc.

^{3}He-A also is the first superfluid, where the gap in the quasiparticle energy spectrum has a node, so that the fermionic quasiparticles are gapless and they resemble in many respects the massless chiral fermions of the Standard Model of electroweak interactions. This establishes the close relation to particle physics and cosmology, which constitutes a successful example of the unity of physics. The fundamental links between cosmology and particle physics, in other words, between macro- and micro-worlds, have been already well established: there is a unified system of laws governing all scales from subatomic particles to the Cosmos and this principle is widely exploited in the description of the physics of the early Universe, baryogenesis, cosmological nucleosynthesis, etc. The connection of these two fields with the third ingredient of the modern physics – condensed matter – is now rapidly developing.

In particular, the counterpart of Abrikosov vortices in RQFT are Nielsen–Olesen cosmic strings that are believed to have precipitated in the phase transitions of the early Universe which probably have some cosmological implications. This analogy allows us to realize the 'cosmological' experiments in the laboratory in which the 'cosmic strings' are created in a rapid phase transition. Theory of the vortex formation under rapid thermal quench is discussed in Chap. 4.

There are the fermionic systems in high energy physics which exhibit Cooper pairing similar to that in unconventional superconductors and superfluids. Among them are the superfluid phases of neutron matter and the superconducting phases of proton matter in neutron stars; at higher density of quark matter the so-called color superconductivity can arise in which the Cooper pairs are colored: they have an $SU(3)$ color charge. Vortices in superfluid and superconducting matter of neutron stars are discussed in Chaps. 5 and 6.

Thermal and kinetic properties of superconductors and Fermi superfluids are mostly determined by fermionic quasiparticles – particles and holes hybridized by the superconducting coherence. They serve as the counterpart of the elementary particles in RQFT. In superconductors with gap Δ (analogue of mass in RQFT), at temperatures well below the superconducting transition temperature, $T \ll T_c$, the quasiparticles in bulk superconductor are exponentially frozen out as $\exp(-\Delta/T)$. Only those quasiparticles who live in the vortex core survive, since they have low energy states in the core potential. These low energy states bound to the vortex core are the counterpart of the fermion zero modes in the core of cosmic strings. They induce the mass, electric charge, linear and angular momenta of the vortex. At low T fermion zero modes play the major role in the transport and thermodynamics of superconductors.

Some of the unconventional superconductors are gapless in the same manner as superfluid ^{3}He-A. An example is provided by the superconductivity of cuprates. The quasiparticles there are also very similar to the massless fermions in RQFT. The contribution of gapless fermions to the transport and thermodynamics is more pronounced since it exhibits the power law behavior, T^n, instead of $\exp(-\Delta/T)$. Nevertheless, the effect of vortices is also crucial: in the presence of vortices the fermionic density of states is enhanced and the exponent n becomes effectively smaller.

In Chap. 7, written by Kopnin, the influence of the fermion zero modes on the vortex dynamics is discussed. The Landau phenomenological picture of the Fermi liquid in terms of the elementary quasiparticles is extended to the fermionic matter living within the vortex core, where the elementary quasiparticles have an exotic energy spectrum characterizing the fermion zero modes. The Boltzmann equation for these fermion zero modes describes the momentum and energy exchange between the core fermions and quasiparticles outside the core. This exchange leads to an extra forces acting on a moving vortex: in addition to the familiar Magnus force, there are spectral flow and Iordanskii forces. The first of the two reflects the physics of the spectral flow in the vortex core, which is essentially the same as the physics of axial anomaly in RQFT. Actually this Kopnin force, whose temperature dependence has been measured in ^{3}He-A and ^{3}He-B, was the first experimental realization of axial anomaly in condensed matter.

Another extra force acting on a vortex – the Iordanskii force – comes from the elastic scattering of external fermions on a vortex. It represents the analogue of Aharonov–Bohm effect which takes place even in electrically neutral superfluid: the vortex line plays the role of the magnetic flux tube, while the neutral quasiparticles play the role of electrons. In RQFT this is completely analogous to the gravitational Aharonov–Bohm effect experienced by the so called spinning cosmic string. General discussion of the Iordanskii force and the analogue of Aharonov–Bohm effect is in Chaps. 8–10.

The phenomenological theory of the influence of fermion zero modes on vortex dynamics is supplemented in Chap. 11 by numerical simulations in conventional and unconventional superconductors in the frequency region beyond the Kopnin theory.

The following three chapters (12–14) are devoted to fermion zero modes in nodal superconductors – in such superconductors where the fermionic quasiparticles are gapless (massless) even in the absence of vortices. These are the high-temperature superconductors with the d-wave pairing, which have lines of nodes in the energy gap; some heavy fermionic materials; and possibly some others including the organic ones. The role of the Abrikosov vortices in these unconventional superconductors is to provide the topologically nontrivial background for the massless fermions. Such problem requires accurate consideration, because as in RQFT, the massless fermions in the presence of the external fields always lead to a variety of the anomalies in the low frequency behavior. In particular, in the superconductors with the lines of nodes the vortices induce the nonanalytic (square-root) behavior of the electronic density of states as a function of magnetic field. This leads to the nonanalytic behavior of the specific heat, thermal conductivity, and other thermodynamic and transport properties, which is now under intensive experimental investigation.

In Chap. 15 the general statistics of fermionic levels in the vortex core is considered. Because of peculiar properties of fermion zero modes the disorder leads to novel types of random-matrix ensemble, which is essentially different from three traditional Wigner–Dyson classes.

Finally the last five chapters (16–20) describe some selected experiments related to the discussed topics. Scanning tunneling spectroscopy experiments on the vortex core structure is discussed in Chap. 16. They reveal a splitting of the vortex core, which is attributed to the quantum tunneling between different pinning sites. However, it cannot be excluded that the core really splits in two, as it happens with the vortex in ^{3}He-B which spontaneously splits into a pair of half-quantum vortices. Chapter 17 provides the first evidence of the electric charge of the vortex core. Chapter 18 discusses the neutron experiments on vortex lattice, which can provide the information on the symmetry and structure of the order parameter. In Chap. 19 the spectral flow phenomenon in the vortex core is extended to the vortex dynamics induced by a temperature gradient, and the corresponding experiments are discussed. Chapter 20 is devoted to the nonlinear effects related with fermion zero modes in the vortex core.

Part I

Vortices in Superconductors, Superfluids, Neutron Stars, and QFT

2 Type II Superconductors and Vortices from the 1950s to the 1990s

A.A. Abrikosov

Summary. This chapter addresses the following topics: The history of the discovery of Type II superconductors and their usual magnetic properties: two critical fields and the mixed state in between, with a partial penetration of the magnetic field in the form of a lattice of quantum vortices. The vortex lattice, as an unavoidable feature of a gauge field theory. Imaging: neutron diffraction, magnetic powder decoration, STM studies, electron holography. Moving vortices and pinning: types of pinning, vortex glass, melting of the vortex lattice. Other types of vortices: Pearl's vortices in thin films placed in a normal field, surface vortices above H_{c2}, Josephson vortices. Resistive state in thin wires; phase-slip centers as vortices in the space-time plane.

2.1 Preamble

The year 1997 was a double jubilee: it was simultaneously 45 years since the discovery of Type II superconductors by the late Nikolay V. Zavaritskii and myself, and 40 years since my publication about the vortex lattice. The first date is not a common knowledge, since the papers by Zavaritskii and myself were published in the Russian journal "Doklady Akademii Nauk SSSR" in 1952, when it was not translated into English. Therefore in textbooks, and elsewhere, one only ever reads that "There exist two types of superconductors...," etc. However, in the citation of the Fritz London award, which I received in 1972, it was written "...for theoretical work in low temperature physics, especially the discovery of Type II Superconductivity". The diploma was signed by John Bardeen and the Award Committee consisted of outstanding physicists: Douglas Scalapino, Michael Tinkham, Pierre Hohenberg, Ernest Lynton and Frederick Reif. The history of my work on this topic I described in my Fritz London address, and it was published in "Physics Today" (1973) [1] and in the LT13 Proceedings. Since that time 27 years have passed, and it can hardly be expected that anybody remembers its contents.

2.2 History

In 1950 Vitalii Ginsburg and Lev Landau (GL) published their famous paper [2] on the theory of superconductivity. The most important properties which were explained by the new theory were the origin of the positive surface energy at the boundary between the normal and superconducting layers, the critical magnetic field of thin superconducting films and the maximal critical current density. The previous theory proposed by Fritz London and Hans London (1935) [3] could not explain these properties. The positive surface energy was introduced phenomenologically by Landau in his theory of the so-called intermediate state [4]; London's theory predicted a negative value. After the GL theory, N. Zavaritskii, a young experimentalist from the P.L. Kapitza Institute for Physical Problems, performed experiments with thin films in order to check the predictions of the GL theory. He was successful: the dependence of the critical field on thickness and temperature exactly fitted the theoretical predictions [5].

In order to obtain more uniform films he started to evaporate the metal on a glass substrate which he kept at helium temperature. And here a strange thing happened: the critical field did not follow the predictions of the GL theory. Since Zavaritskii was my university mate, I often discussed his results with him. The GL theory looked beautiful and firm; so we wanted to find an explanation within its framework. And this idea was correct. The theory included a dimensionless parameter κ (now called the Ginsburg–Landau parameter), characterizing the properties of the substance. All data on the surface energy existing at that time, for superconductors consisting of pure elements, showed that κ was small. Therefore all the calculations in the GL paper were done under this assumption. Moreover, in the same paper it was shown that if κ exceeded some critical value $1/\sqrt{2}$ the surface energy would become negative, and this was considered by the authors as completely unphysical. We asked ourselves: what will happen if $\kappa > 1/\sqrt{2}$? I calculated the critical field for this case, and the results fitted Zavaritskii's data on low-temperature films. After this we published two separate papers (1952) [6], where we expressed the idea that, apart from usual superconductors with a small κ and positive surface energy, there exist Type II superconductors (we called them "second group") with $\kappa > 1/\sqrt{2}$ and a negative surface energy.

After that I became interested in properties of bulk Type II superconductors. Since the surface energy was negative, the phase transition to the normal state in magnetic field had to be of the second order, contrary to Type I superconductors. The critical field could be defined as an eigenvalue of the GL equation for the order parameter Ψ corresponding to the existence of an infinitesimal nucleus of the superconducting phase in a normal medium. The value was

$$H_{c2} = H_{cm}\sqrt{2} \tag{2.1}$$

and at $\kappa > 1/\sqrt{2}$ it exceeded H_{cm} – the thermodynamic critical field, at which the first order transition happened in Type I superconductors. The distributions of the magnetic field and currents in the sample at fields slightly smaller than H_{c2} were defined as a linear combination of the nuclei, which

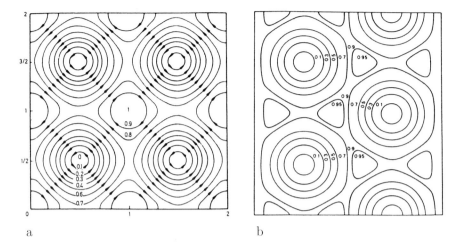

Fig. 2.1. Fields and currents in a vortex lattice in the vicinity of H_{c2} (a cut normal to the applied magnetic field). The solid lines are lines of current and at the same time lines of constant magnetic field. In an isotropic medium the vortex lattice is triangular (**b**); in a tetragonal single crystal with the field along the main axis the lattice may be quadratic (**a**).

corresponded to the minimal free energy. This formed a periodic pattern (Fig. 2.1 a,b).

However, this method failed at lower fields, when the order parameter was not small. After many unsuccessful attempts I carefully analyzed the solution I obtained in the vicinity of H_{c2} and found that it contained points in the plane normal to $\boldsymbol{H}$ (actually lines along $\boldsymbol{H}$), where the complex order parameter Ψ vanished, and its phase changed by 2π along a contour surrounding such a point. Since I did not assume such a behavior beforehand, I wondered what could be the cause and came to the conclusion that this was a necessary property of the equations. If the magnetic field has no systematic variation and remains constant on the average, the vector potential, entering the GL equations, has to grow. The compensations of this growth happens in the following way. Due to gauge invariance, the vector potential enters in the equations as a combination with the gradient of φ – the phase of $\Psi \equiv |\Psi|e^{i\varphi}$:

$$\boldsymbol{A} - (\hbar c/2e)\nabla\varphi \qquad (2.2)$$

(in the original paper I wrote e, instead of $2e$, since at that time the Cooper pairs were not yet discovered). Let us, to be definite, select $\boldsymbol{A}$ along the y direction, and then $H_z = \partial A_y/\partial x$. As usual, we introduce cuts in the complex plane of Ψ in order to make the phase uniquely defined (Fig. 2.2 for a square lattice). If we follow the left shore of such a cut, the singular part of the phase will acquire $-\pi$ at every semicircle, and hence, it will be $-\pi/a$, if a is the period. Following along the right shore we get the singular part equal

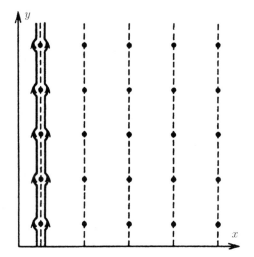

Fig. 2.2. Square vortex lattice in a perpendicular plane (considered as a plane of the variable $x + iy$). The dashed lines are the cuts permitting a unique definition of the phase of the complex order parameter. If one follows the left shore of a cut, the phase changes by $-\pi$ at every semicircle; at a distance y this will sum up to $-\pi y/a$, where a is a period. Similarly we get $\pi y/a$ for the right shore. Therefore the gradient of the phase has a discontinuity $\Delta(\partial\varphi/\partial y) = 2\pi/a$ at every cut.

to $\pi y/a$. Hence we get jumps of $\partial\varphi/\partial y$ at every such cut: $\Delta\partial\varphi/\partial y = 2\pi/a$. This compensates the growth of $\boldsymbol{A}$, if

$$A_y(a) - A_y(0) \equiv \int_0^a H\,dx = a\bar{H} \equiv aB = (\hbar c/2e)(2\pi/a) = \Phi_0/a\,,$$

where B is the magnetic induction equal to the average field, and

$$\Phi_0 = \pi\hbar c/e = 2.05 \times 10^{-7}\,\text{Oe}\,\text{cm}^2 \qquad (2.3)$$

is known as the flux quantum. From here we see that the magnetic flux per period, i.e. Ba^2 equals Φ_0. Afterwards it was realized that this compensation mechanism happens in all gauge field theories now forming the basis of the "standard model" of elementary particles.

When the field decreases, the growth of the vector potential becomes slower, and hence these singular points become more distant. In the limit we can consider them as isolated and analyze one such point separately. This is the region, where the magnetic field is concentrated, surrounded by a circular current – the quantum vortex. Figure 2.3 represents the variation of the field and the magnitude of the order parameter in a vortex. The absolute value of the order parameter increases from zero to the value in absence of magnetic field at the "coherence length" ξ – this is what is known as the vortex core. The magnetic field decreases from its maximum value to zero at the distance

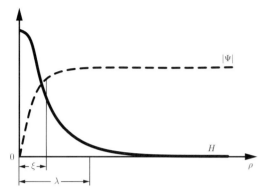

Fig. 2.3. Variation of the magnetic field and the magnitude of the order parameter, $|\Psi|$, in an isolated vortex. The magnetic field decreases from the maximum value at the vortex axis to zero in the bulk of the superconductor at a distance λ – the magnetic penetration depth. The order parameter grows from zero to its value in the bulk superconductor at a distance ξ – the coherence length.

λ – the penetration depth. The relation between these two length scales is $\lambda/\xi \sim \kappa$. Each vortex carries a magnetic flux equal to the flux quantum Φ_0. The field, at which the vortices can first appear, is the first critical field

$$H_{c1} = H_{cm} \left(\kappa\sqrt{2}\right)^{-1} (\ln \kappa + \text{const}) . \qquad (2.4)$$

At higher fields, when many vortices are present, they form a regular lattice, as mentioned previously. In an isotropic substance the most stable is a triangular lattice (Fig. 2.1b), but in a crystal, if the field is directed along a 4th order axis, the vortex lattice is likely to be quadratic (Fig. 2.1a). Figure 2.4 represents the magnetization curves for various values of κ. The region between the critical field H_{c1} and H_{c2} with a partial penetration of the magnetic field I called the mixed state.

I found this all in 1953, but Landau did not agree with my concept, and approved it only after Feynman's work of 1955 [7] on vortices in rotating superfluid helium. So my paper was published only in 1957 [8]. However, even after that it did not attract any attention, and only after the discovery at the beginning of 1960s of superconducting compounds and alloys with very high critical fields did my paper became popular. I owe this in a great deal to Bruce Goodman, who had his own concept of superconducting alloys and published a special paper comparing our concepts; his conclusion was that mine was the true one. This noble action had no analogs either before or after.

After the appearance of the microscopic theory by J. Bardeen, L.N. Cooper and J.R. Schrieffer [9] (1957), L.P. Gor'kov [10] showed that the GL theory is the limit of the BCS theory in the vicinity of T_c and found the microscopic expressions of its phenomenological parameters. A subsequent analysis of the

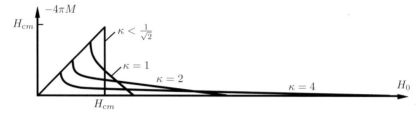

Fig. 2.4. Magnetization ($-4\pi M$) of a cylindrical sample in a longitudinal field as function of the applied magnetic field for different values of the GL parameter κ. For $\kappa < 1/\sqrt{2}$ (Type I) superconductivity makes the sample an absolute diamagnet (no field penetration) up to the thermodynamical critical field H_{cm}, where the phase transition to the normal state takes place. For $\kappa > 1/\sqrt{2}$ (Type II) the Meissner effect continues up to the field H_{c1}. At higher fields the sample is in the mixed state with a vortex lattice, and the magnetization decreases with H, until at H_{c2} the transition to the normal state takes place. The larger is κ, the higher is H_{c2}, and the lower is H_{c1}.

dependence of these parameters on the concentration of impurities showed, particularly, that the value of the GL parameter κ can be increased by reduction of the mean free path of the electrons, ℓ. In the case, when $\ell \ll \xi$ (we recall that ξ is the coherence length, or the size of the Cooper pair in the BCS theory), $\kappa \sim \lambda_L/\ell$, where λ_L is the penetration depth in a clean material. Therefore κ is proportional to the impurity concentration. It is no surprise that the first Type II superconductors, observed *de facto* by Kasimir–Jonker and De Haas in 1935 [11], were superconducting alloys. At that time there was no idea about Type II superconductivity, and the authors explained the gradually changing magnetization by inhomogeneities with different critical fields. The same explanation was used by L.V. Schubnikov et al. in 1937 [12], who managed to reach a great perfection in the quality of their samples of superconducting alloys.

The BCS theory had some other important consequences. Among them were the localized electronic states in the vortex core obtained from the BCS theory by C. Caroli et al. (1965) [13]. For low-temperature Type II superconductors the energy level spacing was so small that only the density of states was important, and it was exactly that of the normal metal. This was described as a "normal core" and led to the prediction of a specific heat in the mixed state, linear in field and temperature

$$C(T) - C(O) = \gamma T(B/H_{c2}) ; \qquad (2.5)$$

later this was confirmed experimentally.

2.3 Imaging

Even after publication of the paper by Bruce Goodman, where it was shown that my theory fits the experimental magnetization curves, a lot of doubt remained because people found it difficult to imagine the existence of some periodic structure incommensurate with the crystalline lattice. The first structural probe was the neutron diffraction experiment by D. Cribier et al. (1964) [14]. Even this did not convince some skeptics. Then came the decoration experiment by U. Essman and H. Träuble (1967) [15] who used very fine ferromagnetic powders (Fig. 2.5). After that nobody could deny the existence of a vortex lattice. Both of these methods are used even now, mostly for high-T_c superconductors.

The methods described above have certain disadvantages. Both neutrons and magnetic powders respond to the magnetic field, which varies at distances of the order of the magnetic penetration depth λ. In superconductors with large κ's the coherence length ξ is small but the penetration depth λ is usually large. If the vortex array is dense, their magnetic fields overlap, and the resolution becomes poor. Practically this happens starting from the fields only slightly larger than H_{c1}, and this means that in most of the mixed state the vortex lattice cannot be resolved by these methods. Another disadvantage is that these methods apply only for a static lattice and the dynamics cannot be studied.

A real breakthrough was achieved by H.F. Hess et al. (1988) [16] through the application of scanning tunneling microscopy (STM). In this method the $I(V)$ characteristic and its derivative $dI/dV = G(V)$ are measured at every point at the surface of the sample. The latter quantity is proportional to the density of states, which depends on the order parameter, i.e., varies at

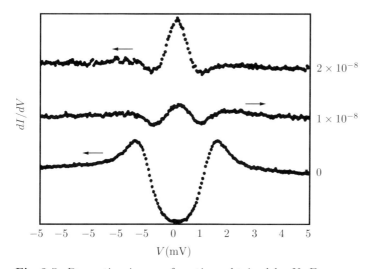

Fig. 2.5. Decoration image of vortices obtained by U. Essmann and H. Träuble [15] on Pb+4% in using a very fine magnetic powder. The ferromagnetic particles are attracted to regions with magnetic field, where the vortices emerge from the sample in the mixed state.

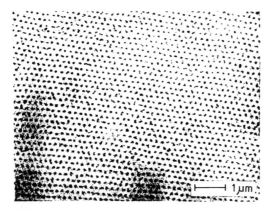

Fig. 2.6. Tunneling conductance $G(V) = dI/dV$ curves as function of applied voltage obtained in NbSe$_2$ by scanning tunneling microscopy (STM) [16]. The lowest curve is taken with the tip far from the vortex core. The dip in the vicinity of $V = 0$ corresponds to the superconducting gap, i.e. absence of exited states of electrons with energies less than Δ. The upper curve is taken with the tip above the vortex core. The maximum around $V = 0$ describes electronic excited states localized in the vortex core. The central curve corresponds to some displacement from the axis.

distances of the order ξ. Without vortices it has a dip around zero voltage corresponding to the superconducting energy gap, and this shape is seen far from the vortex core. If the tip is above the vortex, a maximum of $G(V)$ appears, corresponding to the contribution of localized states (Fig. 2.6). Therefore, this method permits the study not only the geometry of the vortex lattice (Fig. 2.7), but also of the states in the core. Nevertheless, since this method requires separate measurements at every point, it is not convenient for studies of the vortex dynamics (recently, however, such measurements were performed by applying fast automatic registration).

Electron microscopy provided an excellent tool for observing vortex dynamics (see the book by A. Tonomura, 1993 [17]). Two observation methods are used. One is based on the so-called Aharonov–Bohm effect. The quasiclassical wave function of an electron passing through a magnetic field acquires a phase factor $\exp[-\mathrm{i}(e/c\hbar) \int \boldsymbol{A} d\boldsymbol{l}]$, where $\boldsymbol{A}$ is the vector potential, and the integral is along the electron trajectory. This is used in the electron holography, which provides patterns produced by interference of the original beam and the one which has passed through the region under investigation, in our case, the vortex array. This can be done in the "profile mode", where the field outside the superconductor is observed, and so the entrance of vortices can be seen, or in the transmission mode. Another method is called Lorenz microscopy; it registers the deflection of the electrons under the influence of the Lorentz force. Interference with the original beam achieved by suitable electron optics, producing controlled defocusing, gives a direct image of the

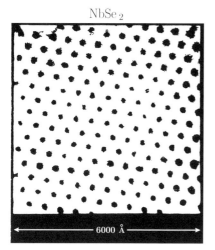

Fig. 2.7. Vortex lattice structure in NbSe$_2$ measured by STM [16].

vortices in transmission micrographs (Fig. 2.8). This method has a low resolution but it is by now the best method for dynamic imaging. The latter is very useful for studying the processes of vortex pinning and melting of the vortex lattice (next section).

2.4 Pinning and Melting of the Vortex Lattice

According to the theory, in the mixed state the external magnetic field penetrates the bulk Type II superconductor in the form of vortices forming, at low temperatures, a regular triangular lattice. However, there exist some departures from this ideal theoretical picture. First of all, at sufficiently high

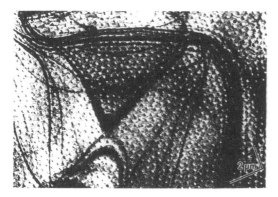

Fig. 2.8. Image of the vortex lattice in a thin Nb film obtained by Lorentz microscopy [17].

temperatures the lattice can melt. The second departure occurs in substances with defects. As it was shown by A. Larkin and Yu. Ovchinnikov [18], the long range order of the vortex lattice disappears in such substances and the array turns into a so-called "vortex glass". I will not address this in detail; for the moment let us imagine an ideal vortex lattice.

For practical applications we are interested in large critical fields. According to (1), for this we need substances with large values of κ. However, according to (4), the first critical field in this case becomes very small. Hence, in order to make superconducting magnets we have to use Type II superconductors in the mixed state. Formally they are superconducting, because the order parameter is finite. However, as was first shown by J. Bardeen and M. Stephen (1965) [19], a finite resistivity actually exists. Indeed, if a current is flowing, there is a Lorentz force acting on every vortex. Under the influence of this force the vortex lattice will start to move perpendicular to the field and to the current. A moving magnetic array will lead to the appearance of an electric field, parallel to the current, and hence will cause a resistance.

In order to prevent this from happening the vortex lattice must be "pinned", and this can be achieved by defects. Two possibilities exist: one, which is called "strong pinning" is achieved by such defects as pores, or non-superconducting regions in the material, having sizes larger than ξ. Such defects can stop the motion at finite temperatures, due to a small probability of thermal excitations assisting the vortices to jump over potential barriers. The resistance in this case, even at finite temperatures, will be small. If it is so small that a supercurrent excited in a ring or in a hollow cylinder will decay, say, only in 100 years, then for practical purposes the metal will be superconducting. Such defects can be introduced artificially, e.g. by bombardment with heavy ions, which produce what is known as "columnar defects".

Another case is that termed "collective pinning". As I said, κ can be enhanced by introduction of electron scatterers, e.g. impurities. The concentration can fluctuate over the sample, and create some random potential slowing down the motion of the vortex lattice. The important point is that every single fluctuation cannot effectively pin the lattice but since there are many of them throughout the sample, and the vortices are held together by elastic forces, such fluctuations can have a considerable pinning effect. This phenomenon is very similar to mechanical friction.

The existence of the vortex lattice is very important for pinning. In high-temperature superconductors the lattice can melt at sufficiently high temperatures, or at sufficiently low densities of vortices, i.e., at small magnetic fields. The pinning becomes very weak, and this limits the practical use of such materials. This trend is enhanced in compounds with the highest T_c, such as Bi-, Tl-, or Hg-based cuprate superconductors, where due to a weak connection between the CuO_2 layers, which are responsible for superconductivity, the vortex lines split into chains of "pancake vortices" (J. Clem, 1991 [20]).

All these issues form a new and rather popular field of "vortex matter physics" which has connections with the physics of dislocations, plasticity, tribology, etc. I will not go further into these matters.

2.5 Other Kinds of Vortices

The vortex lattice also appears in other various problems. The general cause is the appearance in the equations of some growing vector or scalar potential, whereas the physical quantity, i.e., the appropriate field, does not grow on the average. Nature has no other means to compensate for this growth than creating a vortex lattice. It should be stressed that this happens only in quantum problems, since the classical ones can always be described by equations containing only the fields.

One of the further examples are the vortices appearing in a thin superconducting film placed in a perpendicular magnetic field (J. Pearl, 1964 [21]). The difference from the ordinary vortices is that their interaction is defined not only by the magnetic field and current inside the superconductor but also by the distortion of the magnetic field in the surrounding space. This leads to the appearance of an effective penetration depth

$$\lambda_{\text{eff}} = \lambda^2/d ,\qquad (2.6)$$

where λ is the usual penetration depth, and d in the thickness of the film. In order to be a Type II superconductor it must obey the inequality $\lambda_{\text{eff}} > \xi$, or $d < \lambda^2/\xi$. This condition can be fulfilled even for Type I superconductors with $\lambda < \xi$, provided that the film is sufficiently thin.

Another unusual case is that of vortices associated with the "surface superconductivity" discovered by Saint–James and De Gennes (1963) [22] and appearing in magnetic fields

$$H_{c2} < H < H_{c3} = 1.695 H_{c2} .\qquad (2.7)$$

In such fields superconductivity is absent in the bulk of the sample but remains in a surface layer having a thickness ξ. The critical field H_{c3} has its highest value if it is oriented parallel to the surface. A tilt leads to the decrease of this value down to H_{c2} at normal orientation. Simultaneously the current and field distribution in the surface layer change to an array of vortices with axes normal to the surface and the period decreasing with the tilt of the applied magnetic field (I.O. Kulik, 1968 [23]). The existence of such vortices was established indirectly by comparing critical currents for surface superconductivity in samples with a smooth surface, where the critical current was low due to vortex motion, and others with a rough surface, where the current increased considerably due to vortex pinning.

Soon after the discovery of the Josephson effect (B.D. Josephson, 1962 [24]), it was established that the Josephson junction with a supercurrent also has a sort of Meissner effect, i.e., it expels the external magnetic field. Since, however, the supercurrent is weak, the corresponding penetration depth λ_J is large, much larger than the bulk λ. If the junction is sufficiently wide, this is the distance at which the field normal to the junction penetrates into the junction from its edges (Fig. 2.9a). If the field is large enough, vortices will form (Fig. 2.9b).

The last type of vortices that I shall address are the so-called "phase slip centers". Let us imagine a very thin superconducting wire, thinner than λ and ξ, carrying an electric current. For such a thin wire there exists a critical value of the velocity of Cooper pairs, or a critical current density. If the current exceeds this value, resistance, and hence, an electric field, will appear. In an electric field Cooper pairs, encountering no resistance, should be accelerated, i.e., the current density would become even larger. This contradiction is resolved by appearance of the "phase slip centers" [25], which can be viewed as vortices in the (x,t)-plane. At definite points of the wire, the phase of the order parameter is periodically reduced by 2π; at this point the absolute value $|\Psi| = 0$, and hence the supercurrent vanishes (Fig. 2.10). It is compensated by the normal current and so there is indeed a finite resistance.

From these examples one can see that vortices in superconductors present a very general phenomenon. In all these examples I referred only to simple superconductors with an order parameter being a complex scalar and having only minor dependence on momentum. However, there are other possibilities. Nowadays there is little doubt that the order parameter in many high-T_c layered cuprates has d-wave symmetry. Recently evidence was found that the superconducting compound Sr_2RuO_4 has a p-type order parameter reminiscent of superfluid He^3. Quantized vortices in such substances have many distinct features, and besides "ordinary" vortices, other types of sin-

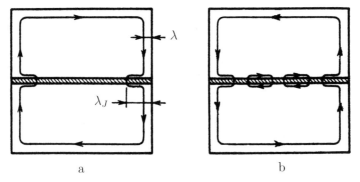

Fig. 2.9. Schematic presentation of a Josephson contact in an external field. (**a**) If the contact is wider than the Josephson penetration depth λ_J and the magnetic field is small, a sort of "Meissner effect" takes place: the inner part of the contact is screened from the magnetic field by a supercurrent. (**b**) With increasing field "Josephson vortices" appear.

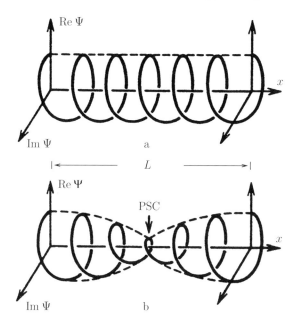

Fig. 2.10. Variation of the order parameter Ψ along a thin superconducting wire carrying a current. (**a**) The current is less than the critical value. The supercurrent density is proportional to the gradient of the phase, and it is constant along the wire. (**b**) The current exceeds the critical value. Then phase slip centers appear permitting a reduction of the phase gradient at certain moments. Since $|\Psi|$ at this moment vanishes, the current is conserved due to appearance of a normal current, and, hence, a resistance.

gularities of the order parameter are possible. Their study is a task for the future.

References

1. A.A. Abrikosov: Physics Today, January 1973, p. 56
2. V.L. Ginzburg, L.D. Landau: Zh. Eksp. Teor. Fiz. **20**, 1064 (1950)
3. F. London, H. London: Proc. Roy. Soc. London, Ser. A **149**, 71 (1935)
4. L.D. Landau: Zh. Eksp. Teor. Fiz. **7**, 19 (1937)
5. N.V. Zavaritskii: Dokl. Akademii Nauk SSSR **78**, 665 (1951)
6. A.A. Abrikosov: Dokl. Akademii Nauk SSSR **86**, 489 (1952); N.V. Zavaritskii, ibid, **86**, 501 (1952)
7. R.P. Feynman: in *Progress in Low Temperature Physics*, ed. by D.F. Brewer (North–Holland, Amsterdam 1955), V. 1, Ch. 11
8. A.A. Abrikosov: Zh. Eksp. Teor. Fiz. **32**, 1442 (1957); Sov. Phys. JETP **5**, 1174 (1957)
9. J. Bardeen, L.N. Cooper, J.R. Schrieffer: Phys. Rev. **106**, 162 (1957); **108**, 1175 (1957)

10. L.P. Gor'kov: Zh. Eksp. Teor. Fiz. **36**, 1918 (1959)
11. J.M. Kasimir–Jonker, W.J. De Haas: Physica **2**, 943 (1935)
12. L.V. Shubnikov et al.: Zh. Eksp. Teor. Fiz. **7**, 221 (1937)
13. C. Caroli et al.: J. Phys. Lett. (France) **9**, 307 (1965)
14. D. Cribier et al.: Phys. Lett. **9**, 106 (1964)
15. U. Essmann, H. Träuble: Phys. Lett. A **24**, 526 (1967)
16. H.F. Hess et al.: Phys. Rev. Lett. **62**, 214 (1988)
17. A. Tonomura: *Electron Holography* (Springer, 1993)
18. A.I. Larkin, Yu.N. Ovchinnikov: Zh. Eksp. Teor. Fiz **65**, 1704 (1973)
19. J. Bardeen, M.J. Stephen: Phys. Rev. A **140**, 1197 (1965)
20. J. Clem: Phys. Rev. B **43**, 7837 (1991)
21. J. Pearl: J. Appl. Phys. Lett. **5**, 65 (1964)
22. D. Saint–James, P.G. De Gennes: Phys. Rev. Lett. **1**, 251 (1963)
23. I. Kulik: Zh. Exp. Teor. Phys. **55**, 218 (1968)
24. B.D. Josephson: Phys. Rev. Lett. **1**, 251 (1962); Rev. Mod. Phys. **36**, 216 (1964)
25. P.W. Anderson: Rev. Mod. Phys. **38**, 298 (1966); W. Little: Phys. Rev. **156**, 396 (1967); J.S. Langer, V. Ambegoakar: Phys. Rev. **164**, 498 (1997)

3 What Can Superconductivity Learn from Quantized Vorticity in ^{3}He Superfluids?

G.E. Volovik, V.B. Eltsov, and M. Krusius

Summary. In ^{3}He superfluids quantized vorticity can take many different forms: It can appear as distributed periodic textures, as sheets, or as lines. In the anisotropic ^{3}He-A phase in most cases the amplitude of the order parameter remains constant throughout the vortex structure and only its orientation changes in space. In the quasi-isotropic ^{3}He-B phase vortex lines have a hard core where the order parameter has a reduced, but finite amplitude. The different structures have been firmly identified, based on both measurement and calculation. What parallels can be drawn from this information to the new unconventional superconductors or Bose–Einstein condensates?

3.1 Unconventional Quantized Vorticity

Soon after the discovery of the ^{3}He superfluids in 1972 it was understood that they represented the first example of unconventional Cooper pairing among Fermi systems, a p-wave state with total spin $S = 1$ and orbital momentum $L = 1$ [1]. This lead to a wide variety of new phenomena, of which one of the most important is the discovery of new vortex structures [2]. These can be studied with NMR spectroscopy [3], when this is combined with a calculation of the order parameter texture [4].

In recent years other unconventional macroscopic quantum systems have been found and have taken the centre stage. Intermetallic alloys such as the heavy fermion metals, the high-temperature superconductors, and the most recent addition, the layered superconductors of Sr_2RuO_4 type, do not fit in the conventional picture of s-wave pairing. Is it possible that unconventional vortex structures, similar perhaps to some of those in the ^{3}He superfluids, might also be present in these new systems?

Current belief holds that the superconducting state in the tetragonal Sr_2RuO_4 material is described by an order parameter of the same symmetry class as that in ^{3}He-A [5,45], an anisotropic superfluid with uniaxial symmetry (where both time reversal symmetry and reflection symmetry are spontaneously broken). Recent advances in optical trapping and cooling of alkali atom clouds to Bose–Einstein condensates have produced Bose systems which are also described by a multi-component order parameter: The spinor representation of the hyperfine spin manifold $F = 1$, for instance, would allow the presence of similar vorticity as in ^{3}He-A.

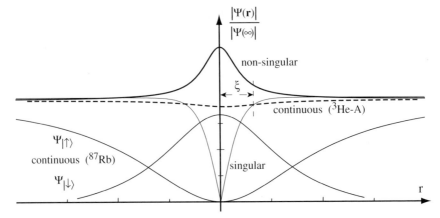

Fig. 3.1. Nomenclature of vortex-core structures: (i) The *singular* vortex has a *hard* core whose radius is comparable to the superfluid coherence length $\xi(T,P)$ and where the order–parameter amplitude vanishes in the centre. (ii) The *non-singular* vortex has a *hard* core in which the order parameter has a finite amplitude everywhere within the core. In principle, this amplitude can even be larger within the core than far outside. (iii) The *continuous* vortex has an almost constant order–parameter amplitude throughout the *soft* core, whose size is much larger than $\xi(T,P)$. Within the soft core primarily the orientation of the order parameter changes. Examples of *continuous* vortices are the doubly quantized vortex in ^{3}He-A and the vortex in a two-component Bose–Einstein condensate, which has a wide inflated core. The latter vortex is formed from the condensate fraction $\Psi_{|\uparrow\rangle}$, while the soft core is filled by the superfluid component $\Psi_{|\downarrow\rangle}$.

Thus the existence of unconventional vorticity has moved into the centre of interdisciplinary debate: To what extent will reduced symmetry influence the structure of quantized vorticity? Vortex lines are defects of the order parameter field, which carry phase winding and circulation of the respective supercurrent. The conventional structure is built around a narrow *singular hard vortex core*: The order parameter vanishes in the centre of the core. By now it has been thoroughly verified that quantized vorticity can take many other forms. Some of the different core structures are schematically listed in Fig. 3.1. Differences exist in the nomenclature of the ^{3}He literature and the current theoretical discussion of vortex line structure in unconventional superconductors, which we try to unify here.

(i) If the vortex has a *hard* core, whose radius is comparable to the temperature- and pressure-dependent superfluid coherence length $\xi(T,P)$ and where the order–parameter amplitude vanishes in the centre, then we call it *singular*. Conventional vortices in the ^{4}He-II superfluid and in s-wave superconductors are singular. Here : $|\Psi(r)|_{r\to 0} = 0$. To a good approximation such a core can be pictured to be a tube with a diameter comparable to the coherence length and filled with normal-state material.

(ii) If the vortex has a hard core but the order parameter has a finite amplitude everywhere within the core, then we call this a *non-singular* vortex, as is generally done in the literature on unconventional superconductivity. Such a core can be viewed as consisting of a different broken symmetry state than the bulk phase outside the core. Vortices in ^{3}He-B are non-singular: the core of the vortex is made up of some non-B-phase components of the order parameter, either the axial (ie. ^{3}He-A) or axiplanar state (Sect. 3.8). The length scale, which determines the core radius, is $\xi(T,P) \sim 10$–100 nm.

(iii) The *continuous* vortex has an almost constant order–parameter amplitude throughout the *soft* vortex core. By the soft core we mean a core whose size is much larger than $\xi(T,P)$. Within the soft core primarily the orientation of the order parameter changes. The larger the soft core diameter, the smoother the distribution of the order–parameter amplitude. Examples of continuous vortices are the *doubly quantized vortex* in ^{3}He-A (Sect. 3.3) and the vortex in a two-component Bose–Einstein condensate (Sect. 3.4). In ^{3}He-A the length scale, which determines the soft-core radius, is the healing length of the dipolar spin–orbit interaction: $\xi_D(T,P) \sim 10$–40 μm.

(iv) A vortex with a *composite* core has an onion structure: It has a hard core (with radius $\sim \xi$), which can be either singular or non-singular, but is embedded within a soft core (with radius $\sim \xi_D$). (a) The *singly quantized vortex* in ^{3}He-A is the prime example (Sect. 3.5). It has a hard non-singular core within a large soft core. The superfluid circulation is generated by the soft core, but the hard core is needed to satisfy the boundary conditions on the orbital part of the order–parameter field, such that it becomes continuous with respect to the bulk fluid. (b) At a high enough magnetic field, the singly quantized vortices in ^{3}He-B have composite cores. Here the vorticity is concentrated in the hard core, while the soft core supports an inhomogeneous order–parameter distribution where the spin–orbit interaction is not minimized. This soft core is a deformation in the order–parameter texture which occurs primarily in the spin part. (c) The *spin-mass vortex* in ^{3}He-B is a composite defect with a narrow hard core, around which the superfluid circulation is trapped. This core is embedded within a planar domain-wall-like soliton defect where the spin–orbit interaction is not minimized and which supports a spin current.

The distinction between non-singular and continuous vortices, and also the existence of a vortex with a composite core rely upon the presence of two length scales, which in turn are determined by two energy scales. In superfluid ^{3}He the two relevant energy scales are the weak spin–orbit interaction and the two orders of magnitude larger superfluid condensation energy. These fix the soft and the hard core sizes, respectively, and as a result the soft core is typically two orders of magnitude larger in diameter than the hard core. If the two scales become comparable in magnitude (as typically occurs in superconductors), then the difference between the continuous and non-singular vortices is washed out.

An understanding of the various structures, in which quantized vorticity may appear, has led to new insight in the physics of macroscopic quantum systems. This new understanding now promises to bring surprising rewards. The discovery of gap nodes in the spectrum of quasiparticle excitations is generally taken to be a signal for unconventional pairing in Fermi systems. An important observation from recent years is the fact that in the vicinity of these gap nodes the energy spectrum is linear and the system acquires all the attributes of relativistic quantum field theory: the analogues of Lorentz invariance, gauge invariance, general covariance, etc. are all present. Therefore fermion superfluids and superconductors on one hand and quantum field theory on the other hand show surprising conceptual similarity. This makes it possible to treat the condensed-matter quantum systems as laboratory models to study physical principles which might also be effective in high energy physics or cosmology. The first examples of such work have been seen in "cosmological" laboratory experiments. For instance, it was recently demonstrated that quantized vortex lines, or linear topological defects as they are known in field theory, are produced in quench-cooled transitions from the normal to the superfluid/superconducting state [7]. This process has been suggested to mimic the production of cosmic strings in the Big-Bang expansion. A second effort was related to the dynamics of vortex lines and was exploited to explain the matter-antimatter asymmetry in the Early Universe [27], if it is assumed to result from the axial anomaly of relativistic field theory. Relativistic quantum field theory may just have found itself an unexpected accomplice!

3.2 Special Features of ^{3}He Superfluids

Superfluid ^{3}He has been blessed with the most ideal properties among the dense coherent quantum systems, approaching those of the alkali atom clouds in Bose–Einstein condensed states: (i) There are no bulk impurities since all alien particles are expelled to the container walls. (ii) The superfluid coherence length $\xi(T, P)$, in addition to being a function of temperature T, also depends on the applied pressure P. By choosing the pressure, the density and interactions can be varied and $\xi(0, P)$ decreases from 65 nm at zero pressure to 12 nm at the liquid-solid transition ($P = 34.4$ bar). (iii) For the best wall materials, surface roughness can be reduced close to the level of $\xi(T, P)$. Experimentally this has important consequences when the container walls approach ideal solid boundaries. (iv) Being an isotropic liquid and a Fermi system, liquid ^{3}He is theoretically more tractable than either superconductors or liquid ^{4}He. All complexity and anisotropy is simply derived from the order parameter, with Cooper pairs in p-wave states [9]. In practice this means

3.3 Continuous Vortices, Skyrmions and Merons

One of the most striking results emerging from ^{3}He research is the existence of vortices with perfectly continuous singularity-free structure. In such a vortex the order–parameter amplitude remains constant throughout the whole structure while its orientation changes in a continuous manner. In the ^{3}He-A phase with axial anisotropy (and Cooper pairs in $L = S = 1$ states), the orbital order–parameter orientation is denoted by the symmetry axis $\hat{l}$ which points in the direction of the nodes of the superfluid energy gap. At very low magnetic fields the vorticity is distributed over the entire primitive cell of the vortex array and thereby takes the form of a periodic order–parameter texture. At higher fields a region of concentrated vorticity is formed, the soft vortex core. An integral number of phase windings is reached along a closed path which encircles the boundary of the unit cell or the soft core.

The simplest possible vortex structure with continuous vorticity in ^{3}He-A is the doubly quantized vortex line. In Fig. 3.2 the orientational distribution of $\hat{l}$ within the soft vortex core is depicted with arrows. By following their flow, it is noted that all possible 4π directions of the radius vector of the unit sphere are present here. This topology of order–parameter orientations ensures two quanta of circulation ($N = 2$).

The 4π topology of $\hat{l}$ orientations within the soft core is known as a *skyrmion*. It can be divided into a pair of *merons* [10–12] ($\mu\epsilon\rho o\sigma$ means fraction [10]), which in the ^{3}He literature are called Mermin–Ho vortices. In the complete skyrmion the $\hat{l}$ vector sweeps the whole unit sphere while each meron, or Mermin–Ho vortex, covers only the orientations in one hemisphere and therefore carries one quantum of vorticity ($N = 1$). The meron covering the northern hemisphere forms a circular 2π Mermin–Ho vortex, while the meron covering the southern hemisphere is the hyperbolic 2π Mermin–Ho vortex. The centers of the merons correspond to minima in the potential of trapped spin-wave states, which generate the satellite peak in the NMR absorption spectrum and make the soft cores observable in NMR measurements. The satellite from the doubly quantized vortex was first detected in rotating ^{3}He-A in 1982 [13], but it was only recently that single-vortex sensitivity was reached and the quantization number $N = 2$ was verified directly [14].

Skyrmions and merons are popular structures in physics: For instance, the double-quantum vortex, in the form of a pair of merons similar to that in ^{3}He-A, is also discussed in the quantum Hall effect where it is formed by pseudo-spin orientations in the magnetic structure [15]. In QCD merons are suggested to produce the color confinement [12].

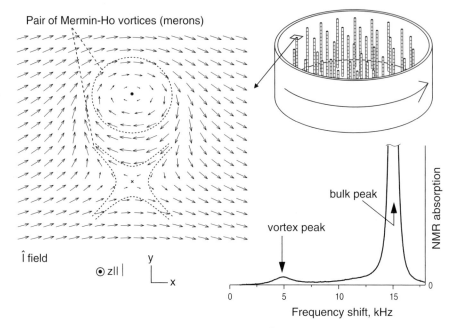

Fig. 3.2. The doubly quantized vortex line in ^{3}He-A with continuous structure in the soft vortex core: (*Top right*) The rotating container with quantized vortex lines. The pillars depict the soft vortex cores, with a diameter of roughly 80 μm $\gg \xi(T, P)$. Each soft core is encircled by a persistent superfluid circulation of two quanta $2\kappa = h/m_3 = 0.13\,\text{mm}^2/\text{s}$. (*Bottom right*) NMR spectroscopy of topological defects in ^{3}He-A. In an external magnetic field, which exceeds the equivalent of the spin–orbit interaction, $H > H_\text{D} \approx 3\,\text{mT}$, the spin $\hat{d}$ and orbital $\hat{l}$ axes are not aligned parallel in the soft core. Spin-orbit interaction exerts an extra torque on the spin precession in nuclear magnetic resonance which shifts the NMR frequency [1]. This torque has a different value within and outside the soft core, and thus gives rise to a satellite absorption peak. Both the frequency shift and the absorption intensity of the satellite are characteristic of the order–parameter texture in the soft core [3]. (*Left*) The orientational distribution of the orbital quantization axis $\hat{l}$ in the soft core, depicted in terms of the projection of $\hat{l}$ on the plane perpendicular to the vortex axis. The $\hat{l}$ orientations cover a solid angle of 4π and the distribution is continuous everywhere. This gives rise to a superfluid circulation of two quanta around the soft core.

In superconductors continuous vortices have been discussed in [16] within a model where the spin–orbit coupling between the electronic spins and the crystal lattice is small and the spin rotation group $SO(3)_S$ is almost exact. The vortex has essentially the same topology as in ^{3}He-A except that instead of the orbital momentum $\hat{l}$ it is the spin orientations $\hat{s}$ of the Cooper pairs which cover a solid angle of 4π in the soft core.

3.4 Transformation from Singular to Continuous Vortex

From the topological point of view, all vortices with the same winding number N can be transformed to each other without changing the asymptotic behaviour of the order parameter, simply by reconstruction of the vortex core. To obtain the continuous vortex in ^{3}He-A (Fig. 3.2) we can start with a $N = 2$ *singular phase vortex*, which has a hard core the size of the coherence length and a uniformly oriented orbital momentum axis $\hat{l}$ along $\hat{x}$. This is a pure phase vortex for the orbital component $L_x = +1$ of the order parameter, which vanishes only on the vortex axis while the other components are zero everywhere. If we now allow for the presence of the other components $L_x = 0$ and $L_x = -1$, then we might fill the hard core with these components, such that the core becomes non-singular. The non-singular core can expand further to form a continuous soft core, which is much larger in radius than the coherence length. Its size is limited by some other weaker energy scales, for instance in ^{3}He-A by the tiny spin–orbit interaction. Within the core $\hat{l}$ sweeps through all possible orientations and the topology becomes that of a skyrmion.

This transformation between continuous and singular vorticity has a direct application to Bose–Einstein condensates. Suppose a mixture of two sub states can be rotated in a laser manipulated trap. Here one starts with a single Bose-condensate which we denote as the $|\uparrow\rangle$ component. Within this component a pure $N = 1$ phase vortex with a singular core is created. Next the hard core of the vortex is filled with the second component in the $|\downarrow\rangle$ state. As a result the core expands and a vortex with continuous structure is obtained. Such a skyrmion has recently been observed with ^{87}Rb atoms [17]. It can be represented in terms of the $\hat{l}$ vector which is constructed from the components of the order parameter as

$$\begin{pmatrix} \Psi_\uparrow \\ \Psi_\downarrow \end{pmatrix} = |\Psi_\uparrow(\infty)| \begin{pmatrix} e^{i\phi} \cos \frac{\beta(r)}{2} \\ \sin \frac{\beta(r)}{2} \end{pmatrix}, \quad \hat{l} = (\sin\beta\cos\phi, -\sin\beta\sin\phi, \cos\beta) \ . \quad (3.1)$$

Here r and ϕ are the cylindrical coordinates with the axis $\hat{z}$ along the vortex axis. The polar angle $\beta(r)$ of the $\hat{l}$ vector changes from 0 at infinity, where $\hat{l} = \hat{z}$ and only the $|\uparrow\rangle$ component is present, to $\beta(0) = \pi$ at the axis, where $\hat{l} = -\hat{z}$ and only the $|\downarrow\rangle$ component is present (Fig. 3.1). Thus the vector $\hat{l}$ sweeps the whole unit sphere. As distinct from ^{3}He-A, where the orbital part of the order parameter has three components and the vortex has $N = 2$ circulation, this mixture of Bose–condensates has two components and the continuous vortex has $N = 1$ phase winding. Various schemes have recently been discussed by which a meron with phase winding can be created in a Bose-condensate formed within a three-component $F = 1$ manifold [18–20].

3.5 Vortex with Composite Core

The singly quantized vortex in ^{3}He-A has a composite core: It is a $N = 1$ vortex with a non-singular hard core, but where the superfluid circulation is generated by the soft core. Thus this $N = 1$ vortex is not a simple $U(1)$ phase vortex, which would have circulation trapped around a hard core in an otherwise homogeneous $\hat{l}$ texture. In fact, the function of the hard core is only to provide the topological stability of the soft core, which with its 2π orientational distribution of the $\hat{l}$ field produces all the vorticity. This vortex structure has the lowest energy in a magnetic field ($H > 3\,\mathrm{mT}$) at low rotation velocity ($\Omega \lesssim 1\,\mathrm{rad/s}$) [21]. Thus it can be created by cooling slowly through T_c in rotation at low Ω, which secures the equilibrium vortex state in an adiabatically slow transition. In contrast, if one starts to rotate the fluid when it is already in the superfluid state below T_c, then the $N = 1$ vortex is generally not formed, because the $N = 2$ vortex has lower critical velocity and is created first. The larger critical velocity of the $N = 1$ vortex reflects the much larger energy barrier involved in the creation of the hard vortex core (Sect. 3.9.3). Experimentally the singly and doubly quantized vortex lines can be distinguished by their very different NMR absorption satellites.

3.6 Vortex Sheet

3.6.1 Vortex-Sheet Structure in ^{3}He-A

In ^{3}He-A the double-quantum vortex line is not the only unconventional vortex structure with perfect continuity in the order parameter amplitude. The most unusual continuous structure is the *vortex sheet* [22], with planar topology (Fig. 3.3). It consists of a folded domain-wall-like structure, a soliton sheet, which separates regions with opposite orientations of the $\hat{l}$ vector. Within this meandering sheet the continuous Mermin–Ho vortex lines with $N = 1$ are confined. These vortices form a chain of alternating circular and hyperbolic merons, which are similar to the two constituents of the soft core in Fig. 3.2. Each meron represents a kink in the soliton structure and is thus topologically trapped within the soliton: it cannot exist as an independent object in the bulk liquid outside the soliton. An analogous example are Bloch lines within a Bloch wall in ferromagnetic materials. The sheet is attached along two connection lines to the lateral boundaries. It is through these connection lines that merons with $N = 1$ vortex quanta can enter or leave the sheet.

3.6.2 Vortex Sheet in Rotating Superfluid

The vortex sheet is well known from classical turbulence as a thin interface across which the tangential component of the flow velocity is discontinuous.

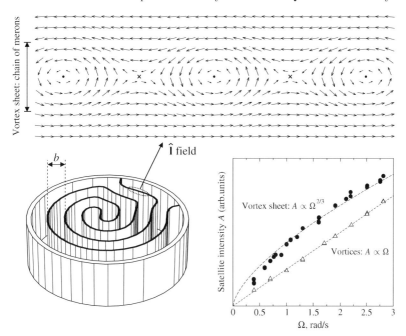

Fig. 3.3. A vortex sheet in ^{3}He-A: (*Bottom left*) The macroscopic configuration of the equilibrium vortex sheet consists of a single continuously folded sheet which fills the rotating container evenly. In an axially oriented external magnetic field the meander forms a double spiral which mimics the original Onsager suggestion of concentric coaxial cylindrical surfaces. (*Top*) An orbital order–parameter texture in the vortex sheet. The arrows depict the projection of $\hat{l}$ in the xy-plane (perpendicular to the rotation axis Ω). This texture consists of an alternating chain of circular (centre marked with black dot) and hyperbolic (centre marked with cross) merons. If one follows a path along the centers of merons, the $\hat{l}$ orientation winds continuously: thus the vorticity is continuously distributed along the sheet. Outside the sheet both $\hat{l}$ and the counterflow velocity $v = v_\mathrm{n} - v_\mathrm{s}$ are oriented parallel to the sheet, but in opposite directions on the two sides of the sheet. (*Bottom right*) The integrated NMR absorption in the satellite peak of continuous vortex lines in Fig. 3.2 follows a linear dependence on the rotation velocity Ω or, equivalently, the number of vortex lines in the equilibrium state of rotation. In contrast, if a soliton sheet preexists in the container and rotation is slowly increased, then a satellite peak with a nonlinear dependence on Ω is grown. This is the signature of the vortex sheet as a rotating state of superfluid.

Within this sheet vorticity approaches infinity while the width of the sheet approaches zero [23]. Historically in superfluids, vorticity was first suggested to be confined to sheets, when Onsager [24] and London [25] described the superfluid state of ^{4}He-II under rotation. It soon turned out, however, that in ^{4}He-II a vortex sheet is unstable with respect to break-up into separated quantized vortex lines. Nevertheless, a calculation on the sheet spacing by

Landau and Lifshitz [26], who did not impose a quantization requirement, happens to be exactly to the point for the vortex sheet in ^{3}He-A. Here the vortex sheet proved to be stable owing to the topological confinement of the vorticity within the topologically stable soliton sheet [22].

The equilibrium state of the vortex sheet in a rotating vessel is constructed by considering the kinetic energy from the flow between the folds and the surface tension σ from the soliton sheet. It is then concluded that the distance between the parallel folds has to be $b = (3\sigma/\rho_{s\|})^{1/3} \, \Omega^{-2/3}$. This is somewhat larger than the inter-vortex distance in a cluster of vortex lines. The areal density of circulation quanta has approximately the solid-body value $n_v = 2\Omega/\kappa$. This means that the length of the vortex sheet per two circulation quanta is $p = \kappa/(b\Omega)$, which is the periodicity of the order–parameter structure in Fig. 3.3 ($p \approx 180\,\mu$m at $\Omega = 1$ rad/s). The NMR absorption in the vortex-sheet satellite is proportional to the total volume of the sheet which in turn is proportional to $1/b \propto \Omega^{2/3}$. This nonlinear dependence of the satellite absorption on rotation velocity is one of the experimental signatures. Locally the equilibrium vortex sheet corresponds to a configuration with accurately equidistant layers. This is manifested by Bragg reflections of spin waves between the folds of the sheet, which produces a characteristic frequency shifted absorption in the observed NMR line shape [27].

3.6.3 Vortex Sheet in Superconductor

The vortex sheet has also been discussed in unconventional superconductors [28,29]. Similar to superfluid ^{3}He-A the vorticity is trapped in a domain wall which separates two domains with opposite orientations of the $\hat{l}$-vector [30]. But, unlike the case of ^{3}He-A, the trapped kink is not a meron but a singular vortex with the fractional winding number $N = 1/2$ (in contrast to isolated vortex lines which are singly quantized). If there are many trapped fractional vortices, then they form a vortex sheet, which as suggested in [29], can be responsible for the flux flow dynamics in the low-temperature phase of the heavy-fermion superconductor UPt$_3$. In superconductors a vortex-sheet-like structure could also appear dynamically, when it is topologically not stable. This state would correspond to a smectic phase of the flowing vortex matter, as was argued in the case of eg. NbSe$_2$ [31].

3.7 Fractional Vorticity and Fractional Flux

An unusual object is the *half-quantum vortex*. Here the order–parameter phase changes by π on circling once around this line. The change of sign of the order parameter can be compensated by some extra degree of freedom, which usually is the spin. Such a linear structure was originally predicted to appear in ^{3}He-A [32], but has not yet been observed there experimentally. Later it was also predicted to appear in unconventional superconductors [33].

Some years ago it was discovered as the intersection line of three grain boundary planes in a thin film of YBa$_2$Cu$_3$O$_{7-\delta}$ [34]. The half-quantum vortex has also been suggested to exist in Bose–Einstein condensates with a hyperfine spin $F = 1$ [35].

Based on the ^{3}He-A example more possible structures of the fractional vortices in unconvetional superconductors can be predicted. In ^{3}He-A the discrete symmetry, which supports the half-quantum vortex, arises from the changes in sign when the spin axis $\hat{\boldsymbol{d}}$ and the orbital axis $\hat{\boldsymbol{e}}_1 + i\hat{\boldsymbol{e}}_2$ are taken once around the line and rotate into $-\hat{\boldsymbol{d}}$ and $-(\hat{\boldsymbol{e}}_1 + i\hat{\boldsymbol{e}}_2)$. When both of these axes change sign, then the order parameter returns to its initial value

$$\hat{\boldsymbol{d}} = \hat{\boldsymbol{x}} \cos \frac{\phi}{2} + \hat{\boldsymbol{y}} \sin \frac{\phi}{2} \ , \ \ \hat{\boldsymbol{e}}_1 + i\hat{\boldsymbol{e}}_2 = (\hat{\boldsymbol{x}} + i\hat{\boldsymbol{y}})e^{i\phi/2} \ . \tag{3.2}$$

Here ϕ is the azimuthal angle of the cylindrical coordinate system and the magnetic field is applied along $\hat{\boldsymbol{z}}$ to keep $\hat{\boldsymbol{d}}$ in the xy plane. The spin axis $\hat{\boldsymbol{d}}$ rotates by π on circling around the half-quantum vortex. Thus a "spectator" in ^{3}He-A, who travels around the vortex, would find its spin reversed with respect to the spin of a "spectator" who was at rest. This situation is analogous to the Alice string in particle physics [36] where a particle, which moves around the string in a continuous manner, flips its charge or parity or enters the "shadow" world [37].

In superconductors the crystalline structure must be taken into account. In the simplest representation which preserves tetragonal symmetry, the p-wave order parameter in Sr$_2$RuO$_4$ has the form

$$\Delta(\boldsymbol{k}) = \Delta_0 \, (\hat{\boldsymbol{d}} \cdot \sigma) \, (\sin \boldsymbol{k} \cdot \boldsymbol{a} + i \sin \boldsymbol{k} \cdot \boldsymbol{b}) \, e^{i\theta} \ , \tag{3.3}$$

where $\boldsymbol{k}$ is momentum, θ is the phase of the order parameter, $\boldsymbol{a}$ and $\boldsymbol{b}$ are the elementary vectors in the basal plane of the crystal lattice. Vortices with fractional quantization N can now be constructed in two ways. If the $\hat{\boldsymbol{d}}$-field is sufficiently flexible, the analog of the vortex with $N = 1/2$ in (3.2) becomes possible, where $\hat{\boldsymbol{d}} \to -\hat{\boldsymbol{d}}$ and $\theta \to \theta + \pi$ on circling once around the vortex [38]. Another possibility is the Möbius–strip geometry. Here the crystal axes $\boldsymbol{a}$ and $\boldsymbol{b}$ are twisted continuously by the angle $\pi/2$ on traversing around the closed wire loop [39]. This closed loop traps fractional flux, since the local orientation of the crystal lattice continuously changes by $\pi/2$ around the loop, $\boldsymbol{a} \to \boldsymbol{b}$ and $\boldsymbol{b} \to -\boldsymbol{a}$, which means that the order parameter becomes multiplied by i. The single-valuedness of the order parameter requires that this change must be compensated by a change in the phase θ by $\pi/2$. As a result the phase winding around the loop is $\pi/2$ and $N = 1/4$.

This, however, does not mean that such a Möbius loop in a chiral p-wave superconductor traps $1/4$ of the magnetic flux Φ_0 of a conventional Abrikosov vortex. Because of the breaking of time reversal symmetry in chiral crystalline superconductors, persistent electric currents arise not only due to phase coherence but also due to deformations of the crystal [40]:

$$\boldsymbol{j} = \rho_s \left(\boldsymbol{v}_s - \frac{e}{mc} \boldsymbol{A} \right) + K a_i \nabla b_i \ , \ \ \boldsymbol{v}_s = \frac{\hbar}{2m} \nabla \theta \ . \tag{3.4}$$

The magnetic flux trapped in the loop is obtained from the condition of zero current, $j = 0$ in (3.4). Thus, the trapped flux depends on the parameter K in the deformation current. In the limiting case of $K = 0$ the flux is $\Phi_0/4$ (or $\Phi_0/6$ if the underlying crystal lattice has hexagonal symmetry).

In a nonchiral d-wave superconductor of layered cuprate-oxide structure the order parameter can be represented by:

$$\Delta(\mathbf{k}) = \Delta_0 \left(\sin^2 \mathbf{k} \cdot \mathbf{a} - \sin^2 \mathbf{k} \cdot \mathbf{b}\right) e^{i\theta} . \tag{3.5}$$

The same twisted wire loop which transforms $\mathbf{a} \to \mathbf{b}$ and $\mathbf{b} \to -\mathbf{a}$, produces a change of sign of the order parameter, which must be compensated by a change of the phase θ by π. This corresponds to a circulation of half a quantum $N = 1/2$, i.e. the fractional flux trapped by this loop is $\Phi_0/2$, since the parameter K in (3.4) is exactly zero in nonchiral superconductors. The same reasoning gives rise to the $\Phi_0/2$ flux associated with the intersection line of three grain boundary planes, the tri-crystal line [34]: around this line $\mathbf{a} \to \mathbf{b}$ and $\mathbf{b} \to -\mathbf{a}$. Only the observation of a fractional flux different from $\Phi_0/2$ would indicate the breaking of time reversal symmetry [28,40,41].

3.8 Broken Symmetry in the Vortex Core

In the quasi-isotropic ^{3}He-B phase (with Cooper pairs in the total angular momentum state $J = 0$), the simplest possible vortex is a singular vortex with $N = 1$, where all the order–parameter components are zero on the axis of the vortex core. However, this vortex is never realized. All vortices (and other linear defects) observed in ^{3}He-B have a non-singular hard core. This is a typical situation for superfluids/superconductors with a multi-component order parameter. As a rule the superfluid/superconductor does not tolerate a full suppression of the superfluid fraction in the core, if there is a possibility to escape this by filling the core with other components of the order parameter. This rule applies also to high-temperature superconductors with d-wave pairing, where the $N = 1$ vortex supports a nonzero s-wave component in the vortex core [42].

3.8.1 Vortex Core Transition

The first NMR measurement on rotating ^{3}He-B revealed, as a function of temperature and pressure, a first order phase transition in the vortex core structure [43]. This phenomenon in Fig. 3.4 was the first example of a phase transition in the structure of any topological defect. Other transitions have been identified in ^{3}He-A since then, but the B-phase transition remains the most prominent one. This transition separates two vortices with the same topology ($N = 1$), but with a different structure of the hard core. The existence of the transition illustrates that vortices in ^{3}He-B have to be non-singular and have a complex structure of the hard core. Later these two stable vortex core

3 What Can Superconductivity Learn from Quantized Vorticity? 33

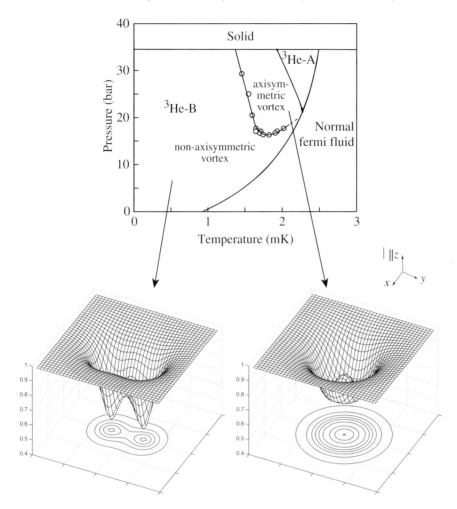

Fig. 3.4. Phase transition in the structure of the singular vortex core in ^{3}He-B: (*Top*) Pressure vs. temperature phase diagram of ^{3}He liquids, with the phase transition line for the B-phase vortex core structure. The exact intercept of the transition line at T_c is unknown. (*Bottom*) Both measurement [45] and calculation [46] confirm that the axi-symmetric core at high temperatures spontaneously reduces its symmetry and transforms to a double-well structure at low temperatures. The magnitude of the square of the order–parameter amplitude has here been plotted in the xy-plane (perpendicular to the rotation axis) for the two core structures. This quantity remains finite throughout the cross section of the core.

structures were theoretically identified as two different stationary minimum energy solutions as a function of pressure in the Ginzburg–Landau temperature regime. Both structures have broken parity in the core.

3.8.2 Ferromagnetic Core

The high-pressure ^{3}He-B vortex has an axi-symmetric core in which the A-phase order–parameter components dominate. These components produce an increase in the order parameter amplitude close to vortex axis which is shown in Fig. 3.4, bottom right. This non-singular vortex with a superfluid core lies lower in energy than the simplest most symmetric solution with a normal core, and displays ferromagnetic spin polarization [44], which is observed in the rotating NMR experiments as a gyromagnetism [45].

Similar, but antiferromagnetic cores have been discussed for high-T_c vortices within the popular $SO(5)$ model for superconductivity and antiferromagnetism [47]. Here it has been established that in the Ginzburg–Landau regime, in certain regions of the parameter values, a solution with normal vortex cores is unstable with respect to non-singular antiferromagnetic cores [48]. Also non-singular vortex cores in the heavy fermion superconductor UPt$_3$ could possibly explain why a three times larger flux-flow resistivity is observed parallel to the $\hat{c}$ axis compared to the perpendicular directions [49].

3.8.3 Asymmetric Double Core

The low-temperature B-phase vortex has a non-axisymmetric core, i.e. the axial $U(1)$ symmetry of the ferromagnetic core is spontaneously broken to create a dumbbell-like double core (Fig. 3.4, bottom left). It can be considered as a pair of half-quantum vortices connected by a non-topological wall [2,46,51]. The separation of the half-quantum vortices increases with decreasing pressure and thus the two-core structure is most pronounced at zero pressure [52]. The asymmetry of the core has been verified by direct observation of a Goldstone mode which is a direct consequence from the loss of axial symmetry: the deformed vortex core can become twisted in the presence of a special type of B-phase NMR mode, which is then detected as a reduction in rf absorption [50].

Related phenomena are also possible in superconductors. In [53] the splitting of the vortex core into a pair of half-quantum vortices has been discussed in heavy-fermion superconductors. In fact, a vortex-core splitting may have been observed in high-T_c superconductors [54]. These observations were interpreted as the tunneling of a vortex between two neighbouring sites in the potential wells created by impurities. However, the phenomenon can also be explained in terms of vortex-core splitting.

3.9 Vortex Formation by Intrinsic Mechanisms

In addition to vortex structure, the second most important question becomes the creation of the different forms of vorticity. Vortex formation by intrinsic mechanisms is a topic which has been discussed for decades in superfluidity, but which has been notoriously difficult. The problems are caused by interference from extrinsic effects, primarily from remanent vorticity trapped on rough surfaces. One of the major developments from the last ten years has been the emergence of reliable measurements on critical velocities in ^{4}He-II. These have been performed by monitoring the superflow through sufficiently small sub-micron-size orifices [55,56]. Intrinsic vortex formation has thereby become an important phenomenon in superfluids – quite unlike in superconductors, where typically vortices appear due to different types of extrinsic effects at the lower critical field H_{c1}. However, it is useful even in the case of superconductors to keep in mind the more ideal properties of vortex formation. Also the process of unpinning of vortices, which is important for the problem of vortex creep in superconductors, can be discussed in terms of vortex nucleation: The motion of a vortex from the pinning site is equivalent to creation of a vortex loop which annihilates the pinned part of the vortex line.

3.9.1 Nucleation Barrier

Many vortex phenomena, not only nucleation, but also the unpinning of remanent vorticity, involve energy barriers which are usually overcome by thermal activation. At the lowest temperatures quantum tunneling has been suggested as a possible mechanism, where a macroscopic amount of matter is assumed to participate coherently in a tunneling process. Vortex nucleation in orifice-flow of ^{4}He-II displays a characteristic low-temperature plateau in the temperature dependence of the critical velocity which has been claimed to support the quantum tunneling concept [57]. This question is also discussed in superconductors in the context of unpinning and creep. However, firm proof for such interpretation is still missing.

In ^{3}He-B the critical velocity was measured in rotating experiments with single-vortex resolution in the early 1990's. This proved to be a more straightforward measurement than in ^{4}He-II. The measured temperature dependence of the critical velocity resembles that of the superfluid energy gap $\Delta(T,P)$ [58], which at the lowest temperatures also approaches a temperature-independent plateau. The explanation here, however, does not involve quantum tunneling, but the superflow instability. This phenomenon resembles a second order transition where the energy barrier goes to zero as a function of the scanned variable, in this case the superflow velocity.

When a cylinder with superfluid ^{3}He-B is slowly accelerated to rotation, the state with one single vortex line becomes energetically favorable when the superflow at the circumference exceeds the Feynman critical velocity $v_{c1} = \kappa/(2\pi R) \ln(R/r_c)$ [59]. With a container radius R of a few millimeters and a circulation quantum $\kappa = h/(2m_3) = 0.066$ mm^2/s, this velocity is only 10^{-2} mm/s. Above this velocity remanent vorticity, which has been trapped on the cylinder wall, may start to unpin and then to expand to rectilinear vortex lines. It is the equivalent of H_{c1} in superconductors. However, if we exclude extrinsic mechanisms of vortex formation, then the vortex-free state will persist metastably to much higher velocities because of the nucleation barrier.

In ^{3}He-B the nucleation barrier is practically impenetrable. The argument is the following: The vortex is nucleated on the wall as a segment of a vortex ring. The radius of a ring sustained by superflow at the velocity v_s can be written as $r_\circ = (\kappa/4\pi v_s) \ln(r_\circ/r_c)$, where the vortex-core radius r_c is of the order of the superfluid coherence length $\xi_{3He} \sim 10 - 100$ nm. The energy of a ring is $E(v_s) = \frac{1}{2}\rho_s \kappa^2 r_\circ \ln(r_\circ/r_c)$, where $\rho_s \sim m/a^3$ is the superfluid density, and a the interatomic distance. This energy constitutes the nucleation barrier. On dimensional grounds we may write $E(v_s)/k_B T \sim (r_\circ/a)(T_F/T) \ln(r_\circ/r_c)$, where $T_F = \hbar^2/2ma^2 k_B \sim 1$ K is the degeneracy temperature of the ^{3}He quantum fluid. Assuming that $r_\circ > r_c$, we find $E(v_s)/k_B T > 10^5 \ln(r_\circ/r_c)$. Such a barrier height in ^{3}He-B is inaccessible at all temperatures below T_c. In contrast, for ^{4}He-II, $\xi_{4He} \sim a$, and the barrier is low, $E(v_s)/k_B T > \ln(r_\circ/r_c)$. It can be thermally overcome, except at the lowest temperatures below 0.2 K.

3.9.2 Vortex Formation in a Hydrodynamic Instability

The huge barrier in ^{3}He-B means that the vortex formation mechanism cannot be thermal activation. When the superflow velocity is increased (by increasing the rotation) a threshold v_{cb} is finally reached, above which homogeneous flow loses local stability. This occurs when the energy density of the superflow, $\rho_s v_s^2/2$, exceeds the energy responsible for the topological stability of a vortex, which is of order $\rho_s (\kappa/2\pi r_c)^2$. An order of magnitude estimate of the maximum velocity is thus $v_{cb} \sim \kappa/2\pi r_c$. At this velocity, the radius $r_\circ$ of the nascent ring becomes comparable to r_c and the nucleation barrier goes to zero. The instability inevitably leads to the creation of a vortex when no other mechanism is available at lower v_s.

In ^{4}He-II well below T_λ, the estimate of v_{cb} in the form $\kappa/2\pi\xi_{4He}$ agrees in order of magnitude with the Landau velocity, defined by the roton gap Δ_r and momentum p_r as $v_L = \Delta_r/p_r \sim 60$ m/s [59]. However, this limiting velocity is not observed directly in the measurements of orifice flow, by extrapolating the thermally activated process to $T \to 0$. The reason is that the measured quantity is the average flow velocity through the aperture and not the local critical velocity at the nucleation site. On the circumference of the orifice, the local velocity will be enhanced from the average value by surface roughness, in particular when the superflow is deflected around any sharp protuberances which match the length scale $r_\circ$ of the evolving vortex half ring. The most effective of such excrescences on the circumference will then selectively become the nucleation centre [58]. For this reason the measured critical velocity is expected to be roughly a factor of $\lesssim 10$ smaller than the ideal limiting value (Fig. 3.5).

In ^{3}He-B, the estimate of $v_{cb} \sim \kappa/2\pi\xi_{3He}$ is smaller by 3 orders of magnitude than in ^{4}He-II, but again comparable to the appropriate Landau limit,

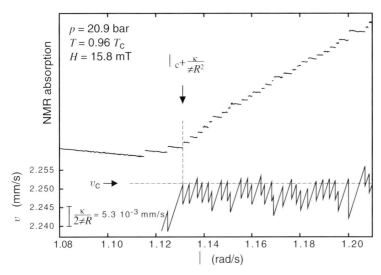

Fig. 3.5. Single-vortex formation as a function of rotation velocity $\Omega(t)$, for ^{3}He-B in a cylindrical container. *Top*: The vertical axis denotes the height of a NMR absorption peak, where the intensity increases per each newly added vortex line by a constant amount. Vortex formation starts with a first step increase at $\Omega_c = 1.115\,\mathrm{rad/s}$, but the critical threshold is identified from the third step (dashed vertical line) where the maximum flow velocity v_c reaches a stable value (dashed horizontal line in the plot at bottom). *Bottom*: The corresponding superflow velocity $v_s(R)$ at the cylinder wall. Each step increase in the upper plot corresponds to a drop by $\kappa/(2\pi R)$ in the velocity $v_s(R)$, the equivalent of one circulation quantum $\kappa = h/(2m_3) = 0.066\,\mathrm{mm}^2/\mathrm{s}$. The average of the maximum velocities v_s defines the critical velocity v_c. The scatter from the average is probably of experimental origin. The rotation here is increased at a constant slow rate $(d\Omega/dt = 2 \cdot 10^{-4}\,\mathrm{rad/s}^2)$, the sample is contained in an epoxy-resin-walled cylinder with radius $R = 2\,\mathrm{mm}$ at a pressure $P = 20.9\,\mathrm{bar}$, in a magnetic field $H = 15.8\,\mathrm{mT}$, and temperature $T = 0.96\,T_c$. (From [58])

defined by the energy gap and the Fermi momentum as $\Delta(T)/p_\mathrm{F}$. In this case the velocity v_{cb} is also known from direct calculations of the stability limit of homogeneous superflow [60]. As in orifice flow, the measured average velocity at vortex formation is smaller than the calculated bulk v_{cb} and depends on the surface roughness. The principal difference from ^{4}He-II is the much larger length scale $r_\circ \sim \xi_{3\mathrm{He}}$, which means that experimentally the influence of surface roughness is less prominent, remanent vorticity can be avoided in the presence of sufficiently smooth surfaces, and stable periodic vortex formation can be investigated with bulk liquid flowing past a flat wall. This is in stark contrast to ^{4}He-II, where the coherence length is of atomic size, surface roughness generally provides an unlimited source of trapped remanent vorticity, and intrinsic nucleation can only be observed in flow through a sufficiently small orifice.

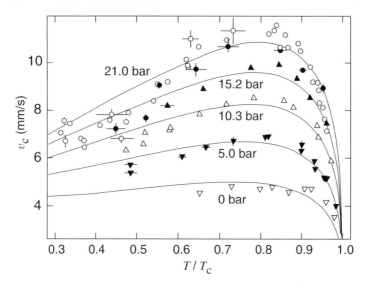

Fig. 3.6. Measurement of critical flow velocity v_c in ^{3}He-B vs. temperature for pressures $P \leq 21$ bar. The solid curves have been fitted to the data with $v_c = v_{cb}(\xi/d)^\chi$, where the fitting parameters d and χ characterize the surface roughness on the cylinder wall, $v_{cb}(T, P)$ is the calculated bulk liquid superflow instability [60] and $\xi(T, P) = \xi(0, P)[\Delta(0)/\Delta(T)]$ the superfluid coherence length. The roughness is modeled by the protuberance with height d and apex angle $\pi/(1-\chi)$ which acts as the nucleation centre. The fit gives $d = 3.1\,\mu$m and $\chi = 0.45$ (apex angle $\approx 30°$). The measurements have been performed with an epoxy-resin-walled container with radius $R = 3.5$ mm. (From [58])

Owing to the excessively high nucleation barrier, in ^{3}He-B it must be the velocity v_{cb} of superflow instability which becomes the appropriate velocity of vortex formation. The process then corresponds to a classical instability, which occurs at the pair-breaking velocity. Thus in the case of ^{3}He-B, the reason for a plateau in $v_c(T)$ in the $T \to 0$ limit is that the characteristic physical quantities, such as the gap amplitude $\Delta(T)$, which determine the instability velocity, become temperature independent. Consequently not only quantum tunneling, but also the intrinsic instability provides an explanation for the low-temperature plateaus which are observed in many different systems, including the case of ^{4}He-II at the lowest temperatures.

3.9.3 Formation of Continuous Vortex Lines: Dependence of Critical Velocity on Core Size

The order-of-magnitude expectation for the bulk superflow instability is $v_c \sim \kappa/(2\pi r_c)$, where r_c is the core size of the emerging vortex. For ^{4}He-II and ^{3}He-B it is the size of the hard core, which is on the order of the superfluid coherence length. For the continuous vortex in ^{3}He-A the length scale of the

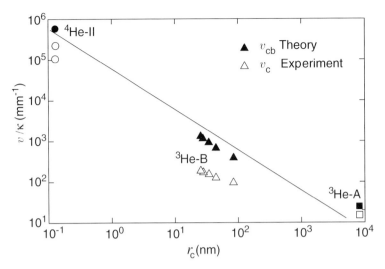

Fig. 3.7. Theoretical v_{cb}/κ and maximum experimental $v_{c,\mathrm{max}}/\kappa$ plotted for 3 superfluids as a function of their core size r_c. For r_c we use 0.1 nm in ^{4}He-II, in ^{3}He-B it is the superfluid coherence length $\xi(T,P)$, and for continuous vortices in ^{3}He-A in a magnetic field the spin–orbit healing length ξ_D. In ^{4}He-II we use the Landau limit v_L for v_{cb}; in ^{3}He-B v_{cb} is the calculated maximum superflow velocity [60], and in ^{3}He-A it corresponds to the helical textural instability [61,62]. For $v_{c,\mathrm{max}}$ in ^{4}He-II the measured value in Ref. [57] is used, for ^{3}He-B data from [58], and for ^{3}He-A from [61]. The line is a guide for the eye, but it obeys the relation $v_c/\kappa \propto 1/r_c$. (From [58])

soft core structure is much larger: It is the healing length $\xi_D \gtrsim 10\,\mu\mathrm{m}$ of the spin–orbit coupling (Fig. 3.2). The same length scale also applies to the structure of the soliton and vortex sheets. Because of this long length scale the measured critical velocities in the A phase are two orders of magnitude lower than in the B phase, as shown in Fig. 3.7. This explains why the continuous $N = 2$ vortex is formed in the A liquid when it is accelerated into rotation, rather than the composite $N = 1$ vortex with a hard core, although the latter might be energetically preferred.

At the container wall the boundary condition on the order parameter requires that the orbital quantization axis $\hat{l}$ has to be oriented along the surface normal. Therefore the centre of vortex formation must evolve outside a surface layer with a width $\sim \xi_D$, out of the influence of surface roughness. This has been recently experimentally confirmed [14]. Here the only mechanism available is an instability of the order–parameter texture.

An example of a critical process, which exemplifies the fact that the order-parameter texture is involved, is shown in Fig. 3.8. Here the vertical axis is the peak height of the satellite absorption in Fig. 3.2, which is proportional to the number of vortex lines, and the horizontal axis is the external drive. Initially

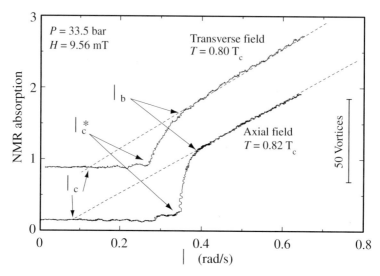

Fig. 3.8. Two examples of vortex-line formation in a transition of the order–parameter texture in ^{3}He-A as a function of the applied rotation Ω. The texture transition occurs at Ω_c^*, where a large number of vortex lines are simultaneously formed. After the transition in the transformed texture the regular periodic vortex formation process sets in, where the flow is limited by a constant critical velocity $v_c = \Omega_c R$. This process gives a linear dependence for the number of vortex lines $\mathcal{N}$ as a function of Ω, with the slope $d\mathcal{N}/d\Omega = 2\pi R^2/\kappa$.

on increasing Ω, no vortex lines are formed. When the critical velocity Ω_c^* is reached, a large number of vortex lines are suddenly formed simultaneously. During further increase of Ω the system recovers and a characteristic linear slope is retrieved. The linear dependence represents a reproducible periodic process where one vortex at a time is formed at a constant critical velocity, similar to that for B-phase vortices in Fig. 3.5. The extrapolation of the linear dependence back to zero vortex number gives for the onset a lower value Ω_c than the actual initially measured Ω_c^* at the sudden jump. This is an example where the global order parameter texture becomes unstable in the increasing superflow and finally a first order transition occurs in the texture to a new configuration with a lower critical velocity. It is also clear evidence for the fact that the value of the critical velocity depends on the global order parameter texture in the rotating container.

Measurements [61] and calculations [62] of the critical velocity in ^{3}He-A show that the maximum limit for the critical velocity is reached with an ordered texture which mimics one where $\hat{l}$ is homogeneously oriented along the superflow v_s. The minimum velocity, in turn, is close to an order of magnitude smaller and is obtained within a soliton sheet where the spin–orbit coupling is broken and the $\hat{l}$ texture is inhomogeneous.

3.9.4 Formation of Vortex Sheet

The dependence of the critical velocity on the core size explains why vorticity with continuous singularity-free structure is formed, when ^{3}He-A is accelerated to rotation. But is it created in the form of lines or sheets? This has turned out to be an interesting question of general validity. The vortex sheet is formed whenever a vertical dipole-unlocked soliton sheet is present in the container, while rotation is started. Here the critical velocity is the lowest possible, ie. the energy barrier for adding more merons into the sheet vanishes at lower velocity than for any other type of vortex structure.

The critical velocity of the vortex sheet is made up of several contributions. First of all there is the low critical velocity at the connection lines between the sheet and the cylinder wall, where the spin–orbit coupling is broken. Some low superflow velocity is required even here, since at least the attractive interaction of the emerging meron with its image within the wall has to be overcome. A second contribution is a small, but nonzero resistance of the texture in the soliton connection line to reach the instability limit. These small contributions at the connection line are the only ones which are effective initially when rotation is started and the first circulation quanta enter the sheet.

As the sheet grows, a second contribution becomes effective. When a new meron is added, it experiences repulsion from the meron which already resides in the sheet close to the connection line. The repulsion depends on the distance between the merons in the sheet and the sheet's resistance to change its folding and the distribution of circulation in the container. Thus the critical velocity becomes Ω-dependent. At low Ω, when the merons are rare, this contribution is much below the critical velocity for the formation of isolated vortex lines or skyrmions. It was suggested and later experimentally confirmed that the critical velocity follows the qualitative dependence $v_c^*(\Omega) \propto \sqrt{\Omega}$ as shown in Fig. 3.9. (The star as a superscript marks the fact that v_c^* is obtained from the measured value of Ω_c through the relation $v_c^* = \Omega_c/R$ which is strictly valid only for an axially symmetric distribution of vorticity.)

The vortex sheet has unusual dynamic properties which bring it to the preferred state if the rotation drive is rapidly changing in time [63]. It is expected that analogous features can evolve in anisotropic p-wave superconductors with periodic vortex flow in the unpinned regime [64]. The dynamic response of ^{3}He-A as a function of the frequency and amplitude of the external drive is currently under study [65]. The time proven method to create the vortex sheet is to apply sinusoidally modulated rotation in the form $\Omega = \Omega_1 \sin \omega t$, where the period of modulation is of order $2\pi/\omega \sim 10\,\mathrm{s}$ and the amplitude $\Omega_1 \sim \hbar/(2m_3 \xi_\mathrm{D} R) \sim 0.3\,\mathrm{rad/s}$ exceeds that required for breaking the spin–orbit coupling. To grow the equilibrium vortex sheet, one applies the oscillating rotation until the signal from the vortex sheet is observed [22]. The final

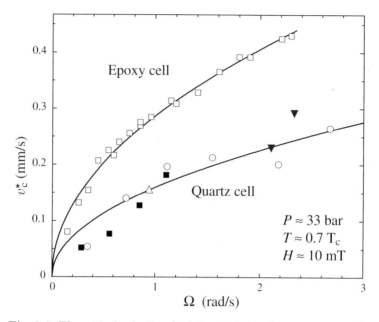

Fig. 3.9. The critical velocity v_c^* of the adiabatically grown vortex sheet as a function of applied rotation Ω: This state consists of a single sheet, which in an axially oriented magnetic field is folded into a double spiral, as shown in Fig. 3.3. The solid curves are fits to the measured data with the dependence $v_c^* \propto \sqrt{\Omega}$. An epoxy container with surface roughness of order 10 μm displays a larger magnitude of v_c^* than a container with fused quartz walls and an order of magnitude smaller roughness ($\lesssim 1$ μm). The difference could arise from a slight pinning of the two connection lines of the sheet along the cylindrical wall which would resist readjustments when new vorticity is added into the sheet during acceleration. The different symbols of data points (quartz cylinder) illustrate the reproducibility of the results from one adiabatically grown vortex sheet to another.

step is to increase Ω from zero to the desired value, using slow acceleration ($d\Omega/dt \lesssim 10^{-3}$ rad/s^2) to ensure adiabatic growth of a single folded sheet.

3.9.5 Vortex Formation in Ionizing Radiation

In the metastable regime of superflow at $v_s < v_c$, vortex formation can be triggered by irradiation with ionizing radiation. Experiments with superfluid ^{3}He-B [7] have shown that quantized vortex lines are formed in the aftermath of a neutron absorption event. According to current belief, vortex formation occurs via the Kibble–Zurek (KZ) mechanism [2,3], which was originally developed to explain the phase transitions of the Early Universe. In this scenario a network of cosmic strings is formed during a rapid non-equilibrium second order phase transition, owing to thermal fluctuations. In ^{3}He the absorption of a thermal neutron causes heating which drives the temperature in a small volume of $\sim 100\,\mu\mathrm{m}$ size above the superfluid transition. Subsequently the heated bubble cools back below T_c with a thermal relaxation time of the order of $1\,\mu\mathrm{s}$. This process forms the necessary conditions for the Kibble–Zurek mechanism within the cooling bubble. It is interesting to note that so far this is the only case of vortex formation in the ^{3}He superfluids which in principle is not confined to the vicinity of the bounding solid walls. In practice the mean free path of thermal neutrons in liquid ^{3}He is only $100\,\mu\mathrm{m}$ and therefore even this process is localized close to the wall.

The real experimental conditions in the neutron irradiation experiment of ^{3}He-B [68] (and also probably in the early Universe) do not coincide with a perfectly homogeneous transition, as is assumed in the KZ scenario: The temperature distribution within the cooling "neutron bubble" is nonuniform, the transition propagates as a phase front between the normal and superfluid phases, and the phase is fixed outside the bubble. These considerations require modifications to the original KZ scenario [69–71] and even raise concerns whether the KZ mechanism is responsible for the defects which are extracted from the neutron bubble and observed in the experiment [72,73]. New measurements demonstrate that a joint defect – the combination of a conventional (mass) vortex and spin vortex – is also formed and is directly observed in the neutron irradiation experiment [74]. This strengthens the importance of the KZ mechanism and places further constraints on the interplay between it and other competing effects.

In superconductors, localized heating can cause the unpinning of vortices from defects in the crystal lattice which can be viewed as a creation of an intermediate vortex ring.

3.10 Vortex Dynamics Without Pinning

As a liquid free of alien impurities, He superfluids do not display bulk pinning. Surface pinning at protuberances on the container wall remains an issue which has been studied in ^{4}He-II [75]. In ^{3}He superfluids vortex core diameters are at least two orders of magnitude larger and pinning correspondingly weaker. Measurements of vortex dynamics have so far not resulted in reliable estimates for pinning parameters. In fact, it has not been settled whether surface pinning plays any observable role in the presence of smooth walls. All indications point in the direction that, even in ^{3}He-B with its smaller vortex core sizes, vortex motion occurs in the limit of weak pinning [21], where collective effects are expected to dominate in pinning. For vortex lines with continuous structure in ^{3}He-A, pinning is expected to be unimportant. The virtual absence of pinning has a number of important consequences:

(i) It allows investigation of vortex dynamics without pinning in the whole temperature range from $T = 0$ to $T = T_\mathrm{c}$. As a result, in ^{3}He-B three topologically different contributions to vortex dynamics have been distinguished from one another, owing to their different temperature behavior [77]. These are (1) the Magnus force with which the flowing superfluid component acts on a vortex line, (2) the Kopnin force which is caused by the spectral flow phenomenon [78] and which acts on the vortex line when it moves with respect to the normal component, and (3) the Iordanskii force which is the analog of the gravitational Aharonov–Bohm effect [79].

(ii) It becomes possible to prepare surfaces with specially prepared pinning sites by micro-fabrication techniques. One can then study vortex dynamics when an array of vortex lines becomes commensurate with a prefabricated lattice of surface pinning sites.

(iii) One can study the trapping and unpinning of circulation from a columnar defect in the order–parameter field – namely a thin wire stretched across the superfluid bath parallel to the rotation axis. This is the Vinen vibrating wire configuration where one quantum of circulation can be trapped around the wire by rotating the container [80]. When rotation is stopped, the trapped circulation can be observed to peel off from the wire as a precessing vortex [81]. With each revolution of the precessing vortex, the phase difference between the ends of the container slips by 2π. This process can be interpreted as a macroscopic manifestation of the ac Josephson effect [82,56]. The corresponding Josephson frequency is remarkably low, approximately 4 mHz.

(iv) Unimpeded by pinning on solid walls, it becomes possible to study interactions of vortex lines with different types of interfaces which can be prepared in the He systems. These interfaces include the free surface of the superfluid bath (i.e. the gas – liquid interface) [83], the superfluid – solid ^{3}He interface [84], the interfaces between ^{3}He and ^{4}He superfluids [21,85], and interfaces between the A and B phases in superfluid ^{3}He [86].

3.11 Conclusion

Although superflow and quantized vortex lines have been the essence of superfluid investigations since the start in the late nineteen forties, nevertheless uniform rotation has not become an important tool for generating vorticity in ^{4}He-II, owing to the uncontrolled release of remanent vorticity. When the construction work of the first rotating nuclear demagnetization cryostat was started in 1978 it was feared that rotation might not prove a useful concept in ^{3}He superfluidity either. However, today we know that the Ω axis is as important in the study of the ^{3}He superfluids as the other experimental parameters H, P, or T, which together control the various order–parameter structures: Measurements on ^{3}He superfluids without access to rotation would be as limited as studies on superconductors without the possibility to turn on the magnetic field!

The central areas so far in superfluid ^{3}He work have been the identification of different topologically stable defect structures in the order–parameter field, the conditions and critical values which control their formation, and phase transitions between different structures as function of the external variables. For these questions the ^{3}He superfluids have been the most ideal system, owing to the large variability in its order–parameter response. Today the largest collection of different theoretically characterized and experimentally identified quantized vortex structures are found in the ^{3}He superfluids. In the coming years ^{3}He work will focus more and more on making use of its ideal text-book-like properties, to employ the ^{3}He liquids as a model system in quantum field theory for the study of such varied questions as the physical vacuum, black hole, or inhomogeneity in the accelerating Universe. For many, questions of this category, working theoretical ^{3}He analogues have already been constructed [87].

References

1. A.J. Leggett: Rev. Mod. Phys. **47**, 331 (1975)
2. M.M. Salomaa, G.E. Volovik: Rev. Mod. Phys. **59**, 533 (1987)
3. V.M. Ruutu, Ü. Parts, M. Krusius: J. Low Temp. Phys. **103**, 331 (1996)
4. J.M. Karimäki, E.V. Thuneberg: Phys. Rev. B **60**, 15290 (1999)
5. M. Rice: Nature **396**, 627 (1998)
6. K. Ishida, H. Mukuda, Y. Kitaoka, K. Asayama, Z. Q. Mao, Y. Mori, Y. Maeno: Nature **396**, 658 (1998)
7. V.M. Ruutu, V.B. Eltsov, A.J. Gill, T.W.B. Kibble, M. Krusius, Yu.G. Makhlin, B. Plaçais, G.E. Volovik, Wen Xu: Nature **382**, 334 (1996); C. Bäuerle, Yu.M. Bunkov, S.N. Fisher, H. Godfrin, G.R. Pickett: Nature **382**, 332 (1996)
8. T.D. Bevan, A.J. Manninen, J.B. Cook, J.R. Hook, H.E. Hall, T. Vachaspati, G.E. Volovik: Nature **386**, 689 (1997)
9. D. Vollhardt, P. Wölfle: *The superfluid phases of He-3* (Taylor & Francis, London, 1990)
10. C.G. Callan, R. Dashen, D.J. Gross: Phys. Lett. **B 66**, 375 (1977)
11. V.A. Fateev, I.V. Frolov, A.S. Schwarz: Nucl. Phys. **B 154**, 1 (1979)
12. J.V. Steel, J.W. Negele: hep-lat/0007006
13. P.J. Hakonen, O.T. Ikkala, S.T. Ikkala: Phys. Rev. Lett. **49**, 1258 (1982)
14. R. Blaauwgeers, V.B. Eltsov, M. Krusius, J.J. Ruohio, R. Schanen, G.E. Volovik: Nature **404**, 471 (2000)
15. S.M. Girvin: Phys. Today **53**, N. 6, 39 (2000)
16. L.I. Burlachkov, N.B. Kopnin: Sov. Phys. JETP **65**, 630 (1987)
17. M.R. Matthews, B.P. Anderson, P.C. Haljan, D.S. Hall, C.E. Wieman, E.A. Cornell: Phys. Rev. Lett. **83**, 2498 (1999)
18. T.-L. Ho: Phys. Rev. Lett. **81**, 742 (1998)
19. T. Isoshima, M. Nakahara, T. Ohmi, K. Machida: Phys. Rev. A **61**, 63610-1 (2000)
20. K.P. Marzlin, W. Zhang, B.C. Sanders: preprint cond-mat/0003273
21. Ü. Parts, J.M. Karimäki, J.H. Koivuniemi, M. Krusius, V.M. Ruutu, E.V. Thuneberg, G.E. Volovik: Phys. Rev. Lett. **75**, 3320 (1995)

22. Ü. Parts, E.V. Thuneberg, G.E. Volovik, J.H. Koivuniemi, V.H. Ruutu, M. Heinilä, J.M. Karimäki, M. Krusius: Phys. Rev. Lett **72**, 3839 (1994); E.V. Thuneberg: Physica B **210**, 287 (1995); M.T. Heinilä, G.E. Volovik, Physica B **210**, 300 (1995); Ü. Parts, V.M. Ruutu, J.H. Koivuniemi, M. Krusius, E.V. Thuneberg, G.E. Volovik: Physica B **210**, 311 (1995)
23. P.G. Saffman: *Vortex dynamics* (Cambridge Univ. Press, Cambridge, UK, 1997; H. Lamb: *Hydrodynamics* (Dover Publ., New York, 1932)
24. L. Onsager: unpublished, see F. London, *Superfluids*, Vol II (Wiley, New York), p 151; see also [59]
25. H. London, in *Report of Intern. Conf. on Fund. Particles and Low Temp.*, Vol II, (Physical Society, London, 1946), p. 48
26. L. Landau, E. Lifshitz: Dokl. Akad. Nauk. **100**, 669 (1955)
27. Ü. Parts, J.H. Koivuniemi, M. Krusius, V.M. Ruutu, E.V. Thuneberg, G.E. Volovik: Pis'ma ZETF **59**, 816 (1994) [JETP Lett. **59**, 851 (1994)]
28. M. Sigrist, T.M. Rice, K. Ueda: Phys. Rev. Lett. **63**, 1727 (1989)
29. E. Shung, T.F. Rosenbaum, M. Sigrist: Phys. Rev. Lett. **80**, 1078 (1998)
30. G.E. Volovik, L.P. Gor'kov: Sov. Phys. JETP **61**, 843 (1985)
31. F. Pardo, F. de la Cruz, P.L. Gammel, E. Bucher, D.J. Bishop, Nature **396**, 348 (1998)
32. G.E. Volovik, V.P. Mineev: JETP Lett. **24**, 561 (1976)
33. V. Geshkenbein, A. Larkin, A. Barone: Phys. Rev. B **36**, 235 (1987)
34. J.R. Kirtley, C.C. Tsuei, M. Rupp, J.Z. Sun, L.S. Yu-Jahnes, A. Gupta, M.B. Ketchen, K.A. Moler, M. Bhushan: Phys. Rev. Lett. **76**, 1336 (1996)
35. U. Leonhardt, G.E. Volovik: Pis'ma ZhETF **72**, 66 (2000) [JETP Lett. **72**, 46 (2000)]
36. A.S. Schwarz: Nucl. Phys. **B 208**, 141 (1982)
37. Z.K. Silagadze: preprint hep-ph/0002255; Mod. Phys. Lett. **A14**, 2321 (1999)
38. H.-Y. Kee, Y.B. Kim, K. Maki: Phys. Rev. B **62**, R9275 (2000)
39. G.E. Volovik: Proc. Natl. Acad. Sc. USA **97**, 2431 (2000)
40. G.E. Volovik, L.P. Gor'kov: JETP Lett. **39**, 674 (1984)
41. M. Sigrist, D.B. Bailey, R.B. Laughlin: Phys. Rev. Lett. **74**, 3249 (1995)
42. G.E. Volovik: Pis'ma ZhETF **58**, 457 (1993) [JETP Lett. **58**, 469 (1993)]
43. O.T. Ikkala, G.E. Volovik, P.J. Hakonen, Yu.M. Bunkov, S.T. Islander, G.A. Kharadze: Pis'ma Zh. Eksp. Teor. Fiz. **35**, 338 (1982) [JETP Lett. **35**,416 (1982)]
44. M.M. Salomaa, G.E. Volovik: Phys. Rev. Lett. **51**, 2040 (1983)
45. P.J. Hakonen, M. Krusius, M.M. Salomaa, J.T. Simola, Yu.M. Bunkov, V.P. Mineev, G.E. Volovik: Phys. Rev. Lett. **51**, 1362 (1983)
46. E.V. Thuneberg: Phys. Rev. Lett. **56**, 359 (1986); M.M. Salomaa, G.E. Volovik: *ibid.* **56**, 363 (1986); E.V. Thuneberg: Phys. Rev. B **33**, 5124 (1986)
47. On $SO(5)$ model see e.g. N.A. Mortensen, H. M. Ronnow, H. Bruus, P. Hedegard: Phys. Rev. B **62**, 8703 (2000) and references therein
48. D.P. Arovas, A. J. Berlinsky, C. Kallin, S.-C. Zhang: Phys. Rev. Lett. **79**, 2871 (1997); S. Alama, A.J. Berlinsky, L. Bronsard, T. Giorgi: Phys. Rev. B **60**, 6901 (1999)
49. T.A. Tokuyasu, D.W. Hess, J.A. Sauls: Phys. Rev. B **41**, 8891 (1990); J.A. Sauls: Adv. Phys. **43**, 113 (1994); N. Lütke-Entrup, R. Blaauwgeers, A. Huxley, S. Kambe, M. Krusius, P. Mathieu, B. Plaçais, Y. Simon: to be published
50. Y. Kondo, J.S. Korhonen, M. Krusius, V.V. Dmitriev, Yu. M. Mukharskiy, E.B. Sonin, G.E. Volovik: Phys. Rev. Lett. **67**, 81 (1991)

51. M.M. Salomaa, G.E. Volovik: J. Low Temp. Phys. **74**, 319 (1989)
52. G.E. Volovik: Pis'ma ZETF **52**, 972 (1990) [JETP Lett. **52** , 358 (1990)]
53. I.A. Luk'yanchuk, M.E. Zhitomirsky: Superconductivity Review **1**, 207 (1995)
54. B.W. Hoogenboom, M. Kugler, B. Revaz, I. Maggio-Aprile, O. Fischer, Ch. Renner: Phys. Rev. B **62**, 9179 (2000)
55. E. Varoquaux, O. Avenel: Physica B **197**, 306 (1994)
56. R.E. Packard: Rev. Mod. Phys. **70**, 641 (1998)
57. J.C. Davis, J. Steinhauer, K. Schwab, Yu.M. Mukharsky, A. Amar, Y. Sasaki, R.E. Packard: Phys. Rev. Lett. **69**, 323 (1992); G.G. Ihas, O. Avenel, R. Aarts, R. Salmelin, E. Varoquaux, Phys. Rev. Lett. **69**, 327 (1992); *ibid.* **70**, 2114 (1993)
58. Ü. Parts, V.M. Ruutu, J.H. Koivuniemi, Yu.M. Bunkov, V.V. Dmitriev, M. Fogelström, M. Huebner, Y. Kondo, N.B. Kopnin, J.S. Korhonen, M. Krusius, O.V. Lounasmaa, P.I. Soininen, G.E. Volovik: Europhys. Lett. **31**, 449 (1995); J. Low Temp. Phys. **107**, 93 (1997)
59. R.J. Donnelly: *Quantized vortices in He-II* (Cambridge Univ. Press, Cambridge, UK, 1991)
60. D. Vollhardt, K. Maki, N. Schopohl: J. Low Temp. Phys. **39**, 79 (1980); H. Kleinert, J. Low Temp. Phys. **39**, 451 (1980)
61. V.M. Ruutu, J. Kopu, M. Krusius, Ü. Parts, B. Plaçais, E.V. Thuneberg, W. Xu: Phys. Rev. Lett. **79**, 5058 (1997)
62. J. Kopu, R. Hänninen, and E.V. Thuneberg, Phys. Rev. B **62**, 12374 (2000)
63. R. Blaauwgeers, V.B. Eltsov, M. Krusius, J.J. Ruohio, R. Schanen: in *Superfluid turbulence and quantized vortex dynamics*, ed. C. Barenghi (Springer Verlag, Berlin, 2001)
64. T. Kita, Phys. Rev. Lett. **83**, 1846 (1999)
65. V.B. Eltsov, R. Blaauwgeers, N.B. Kopnin, M. Krusius, J.J. Ruohio, R. Schanen, E.V. Thuneberg: to be published
66. T.W.B. Kibble, J. Phys. A **9**, 1387 (1976)
67. W.H. Zurek, Nature **317**, 505 (1985)
68. V.B. Eltsov, M. Krusius, G.E. Volovik: in *Prog. Low Temp. Phys.*, ed. W.P. Halperin, Vol. XV (Elsevier Science Publ., Amsterdam, 2001), preprint cond-mat/9809125
69. T.W.B. Kibble and G.E. Volovik, JETP Lett. **65**, 102 (1997)
70. J. Dziarmaga, P. Laguna, W.H. Zurek: Phys. Rev. Lett. **82**, 4749 (1999)
71. N.B. Kopnin, E.V. Thuneberg: Phys. Rev. Lett. **83**, 116 (1999)
72. V.M. Ruutu, V.B. Eltsov, M. Krusius, Yu.G. Makhlin, B. Plaçais, G.E. Volovik: Phys. Rev. Lett. **80**, 1465 (1998)
73. I.S. Aranson, N.B. Kopnin, V.M. Vinokur: Phys. Rev. Lett. **83**, 2600 (1999)
74. V.B. Eltsov, T.W.B. Kibble, M. Krusius, V.M.H. Ruutu, G.E. Volovik: Phys. Rev. Lett. **85**, 4739 (2000).
75. S.G. Hedge, W.I. Glaberson: Phys. Rev. Lett. **45**, 190 (1980); W.I. Glaberson, R.J. Donnelly: in *Prog. Low Temp. Phys.*, ed. D.F. Brewer, Vol. IX, p. 81 (Elsevier Science Publ., Amsterdam, 1986)
76. M. Krusius, J.S. Korhonen, Y. Kondo, E.B. Sonin: Phys. Rev. B **47**, 15113 (1993)
77. T.D. Bevan, A.J. Manninen, J.B. Cook, H. Alles, J.R. Hook, H.E. Hall: J. Low Temp. Phys. **109**, 423 (1997)
78. N.B. Kopnin, G.E. Volovik, Ü. Parts: Europhys. Lett. **32**, 651 (1995)

79. M. Stone, Phys. Rev. **B 61**, 11780 – 11786 (2000)
80. W.F. Vinen, Proc. Royal Soc. London **A 260**, 218 (1961)
81. R.J. Zieve, Yu.M. Mukharsky, J.D. Close, J.C. Davis, R.E. Packard: J. Low Temp. Phys. **91**, 315 (1991); Phys. Rev. Lett. **68**, 1327 (1992)
82. T.Sh. Misirpashaev, G.E. Volovik, Pis'ma Zh. Exp. Teor. Fiz. **56**, 40 (1992) [JETP Lett. **56**, 41 (1992)]
83. E.B. Sonin, A.J. Manninen: Phys. Rev. Lett. **70**, 2585 (1993)
84. A.Ya. Parshin: Physica B **210**, 383 (1995)
85. Y. Kondo, J.S. Korhonen, M. Krusius: Physica B **169**, 533 (1991)
86. Ü. Parts, Y. Kondo, J.S. Korhonen, M. Krusius, E.V. Thuneberg: Phys. Rev. Lett. **71**, 2951 (1993)
87. G.E. Volovik: Phys. Rep., to be published (2001); preprint gr-qc/0005091

4 Nucleation of Vortices in Superfluid ^{3}He-B by Rapid Thermal Quench

Igor S. Aranson, Nikolai B. Kopnin, and Valerii M. Vinokur

Summary. We show by numerical and analytical solution of the time-dependent Ginzburg–Landau equation (TDGLE) that vortex nucleation in superfluid ^{3}He-B by rapid thermal quench in the presence of superflow is dominated by a transverse instability of the moving normal-superfluid interface. Exact expressions for the instability threshold as a function of supercurrent density and the front velocity are found. The dynamics of vortex annihilation and the generalization to the case of complex relaxation rate in the TDGLE are considered.

4.1 Introduction

Formation of topological defects under a rapid quench is a fundamental problem of contemporary physics [11]. For homogeneous cooling a fluctuation-dominated formation mechanism has been suggested by Kibble and Zurek (KZ)[2,3]. Typically, cooling is associated with an inhomogeneous temperature distribution accompanied by a phase separating interface which moves through the system as temperature decreases. A generalization of the KZ scenario for inhomogeneous phase transitions in superfluids was later proposed by Kibble and Volovik [4].

Superfluid ^{3}He offers a unique "testing ground" for rapid phase transitions [5]. Recent experiments, where a rotating superfluid ^{3}He was locally heated well above the critical temperature by absorption of neutrons [6], revealed vortex formation under a rapid second–order phase transition. The TDGL analysis was applied to study a propagating normal–superfluid interface under inhomogeneous cooling [34] and the formation of a large supercooled region was confirmed. The fluctuation–dominated mechanism may thus be responsible for creation of initial vortex loops. It is commonly accepted that these initial vortex loops are further inflated by the superflow and give rise to a macroscopic number of large vortex lines filling the bulk superfluid.

In this chapter we study the entire process of the vortex formation in the presence of a superflow using TDGL dynamics. We take into account the temperature evolution due to thermal diffusion. A preliminary account of some results has been published in [8]. We find analytically and confirmed by numerical simulations that the normal–superfluid interface becomes unstable with respect to transverse undulations in the presence of a superflow. These undulations quickly transform into large primary vortex loops which then separate themselves from the interface. Simultaneously, a large number of small secondary vortex/antivortex nuclei are created in the supercooled region by fluctuations, resembling conventional KZ mechanism. The primary vortex loops screen out the superflow in the inner region causing the annihilation of the secondary vortex/antivortex nuclei. The number of the *survived* secondary vortex loops is thus much smaller then that anticipated from the KZ conjecture. In addition to the previously published results in [8] we include the results on vortex annihilation after the quench and the generalization to the case of a complex relaxation rate in the TDGLE.

4.2 Model

We describe the dynamics of vortex nucleation in superfluid ^{3}He-B by the TDGLE for a scalar order parameter ψ:

$$\partial_t \psi = \Delta \psi + (1 - f(\mathbf{r}, t))\psi - |\psi|^2 \psi + \zeta(\mathbf{r}, t). \tag{4.1}$$

Here Δ is the three-dimensional (3D) Laplace operator, and f describes local temperature evolution. Distances and time are measured in units of the coherence length $\xi(T_\infty)$ and the characteristic time $\tau_{\mathrm{GL}}(T_\infty)$, respectively. These quantities are taken at temperature T_∞ far from the heated bubble. Close to T_c, the local temperature is controlled by normal-state heat diffusion and evolves as $f(\mathbf{r}, t) = E_0 \exp(-r^2/\sigma t) t^{-3/2}$, where σ is the normalized diffusion coefficient. E_0 determines the initial temperature of the hot bubble T^* and is proportional to the deposited energy $\mathcal{E}_0$ such that $E_0 = \mathcal{E}_0 / \left[C(T_c - T_\infty) \xi^3(T_\infty)(\pi\sigma)^{3/2} \right]$ where C is the heat capacity. Since the deposited energy is large compared to the characteristic superfluid energy, we assume $E_0 \gg 1$. Representative value of $E_0 \sim 30-50$. The time at which the temperature in the center of the hot bubble drops down to T_c is $t_{\max} = E_0^{2/3}$. The Langevin force ζ with the correlator $\langle \zeta \zeta' \rangle = 2T_f \delta(\mathbf{r} - \mathbf{r}')\delta(t - t')$ describes thermal fluctuations with a strength T_f at T_c (in the following we neglect dependence of T_f on local temperature).

The microscopic values of the Ginzburg–Landau parameters are $\tau(T_\infty) = \pi\hbar/8(T_c - T_\infty)$, the coherence length is $\xi(T) = \xi_0[(7\zeta(3))/(12(1 - T_\infty/T_c))]^{-1/2}$, where $\xi_0 = \hbar v_F/2\pi T_c$. The diffusion constant $\sigma = [48/7\zeta(3)][D\tau_0/\xi_0^2] \sim \ell/\xi_0$ where ℓ is the quasiparticle mean free path. In ^{3}He, σ is large because $\ell \gg \xi_0$. Effective noise strength is $T_f = (3/\pi^2)[3/7\zeta(3)]^{1/2} \mathrm{Gi}^{-1}[1 - (T/T_c)]^{-1/2}$ where $\mathrm{Gi} = \nu(0)\xi_0^3 T_c \sim 10^4$ is the Ginzburg number, $\nu(0)$ is normal density of states.

Approximation of the ^{3}He-B order parameter structure by a complex scalar ignores the complexity of the nine-component ^{3}He specific order parameter [3,4]. We expect that not too close to the $A - B$ transition line the dynamics of vortex nucleation in the ^{3}He-B phase is described fairly well by a scalar order parameter, except the very vicinity of the vortex cores where small amount of ^{3}He phases with different symmetries may exist. Far away from the transition line these phases are confined at the vortex cores.

Equation (4.1) formally coincides with the TDGLE for superfluid ^{4}He near the λ-point [9] and type-II superconductors near T_c. In contrast to ^{3}He, the corresponding equation for ^{4}He and superconductors has much smaller coherence length ξ and relaxation time τ. As a result, the relative quench rate for the parameters of [6] in ^{4}He is 19 orders of magnitude slower then in ^{3}He, and no vortices can be created [10]. Experiments with thin high-temperature superconducting films find also no evidence of vortex nucleation by rapid cooling [11].

4.3 Results of Simulations

We solved (4.1) by the implicit Crank–Nicholson method. The integration domain was equal to 150^3 units of (4.1) with 200^3 mesh points. The boundary conditions were taken as $\partial_z \psi = ik\psi$ with a constant k at the top and the bottom of the integration domain. This implies a uniform superflow $j_s = k|\psi_0|^2$ along the z-axis far away from the temperature bubble. Since the

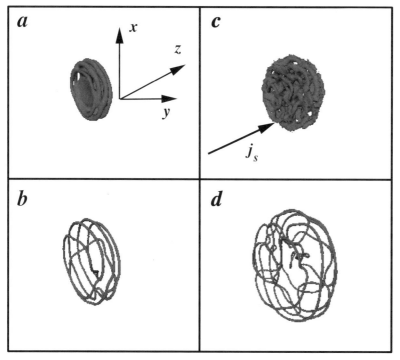

Fig. 4.1. 3D isosurface of $|\psi| = 0.4$ for $\sigma = 400$, $E_0 = 30$ and $k = 0.5$. (**a, b**) $T_f = 0$. Images are taken at times $t = 36, 80$. (**c, d**), $T_f = 0.002$, $t = 24, 80$.

equilibrium value of the order parameter ψ_0 is related to k as $|\psi_0|^2 = 1 - k^2$, it leads to $j_s = k(1 - k^2)$.

Some simulations results are shown in Fig. 4.1. One sees (Fig. 4.1a,b) that without fluctuations (numerical noise only) the vortex rings nucleate upon the passage of the thermal front. Not all of the rings survive: the small ones collapse and only the big ones grow. Although the vortex lines are centered around the point of the quench, they exhibit a certain degree of entanglement. After a long transient period, most of the vortex rings reconnect.

We find that fluctuations have a strong effect at early stages: the vortices nucleate not only at the normal–superfluid interface, but also in the bulk of the supercooled region (Fig. 4.1c). However, later on, small vortex rings in the interior collapse and only larger rings (primary vortices) survive and expand (Fig. 4.1d).

To elucidate the details of nucleation we considered an axi-symmetric version of (4.1) (depending on only r and z coordinates, $\Delta = \partial_r^2 + 1/r\partial_r + \partial_z^2$). The domain was 500^2 with 1000^2 mesh points. We have found that wi-

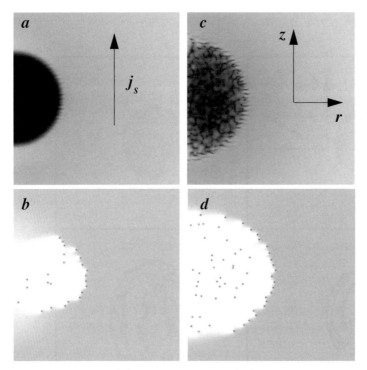

Fig. 4.2. Images of $|\psi|$ for axi–symmetric (4.1) for $\sigma = 5000$, $E_0 = 50$ and $k = 0.5$, black corresponds to $|\psi| = 0$ and white to $|\psi| = 1$; grey ($|\psi| \sim 0.8$) indicates suppression of order parameter by current. Current is along the z-axis. Vortices are seen as black dots. (**a, b**) $T_f = 0$, images are shown for $t = 40, 200$; (**c, d**) $T_f = 0.002$, for $t = 30, 200$

thout thermal fluctuations the vortices nucleate at the front of the normal-superfluid interface (black/grey border in Fig. 4.2a) analogous to the 3D case. The initial instability is seen as a corrugation of the interface. The interface propagates towards the center, leaving the vortices behind. As thermal fluctuations are turned on, the vortex rings also nucleate in the bulk of the supercooled region (black spot in Fig. 4.2c) resulting in the creation of the secondary vortex/antivortex pairs. We have found that the "primary" vortices prevent the supercurrent from penetrating into the region filled with the secondary vortices. One sees that the primary vortices encircle the brighter spots in Fig. 4.2 b,d indicating a larger value of the order parameter and thus a smaller value of the supercurrent. As a result the secondary vortices either annihilate with antivortices due to their mutual attraction or collapse due to the absence of the inflating superflow.

Shown in Fig. 4.3 is the number of vortex rings N vs quench parameter σ and applied current k. At small k N shows threshold behavior while becoming almost linear for larger k values. The deviations from a linear law appear close to the value of the critical current $k_c = 1/\sqrt{3}$ for a homogeneous system, when the vortices nucleate spontaneously everywhere in the bulk.

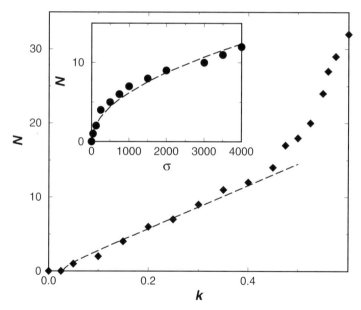

Fig. 4.3. Number of vortex rings N vs k for $E_0 = 50$ and $\sigma = 5000$. Inset: N vs σ for $k = 0.4$ and $E_0 = 50$. Dashed lines show fitting to prediction (4.29).

4.4 Dynamics of Vortex/Antivortex Annihilation

We performed detailed numerical simulations in order to elaborate the statistics of the vortex annihilation and results are shown in Figs. 4.4 and 4.5. Simulations were performed for the quasi-three-dimensional geometry (assuming axial symmetry of the vortex configuration) and the pure two-dimensional geometry.

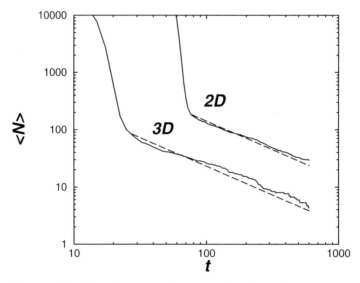

Fig. 4.4. Number of vortices N vs time for $k = 0$ (zero supercurrent). Parameters of simulations: $E_0 = 50, \sigma = 10000$, domain size 1000×1000 for quasi-three-dimensional sample (line 3D), and $E_0 = 50, \sigma = 6000$, domain size 800×800 for the two-dimensional sample. Each line is averaged over 5 independent realization of thermal noise. Dashed lines show the prediction from mean-filed theory $N \sim 1/t$.

For the zero applied current j_s, as one sees from Fig. 4.4, the behavior for both 3D and 2D situations is similar: fast initial relaxation and then slow decay consistent with the dependence $N \sim 1/t$ (in agreement with [12] for the homogeneous quench). This result consistent with the mean-field theory of annihilation based on the assumption that the annihilation rate of vortices is proportional to the local density of antivortices: $dN^+/dt \sim -N^+N^-$ (Vinen's equation [13]). Assuming that $N^+ = N^-$ one readily obtains that $N^\pm \sim 1/t$. Note that [14,15] which claim that long-range interaction between the vortices will result in substantial deviation from the mean-field theory.

If the flow is applied ($k \neq 0$), one has in general $N^+ \neq N^-$. Assuming that $N^\pm$ are uniformly distributed, one obtains from the mean-filed theory exponential relaxation: $N^+ \sim \exp(-\alpha t) + B$, where B is the final number of vortices and α is the relaxation rate $\alpha \sim N^+ - N^-$. This results is in clear

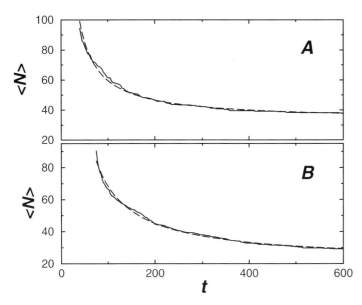

Fig. 4.5. Number of vortices N vs time $k = 0.4$ for the three-dimensional sample (**a**) and $k = 0.3$ for the two-dimensional one (**b**). Other parameters the same as in Fig. 4.4. Dashed lines show the dependence $N^+ = A/t + B$.

disagreement with the results of numerical simulations shown in Fig. 4.5: the relaxation law is the same as in the previous case $k = 0$, with the only difference that N^+ approaches equilibrium value algebraically, $N^+ = A/t + B$. Provided that: (a) the "excess vortices" due to applied current are always expelled to the periphery, see from Figs. 4.1, 4.2; (b) vortices in the bulk do not interact with the supercurrent due to screening by the peripheral vortices, one recovers mean field behavior $N^+ \sim 1/t$ for the bulk vortices.

4.5 Instability of Normal–Superfluid Interface

Our simulations demonstrate that the nucleation happens predominantly at the relatively narrow region at the NS interface where the temperature is close to T_c. Following [3,34,8], we expand the local temperature $1-f$ near T_c. Let us put $x = r_c - r$ where r_c is the radius of the surface at which $T = T_c$ or $f = 1$, i.e., $r_c^2 = (3/2)\sigma t \log(t_{max}/t)$. A positive x is directed towards the hot region. We write $1 - f(r,t) \approx -\alpha(x - vt)$ where $\alpha = -[df/dr]_{f=1} = 2r_c/\sigma t$ is the local temperature gradient and $v = (\alpha \tau_Q)^{-1}$ is the front velocity defined through the quench rate $\tau_Q^{-1} = [\partial f/\partial t]_{f=1}$. We have $v = (3\sigma t - 2r_c^2)/4r_c t$. The front starts to move towards the center at $t > t_* = t_{max}/e$ and disappears at $t = t_{max}$ when the temperature drops below T_c. The front velocity accelerates as the hot bubble collapses. Since the front radius r_c is large compared to the coherence length, it can be considered flat. The coordinates y, z are parallel to the front. In a two-dimensional problem the solution is assumed independent of y. We transform to the frame moving with the velocity v and perform the scaling:

$$\tilde{x} = vx, \ \tilde{z} = vz, \ \tilde{t} = tv^2, \tag{4.2}$$
$$\tilde{\psi} = \psi/v, \ u = v^3/\alpha. \tag{4.3}$$

We drop tildes in what follows. The parameter $u \sim (\sigma^2/t_{\max})/\log^2(t_{\max}/t)$ is of the order 1 in the experiment [6] at the initial time but grows rapidly as the hot bubble shrinks. In our numerical simulations, $u \gg 1$. Equation (4.1) takes the form if we neglect all the term related to the curvature of the hot bubble, i.e. $\Delta = \delta_x^2 + \delta_z^2$:

$$\partial_t \psi = \Delta \psi + \partial_x \psi - \frac{x}{u}\psi - |\psi|^2 \psi. \tag{4.4}$$

4.5.1 Stationary Solutions

Equation (4.4) allows for a family of stationary current-carrying solution

$$\psi = F(x)\exp(ikz) \tag{4.5}$$

with amplitude F satisfying the equation

$$\partial_x^2 F + \partial_x F - \left(\frac{x}{u} + k^2\right) F - F^3 = 0. \tag{4.6}$$

$u \gg 1$ expansion of the stationary solution was studied in [34] and [16].

4.5.2 Linearized Equations

To examine the transverse stability of stationary solution to (4.6) we represent the general solution to 4.4 in the form $\psi = (F + w(x, z, t))\exp(ikz)$, where w is the perturbation. For the function w we derive:

$$\partial_t w = \partial_x^2 w + \partial_z^2 w + 2ik\partial_z w + \partial_x w - \left(\frac{x}{u} + k^2\right) w \tag{4.7}$$
$$- F^2(2w + w^*) - F(2|w|^2 + w^2) - |w|^2 w \ .$$

Separating real and imaginary parts of $w = a + ib$ one has

$$\partial_t a = \partial_x^2 a + \partial_z^2 a - 2k\partial_z b + \partial_x a - \left(\frac{x}{u} + k^2\right) a \tag{4.8}$$
$$- 3F^2 a - F(3a^2 - b^2) - (a^2 + b^2)a$$
$$\partial_t b = \partial_x^2 b + \partial_z^2 b + 2k\partial_z a + \partial_x b - \left(\frac{x}{u} + k^2\right) b \tag{4.9}$$
$$- F^2 b - 2Fab - (a^2 + b^2)b \ .$$

Dropping nonlinear terms and representing the solution to (4.8,4.9) in the form

$$\begin{pmatrix} a \\ b \end{pmatrix} = \begin{pmatrix} A \\ iB \end{pmatrix} \exp(\lambda(q)t + iqz) \tag{4.10}$$

where q is the transverse undulations wavenumber and λ is the growth rate, we obtain ($\chi = kq$, $\Lambda = \lambda + q^2$)

$$\Lambda A + 2\chi B = \partial_x^2 A + \partial_x A - (x/u + k^2) A - 3F^2 A$$
$$\Lambda B + 2\chi A = \partial_x^2 B + \partial_x B - (x/u + k^2) B - F^2 B \tag{4.11}$$

solution to (4.11) can be obtained numerically for arbitrary u and χ and analytically in long-wavelength limit $\chi \ll 1$ or large velocity limit $u \gg 1$.

4.5.3 Long-Wavelength Limit

The eigenvalue Λ for $\chi \to 0$ can be found as an expansion in χ:

$$\Lambda = \chi \Lambda_1 + \chi^2 \Lambda_2 + \dots$$
$$A = A_0 + \chi A_1 + \chi^2 A_2 + \dots \tag{4.12}$$
$$B = B_0 + \chi B_1 + \chi^2 B_2 + \dots$$

In zeroth order in χ second (4.11) coincides with the equation for the stationary solution (4.6), one has $A_0 = 0$, $B_0 = F$. In the first order $B_1 = 0$ and

$$\partial_x^2 A_1 + \partial_x A_1 - (x/u + k^2) A_1 - 3F^2 A_1 = 2F. \tag{4.13}$$

The solution $A_1 = 2u\partial_x F$ is obtained by differentiating (4.6). In the second order to (4.11) one has

$$\partial_x^2 B_2 + \partial_x B_2 - (x/u + k^2 + F^2)B_2 = 4u\partial_x F + \Lambda_2 F . \tag{4.14}$$

A zero mode of (4.14) is F. Equation (4.14) is not self-adjoint, and, therefore, the adjoint zero mode $B^\dagger$ does not coincide with F. The corresponding adjoint operator is of the form:

$$\partial_x^2 B^\dagger - \partial_x B^\dagger - (x/u + k^2 + F^2)B^\dagger = 0. \tag{4.15}$$

One checks that function $B^\dagger = F \exp(x)$ satisfies (4.15). Equation (4.14) has a solution if the solvability condition with respect to the zero mode is fulfilled

$$\int_{-\infty}^{\infty} dx F e^x (4u\partial_x F + \Lambda_2 F) = 0 . \tag{4.16}$$

After integration we obtain $\Lambda_2 = 2u$. Returning to the original notations, we obtain the *exact* result

$$\lambda = q^2(2uk^2 - 1) + O(q^4) . \tag{4.17}$$

The instability occurs above the threshold value $k_v^2 = (2u)^{-1}$ or $k_v^2 \sim \alpha^{2/3}/u^{1/3} \sim \sigma^{-1}\log(t_{\max}/t)$ in the Ginzburg–Landau units. Since it is much smaller than the bulk critical value $k_c = 1/\sqrt{3}$, it can be exceeded for a very small superflow. Equation (4.17) indicates the fact of the instability but does not provide the optimal wavenumber q_c for the most unstable perturbation.

4.5.4 Large u Limit

The eigenvalue Λ vs χ can be derived in the limit of $u \to \infty$ (fast quench rate). In this limit one assumes $\Lambda \sim \chi \sim 1/u$. Substituting $x = \bar{x} - u\gamma - uk^2$, where γ determines the position of the interface. We treat the terms containing Λ, χ and $\bar{x}/u$ as perturbations. For $\varepsilon = 1/u \to 0$ (4.6) is reduced to

$$\partial_{\bar{x}}^2 F + \partial_{\bar{x}} F + \gamma F - F^3 = 0. \tag{4.18}$$

Equation (4.18) possesses a front solution $F(\bar{x} - x_0)$ connecting equilibria $F = \pm\sqrt{\gamma}$ and $F = 0$. Here x_0 is arbitrary constant determining the position of the front fixed by the corresponding solvability condition. Equation (4.18) approximates F obtained from (4.6) fairly in the range of intermediate values of x, but it fails for $|\bar{x}| \gg 1$. For large negative $-\bar{x} \gg u$, the solution should be replaced by its final asymptotics $F = \sqrt{\gamma - \bar{x}/u}$. For large positive x, solution of (4.18) should be matched with true asymptotics of (4.6) given by the expression $F \sim \exp(-\bar{x}/2)\mathrm{Ai}(u^{2/3}(\gamma - 1/4 - \bar{x}/u))$. The matching is possible for $\gamma \to 1/4$ [34]. The matching fixes the value of $\gamma = 1/4 + O(u^{-2/3})$ [16]. The contribution from the tail of F is negligible since for $u \gg 1$ the function $F(x)$ is already small. For $u \to \infty$ (4.11) assume the form:

$$\Lambda A + 2\chi B + \varepsilon \bar{x} A = \partial_{\bar{x}}^2 A + \partial_{\bar{x}} A + \gamma A - 3F^2 A$$
$$\Lambda B + 2\chi A + \varepsilon \bar{x} B = \partial_{\bar{x}}^2 B + \partial_{\bar{x}} B + \gamma B - F^2 B \ . \tag{4.19}$$

For $\varepsilon = 0$ (4.19) have a zero mode: $(A, B) = (0, F(x - x_0))$, similar to (4.11). In addition, (4.19) has an extra zero mode $(A, B) = (F_x(x - x_0), 0)$, which manifest the translation invariance for $\varepsilon = 0$. For any $\varepsilon \neq 0$ the translation invariance is broken down by the perturbation $\sim x/uA, x/uB$ in the l.h.s. of (4.19), and corresponding solvability condition will specify the value of x_0.

In contrast to the case considered in the previous section, the solvability conditions must be fulfilled simultaneously for both zero modes of (4.19). Thus, representing the general zero order solution of (4.19) in the form

$$(A, B) = (a_0 F_x(x - x_0), b_0 F(x - x_0)) \tag{4.20}$$

where a_0, b_0 are arbitrary constants, and performing the integrations with corresponding zero modes, one obtains characteristic equation for Λ:

$$\Lambda^2 + \frac{1}{u}c_1 \Lambda - 4c_2\chi^2 + \frac{d}{u^2} = 0, \tag{4.21}$$

where the coefficients $c_{1,2}, d$ are given in the forms of integrals of F with the corresponding zero modes in the interval $-\infty < \bar{x} < \infty$:

$$c_1 = \frac{1}{u}(i_5/i_2 + i_4/i_1), \quad c_2 = \frac{i_3^2}{i_1 i_2}, \quad d = \frac{i_4 i_5}{u^2 i_1 i_2} \quad (4.22)$$

where

$$i_1 = \int_{-\infty}^{\infty} F^2 e^x dx, \quad i_2 = \int_{-\infty}^{\infty} F_x^2 e^x dx, \quad i_3 = \int_{-\infty}^{\infty} F F_x e^x dx = -1/2 i_1$$

$$i_4 = \int_{-\infty}^{\infty} (x-x_0) F^2 e^x dx, \quad i_5 = \int_{-\infty}^{\infty} (x-x_0) F_x^2 e^x dx. \quad (4.23)$$

For $\bar{x} \to \infty$ the asymptotic behavior of the function F is incorrect, and one has to introduce a cutoff distance $x_c \gg 1$ in order to apply correct asymptotic representation $F \sim \exp(\bar{x})\mathrm{Ai}(\bar{x}/u^{1/3} + \mathrm{const})$ for $\bar{x} > x_c$, see for detail [34]. However, as it was already mentioned, the crossover to this asymptotic behavior occurs when F is already very small [16]. Thus, the contributions to the integrals (4.23) from the asymptotic tail for $x \gg x_c$ can be neglected.

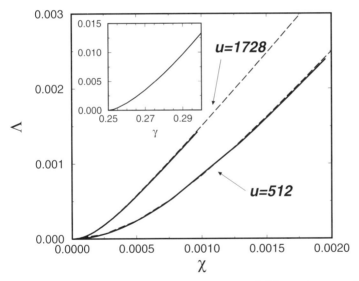

Fig. 4.6. Λ vs χ for $u = 512$ and $u = 1728$, solid lines show results of numerical solution of (4.11), dashed lines represent analytical solution (4.26). Inset: the ratio of integrals $\int_{-\infty}^{\infty} F^4 e^x dx / \int_{-\infty}^{\infty} F^2 e^x dx$ vs. γ.

The constant x_0 is determined from the constraint that (4.19), as original (4.11) always have an eigenvalue $\Lambda = 0$ for $\chi = 0$. It implies that $i_4 = 0$ and fixes the value of x_0 by the condition

$$\int_{-\infty}^{\infty} F_x^2 e^x dx = x_0 i_1. \qquad (4.24)$$

Integrating (4.23) by parts and using that $\int_{-\infty}^{\infty} F^4 e^x dx / \int_{-\infty}^{\infty} F^2 e^x dx \to 0$ for $\gamma \to 1/4$, see Fig. 4.6, one arrives after some algebra [16]:

$$c_1 = \frac{2}{u}, \quad c_2 = 1. \qquad (4.25)$$

Substitution of (4.25) into (4.21) results in

$$\Lambda = \pm\sqrt{1/u^2 + 4\chi^2} - 1/u. \qquad (4.26)$$

Returning to original definitions, we have an explicit expression for the largest eigenvalue of the transverse instability

$$\lambda = \sqrt{1/u^2 + 4k^2 q^2} - 1/u - q^2. \qquad (4.27)$$

Numerical solution of (4.11) demonstrates an excellent agreement with the theoretical expression (4.27), see Fig. 4.6.

4.5.5 Estimate for the Number of Vortices

The number of nucleated vortices is determined by the wavenumber of the most unstable mode. In the case of thermal quench, the normal/superfluid front velocity $u \to \infty$ as time increases, therefore the limit of large u applies. For $u \to \infty$ one has from (4.27) $\lambda = 2|kq| - q^2$. The maximum growth rate is achieved at $q = k$ and is simply k^2. The growth of perturbations near the interface is described by the Fourier integral

$$w \sim \int dq S(q) \exp[\lambda(q)t + iqz], \qquad (4.28)$$

where $w(x, z, t)$ is the perturbation to the interface solution, $S(q)$ is the spectrum of initial perturbation. For the original problem described by (4.1) the velocity of the interface, and, therefore, parameter u, k are certain functions of time, see (4.2). Therefore, instead of expression $\lambda(q)t$ in (4.28) one has to use an integral $\int_0^t \lambda(q(t'))dt'$, valid in the WKB approximation. In the large u limit this time dependence is canceled out trivially and one recovers (4.28).

Taking into account that the thermal noise provides initial perturbations for the interface instability w, and using saddle-point approximation for the integral in (4.28) for $t \gg 1$, one derives $\langle|w|\rangle \sim \sqrt{T_f} \text{Re} \exp[k^2 t + ikz]$. The number of vortices is estimated as $N = r_0 k$, where r_0 is the radius of the front where the perturbations $\langle|w|\rangle$ become of the order of one. The time interval t_0 corresponding to $\langle |w| \rangle = 1$ is $t_0 \sim k^{-2} \log(T_f^{-1})$. Vortices have no time to grow if $t_0 \to t_{\max}$. In this limit $r_0^2 \sim \sigma(t_{\max} - t_0)$ and one arrives at

$$N \sim k r_0 \sim \sqrt{\sigma} E_0^{1/3} \sqrt{(v_s/v_c)^2 - \beta^2 \log(T_f^{-1})/E_0^{2/3}} \qquad (4.29)$$

where $\beta = $ const, while v_s and v_c are the imposed and critical GL superflow velocity, respectively. This estimate is in agreement with the results of simulations, see Fig. 4.3. Equation (4.29) exhibits a slow logarithmic dependence of the number of vortices at the interface on the level of fluctuations and agrees with the results presented in Fig. 4.3. For the experimental values of the parameters our analysis results in about 10 surviving vortices per heating event.

4.6 Generalization

The TDGLE is a reasonable model for ^{3}H-B very close to T_c. Moving away from T_c additional terms must be included, and in general there is no universal description. However, on the qualitative level not too far away from T_c the TDGLE can be generalized by allowing complex relaxation rate of the order parameter:

$$(1 + i\eta)\partial_t \psi = \Delta \psi + (1 - f(\boldsymbol{r}, t))\psi - |\psi|^2 \psi + \zeta(\boldsymbol{r}, t). \qquad (4.30)$$

Close to T_c the parameter $\eta \to 0$, and $\eta \to \infty$ for $T \to 0$. Thus, (4.30) on the qualitative level combines TDGLE at $T = T_c$ and dissipationless Gross–Pitaevskii equation at $T = 0$ [17].

We show that the complex relaxation does not change qualitatively our results on the interface instability. Applying similar transformation as in the Sect. 4.5, we obtain instead of (4.4) the following equation

$$(1 + i\eta)\partial_t \psi = \Delta \psi + (1 + i\eta)\partial_x \psi - \frac{x}{u}\psi - |\psi|^2 \psi. \qquad (4.31)$$

For $\eta \neq 0$ we have to modify the ansatz for stationary solution (4.5) to

$$\psi = F(x) \exp(ikz + i\omega t + i\kappa x) \qquad (4.32)$$

where ω and κ are the frequency and longitudinal wavenumber which will be defined later. One sees that new feature for any $\eta \neq 0$ is the emission of oblique waves from NS interface. Substituting (4.32) into (4.31) one obtains

$$i\omega(1 + i\eta)F = \partial_x^2 F + 2i\kappa \partial_x F \qquad (4.33)$$
$$+ (1 + i\eta)(\partial_x F + i\kappa F) - \left(\frac{x}{u} + k^2 + \kappa^2\right) F - F^3.$$

Fixing $\kappa = -\eta/2$, $\omega = \kappa = -\eta/2$ one derives from (4.33)

$$\partial_x^2 F + \partial_x F - \left(\frac{x}{u} + k^2 + \kappa^2\right) F - F^3 = 0. \qquad (4.34)$$

Thus, one sees for any η (4.34) coincides with (4.6) with the k^2 replaced by $k^2 + \kappa^2$, which can be excluded by the shift of x. Perturbative solution to (4.31) is sought in the form $\psi = (F + w)\exp(ikz + i\omega t + i\kappa x)$. Substituting this ansatz into (4.31) one obtains in the linear order in w:

$$(1 + i\eta)\partial_t w = \Delta w + 2ik\partial_z w + \partial_x w$$
$$- \left(\frac{x}{u} + k^2 + \kappa^2\right) w - F^2(2w + w^*) . \tag{4.35}$$

Separating real and imaginary parts of $w = a + ib$ and representing the solution in the form (4.10), one derives from (4.35):

$$\Lambda A + 2\chi' B = \partial_x^2 A + \partial_x A - \left(\frac{x}{u} + k^2 + \kappa^2\right) A - 3F^2 A$$
$$\Lambda B + 2\chi' A = \partial_x^2 B + \partial_x B - \left(\frac{x}{u} + k^2 + \kappa^2\right) B - F^2 B \tag{4.36}$$

where $\Lambda = \lambda + q^2$ and $\chi' = kq + i\lambda\eta/2$. One sees that the structure of (4.36) is *identical* to (4.11). Therefore, all the results on linear stability can be easily carried over to the case of arbitrary η! In particular, similar to (4.17) for $\eta = 0$ we obtain for the instability threshold $\Lambda = 2u(\chi')^2$ for $\eta \neq 0$, which gives (implicit) condition for λ vs q:

$$\lambda = 2u(kq + i\lambda\eta/2)^2 - q^2 . \tag{4.37}$$

It is easy to check that the threshold for instability is given by the condition $2k^2u = 1$ irrespectively of η, and the growthrate λ near the threshold $2k^2u - 1 \to 0$ is of the form (compare with (4.17))

$$\lambda \approx q^2 \frac{2uk^2 - 1}{1 + i2u\eta kq} . \tag{4.38}$$

Full numerical solution of (4.30) shows close similarity with (4.30), see Fig. 4.7). However, in contrast to the case $\eta = 0$ for nonzero η the shape of the resulting vortex configuration is asymmetric in the direction of applied current. This is due to oblique motion of the vortices with respect to the current direction for $\eta \neq 0$ (see [18] for detail).

4.7 Conclusion

We have found that the rapid normal–superfluid transition in the presence of superflow is dominated by a transverse instability of the normal/superfluid interface propagating from the bulk into the normal region. Our numerical results indicate that the dynamics of vortex/antivortex annihilation in the bulk obeys simple power low $N \sim 1/t$ irrespectively of the dimensionality of the space. We derived analytically exact expressions for the instability threshold and for the growth rate of transverse perturbations in the limit of

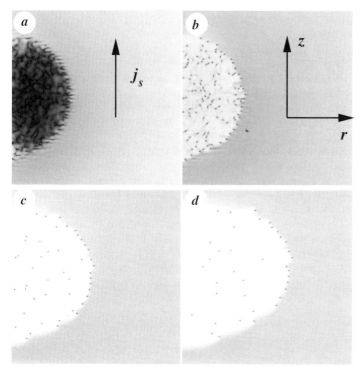

Fig. 4.7. Images of $|\psi|$ obtained by numerical solution of axi-symmetric (4.30) for $E_0 = 50, \sigma = 5000, k = 0.5, T_f = 0.002$. Images are shown for time $t = 30$ (a), $t = 40$ (b), $t = 150$ (c), and $t = 300$ (d).

fast quench. We verified that this scenario remains valid for the TDGLE with complex relaxation rate.

Our results may be useful for interpretation of experiments with ^{3}He [6] and nonlinear optical systems [19] in liquid crystals [20].

We are grateful to V. Eltsov, M. Krusius, G. Volovik, L. Kramer, L. Pismen, V. Steinberg and W. Zurek for stimulating discussions. This research is supported by US DOE, grant W-31-109-ENG-38.

References

1. G.E. Volovik: Physica B **280**, 122 (2000)
2. T.W.B. Kibble: J. Phys. A: Math Gen **9**, 1387 (1976)
3. W. H. Zurek: Nature **317**, 505 (1985); N.D. Antunes, L.M.A. Bettencourt, and W.H. Zurek: Phys. Rev. Lett. **82**, 2824, (1999); J. Dziarmaga, P. Laguna, and W.H. Zurek: Phys. Rev. Lett. **82**, 4749 (1999)
4. T.W.B. Kibble and G.E. Volovik: JETP Lett. **65**, 102 (1997)
5. V.B. Eltsov, M. Krusius, and G.E. Volovik: cond-mat/9809125, to be published

6. V.M.H. Ruutu et al: Nature **382**, 334 (1996); V.M.H. Ruutu et al: Phys. Rev. Lett. **80**, 1465 (1998)
7. N.B. Kopnin and E.V. Thuneberg: Phys. Rev. Lett. **83**, 116 (1999)
8. I.S. Aranson, N.B. Kopnin and V.M. Vinokur: Phys. Rev. Lett. **83**, 2600 (1999)
9. R. J. Donnelly, *Quantized Vortices in He-II* (Cambridge Univ. Press, 1991)
10. R.J. Rivers: Phys. Rev. Lett. **84**, 1248 (2000)
11. R. Carmi and E. Polturak: Phys. Rev. B **60**, 7595 (1999)
12. A. Yates and W. H. Zurek: Phys. Rev. Lett. **80**, 5477 (1998)
13. W.F. Vinen: Proc. R. Soc. London A **242**, 493 (1957)
14. I. Ispolatov and P. Krapivsky: Phys. Rev. E **53**, 3154 (1996)
15. V. V. Ginzburg, L. Radzihovsky, and N. A. Clark: Phys. Rev. E **55**, 1 (1997)
16. I.S. Aranson, N.B. Kopnin and V.M. Vinokur, Phys. Rev. B **63**, 184501 (2001)
17. I.S. Aranson and V. Steinberg: Phys. Rev. B **54**, 13072 (1996)
18. L. M. Pismen: *Vortices in Nonlinear Fields*, (Oxford Science Publications, 1999)
19. S. Ducci,P. L. Ramazza, W. González-Viñas, and F. T. Arecchi: Phys. Rev. Lett. **83**, 5210 (1999)
20. S. Digal, R. Ray, amd A. M. Srivastava: Phys. Rev. Lett. **83**, 5030 (1999)

5 Superfluidity in Relativistic Neutron Stars

David Langlois

Summary. The purpose of this chapter is to give a brief review of superfluidity in neutron stars. After a short presentation explaining why and how superfluidity is expected in the crust and core of neutron stars, consequences on thermal evolution and rotational dynamics are discussed. The second part summarizes a formalism that has been recently developed to describe the hydrodynamics of superfluids or superconductors in the framework of general relativity. As an application, one can compute the oscillations of a two-component relativistic neutron star.

5.1 Introduction

This chapter will discuss superfluidity and superconductivity in a rather extreme environment: neutron stars. Neutron stars, dense stars composed mainly of neutrons, were envisaged by Landau as soon as the neutron was discovered in 1932. In 1934, Baade and Zwicky suggested that supernovae were manifestations of a transition from an ordinary star to a very dense neutron star. In 1967, Hewish and Bell discovered the first "pulsar" (pulsating source of radio), which was soon identified as a rotating neutron star. The following year, two new pulsars, much studied until now, were discovered: the Vela pulsar with a period $P = 89$ ms and the Crab pulsar with $P = 33$ ms. Since then, more than one thousand pulsars have been detected.

Neutron stars are rather impressive objects. They contain a mass of the order of the solar mass confined in a radius $R \sim 10$ km, which implies an average mass density of the order $\rho \sim 10^{14} \mathrm{g/cm}^3$. They can rotate up to several hundred times per second. Due to these extreme conditions, neutron stars are of interest for various branches of physics. First, neutron stars are so dense and so compact that their gravitational field is very strong,

$$\frac{GM}{c^2 R} \sim 0.2, \qquad (5.1)$$

and can be described correctly only with general gelativity. Moreover, binary pulsars have been extraordinary "laboratories" to test the strong field predictions of general relativity, in particular to verify (indirectly) the existence of gravitational radiation.

Second, the magnetic field, typically of the order of 10^{12} G, plays a very important role in the neutron star physics, in particular in the pulsar emission mechanism (see e.g. [1]). Third, neutron stars are unique places where one can find matter in an ultra-dense state. The density is indeed higher than the atomic nuclear density. The corresponding pressure therefore depends essentially on the strong interactions, and neutron star observations are potentially a very rich source of information about the behaviour of strong interactions at high densities, which, until now, remains very poorly known from both theoretical and experimental points of view.

Finally, neutron star physics has also some connection with the low temperature physics. Studying the nucleon interactions, Migdal suggested the possibility that neutron star matter becomes superfluid at sufficiently low temperature. If the temperature of neutron stars, typically $T \simeq 10^6 K$, may seem huge in terrestrial standards, it is in fact smaller than the superfluid critical temperature evaluated as $T_c \sim 10^9 - 10^{10}$ K.

In these notes, we will focus on two aspects of the physics of neutron stars: the property of superfluidity with its consequences on the evolution of a neutron star; the necessity to consider a neutron star as a general relativistic object. Many other interesting aspects of neutron stars will thus be left aside and the curious reader will find some useful information in [2].

The plan will be the following: we will begin with some details about the superfluid properties of the interior of neutron stars. Observable consequences of superfluidity will then be discussed: first, the impact on the thermal evolution of neutron stars, then on their rotational dynamics. In a second part, we will summarize some elements of a formalism that gives a hydrodynamical description of the interior of neutron stars in the framework of general relativity. Finally, as an illustration, oscillations of two-component superfluid neutron stars are considered.

5.2 Superfluidity and Superconductivity

5.2.1 Composition of the Interior of a Neutron Star

Let us start by describing the matter composition inside a neutron star. Because there is a strong density gradient from the exterior to the interior of the star, the composition changes dramatically with the radial distance from the center. Let us progress from the surface of the star towards its interior.

Ignoring here the surface ocean, the first layer of a neutron star is the *outer crust*: it is made of a lattice of nuclei with a gas of relativistic degenerate electrons. The nuclei are richer and richer in neutrons as the density increases (see [3] for a recent review). At some point ($\rho \simeq 4.3 \times 10^{11}$g/cm^3), neutrons begin to leak out of the nuclei. This is called the *neutron drip* transition and this marks the boundary between the outer crust and the *inner crust*. For the latter, one still has a lattice of nuclei immersed in an electron gas, but in addition there is a liquid of free neutrons. The density still increasing, the difference between the neutrons in nuclei and the neutrons outside becomes fainter and fainter until the nuclei simply dissolve (at a density $\rho \sim 2 \times$

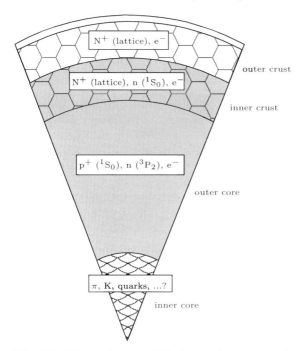

Fig. 5.1. Schematic view of the internal structure of a neutron star.

10^{14}g/cm^3). One has reached the limit of the crust and one now enters the realm of the core.

In the *outer core*, one will find the coexistence of a gas free neutrons, which is by far the main species, with a liquid of free protons, plus a gas of electrons (and muons) which ensures charge neutrality. Finally, the deeper layers of a neutron star, the *inner core*, remain mysterious. Several possibilities can be envisaged, including appearance of hyperons, condensation of pions or kaons, or the transition from hadronic to quark matter.

5.2.2 Energy Gaps and Critical Temperature

Like electrons in a superconductor, the superfluidity of neutrons in neutron stars is due to the pairing of two neutrons near the Fermi surface in momentum space, according to the Cooper mechanism (see [4]). In the inner crust, the neutron Cooper pairs are preferably in a state $^1S_0$. In the outer core, where the density is much higher, the neutron Cooper pairs are of the $^3P_2$ type (see [5] and [6]). The evaluation of the energy gaps give typically values of the order of 1 MeV.

In the same layer, the protons are free and can undergo the same process as neutrons, i.e. form Cooper pairs. Being fewer than the neutrons, they are expected to condense into a $^1S_0$ state [7]. Note that hyperons that may exist in the inner core are also expected to be superfluid.

5.2.3 Various Equations of State

In order to determine the global structure of neutron stars, one needs the effective bulk equation of state of neutron star matter. Inserting this equation of state in Einstein's equations, one obtains immediately, in the case of a *non-rotating star* (see Sect. 6.1), the radial profile of the star, i.e. the radial profile of the energy density, of the pressure and of the metric coefficients. One can also take into account the rotation of the star.

The main problem, however, is that the equation of state is unknown at very high densities. The reason is the lack of experimental data as well as the theoretical and computational challenge to evaluate the interactions of high density matter. Therefore there exist many different equations of state in the literature (see [8] for a recent review) and the hope is to infer from observations of neutron stars some constraints on the high density equations of state (see [9] and [10]).

5.3 Cooling Processes in Neutron Stars

Born with a temperature of the order $10^{11} - 10^{12}$ K, neutron stars cool very rapidly to temperatures of less than 10^{10} K within minutes. The subsequent thermal evolution is more uncertain and could be strongly affected by the presence of superfluidity. The mechanism by which neutron stars cool down is essentially neutrino emission. Neutrinos can be produced in two types of processes:

- Direct URCA processes
  They correspond to the simplest beta reactions
  $$n \to p + e^- + \bar{\nu}_e, \quad p + e^- \to n + \nu_e. \tag{5.2}$$

- Modified URCA processes
  They correspond to beta reactions with the presence of a second nucleon.

Direct URCA reactions are (not surprisingly) more efficient than the modified ones but they can occur only if conservation of energy and momentum is satisfied, which implies some constraint on the species fractions, namely that the proton fraction is sufficiently high. If this is not the case, then direct URCA processes are completely suppressed and cooling occurs only through modified URCA processes and is much slower.

However, taking into account the existence of superfluidity leads to the conclusion that the direct URCA processes could be *partially* suppressed: the neutrino emissivity would be reduced by a factor $\exp(-\Delta/kT)$, where Δ is the superfluid gap energy. The reason is that, before the beta reaction takes place, one needs to break the Cooper pair.

The existence of superfluidity therefore enables us to envisage, between the two extreme scenarios suggested above of *fast cooling* or *slow cooling*, an intermediate scenario of *moderate cooling*, which may be testable in the future by the temperature measurements of neutron stars (see [11] and references therein).

5.4 Rotational Dynamics of Neutron Stars: Glitches

Another important consequence of superfluidity concerns the rotational dynamics of neutron stars, which can be followed extremely precisely via the observation of the radio signal from pulsars. The main feature is an extremely stable periodic signal, the pulsar being analogous to a lighthouse beacon: the radio emission is indeed collimated along the magnetic axis which does not coincide with the rotation axis.

For isolated pulsars, one observes a steady, although tiny, increase of the pulse period, i.e. a steady decrease of the angular velocity. This is interpreted as a loss of angular momentum due to the emission of electromagnetic radiation. Assuming a magnetic dipole, the rotational energy loss is given by

$$\dot{E} = I\Omega\dot{\Omega} = -\frac{B^2 R^6 \Omega^4 \sin^2\theta}{6c^3}, \tag{5.3}$$

where I is the moment of inertia and θ the angle between the magnetic and rotation axes.

In addition to this slow increase of the period, a few pulsars exhibit some rare events named *glitches* at which the period suddenly decreases, i.e. the angular velocity suddenly increases, followed by a slow exponential relaxation with a typical time scale of the order of days to months. The two most famous 'glitching' pulsars are the Vela and Crab pulsars with period changes $\Delta\Omega/\Omega$ of the order of 10^{-6} and 10^{-8} respectively.

The current explanation for these glitches is the following. It is based on the idea that the neutron star interior contains a weakly coupled component that is not directly slowed down by the electromagnetic torque and therefore rotates slightly faster than the star crust. What is directly observable is the angular velocity of the crust, and if, for some reason, there is a rapid transfer of angular momentum from the faster component to the crust, then one should see a sudden increase of the pulsar period. Of course, a natural candidate for the component rotating faster is the *neutron superfluid*.

5.4.1 The Two–Fluid Model

The relaxation after the glitch can also be explained, at least qualitatively, by the simple two–component model of Baym et al. [12]. One considers two components with respective angular velocities Ω_c and Ω_n, and respective moments of inertia, I_c and I_n. The first component corresponds to the crust and to whatever is strongly coupled to it, whereas the second component corresponds to the neutron superfluid weakly coupled to the crust. Note that the pulsar angular velocity Ω which is the directly observable quantity can be identified with the crust angular velocity, so that $\Omega_c = \Omega$. The evolution of the two components is governed by the system

$$I_c \dot{\Omega} = -\frac{I_c}{\tau_c}(\Omega - \Omega_n) - \alpha, \tag{5.4}$$

$$I_n \dot{\Omega}_n = \frac{I_c}{\tau_c}(\Omega - \Omega_n), \tag{5.5}$$

where the first term on the right hand side represents a coupling between the two components, coupling characterized by the timescale τ_c. α corresponds to the electromagnetic torque. One can integrate the above system just after a glitch, characterized by a sudden jump of the crust angular velocity $\Delta\Omega_0 \equiv \Omega(t=0^+) - \Omega(t=0^-)$. The subsequent evolution is then

$$\Omega(t) = \Omega_0(t) + \Delta\Omega_0 \left(Q e^{-t/\tau} + 1 - Q \right), \tag{5.6}$$

where Q is the healing parameter and $\Omega_0(t) \equiv \Omega_0 - (\alpha/I)t$ (Ω_0 being a constant) corresponds to the evolution without glitch, and the characteristic relaxation timescale is given by

$$\tau = \frac{I_n}{I}\tau_c, \tag{5.7}$$

with $I = I_n + I_c$.

5.4.2 Role of the Vortices

The two–fluid model gives an effective view of the rotational evolution, but it is interesting to explore in more details the underlying mechanism responsible for the coupling between the two components. This is where the vortices enter into play.

The neutron star at its birth is a rapidly rotating object. With its cooling, the temperature reaches the critical temperature under which superfluidity appears. However, one essential property of a superfluid is that its flow is *irrotational*. The way the superfluid solves the contradiction between irrotationality and rotation of the star is, like superfluid Helium in laboratories, via the creation of quantized vortices that carry all the angular momentum of the superfluid. The superfluid 'velocity' (in fact, momentum) being the gradient of a phase,

$$\boldsymbol{p}_n = 2m_n \boldsymbol{v}_n = \hbar \boldsymbol{\nabla}\varphi, \tag{5.8}$$

the circulation around any closed path is given by

$$\kappa \equiv \int_C \boldsymbol{dl} \boldsymbol{v}_n = \frac{\hbar}{2m_n}(2\pi N), \qquad (5.9)$$

where N is an integer. Superfluid vortices usually correspond to one unit ($N = 1$) of quantized circulation $\kappa = h/(2m_n)$. One can then easily compute the average density of vortices for a superfluid rotating at uniform angular velocity Ω_n:

$$n_V = 4\frac{\Omega_n m_n}{h}. \qquad (5.10)$$

Putting numbers, one finds

$$n_V = 6.3 \times 10^3 \left(\frac{P}{1\,\mathrm{s}}\right)^{-1} \text{ vortices per cm}^2. \qquad (5.11)$$

In the neutron star core, the existence of a proton superconductor, which is believed to be a type II superconductor, suggests the presence of magnetic vortices, or fluxoids, which carry the magnetic flux going through the neutron star. Note that one must be careful to compute the total energy of a fluxoid in a *rotating* superconducting background [13]. Moreover, because of the 'entrainment effect' by which the neutron superfluid momentum is associated with both neutron and proton currents, neutron superfluid vortices, in the neutron star core, will carry a magnetic flux as well, thus implying a strong coupling between the core superfluid and the crust [14,15].

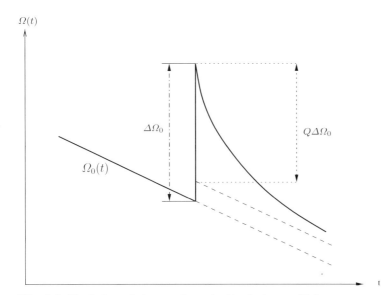

Fig. 5.2. Evolution of the angular velocity during a glitch

Now, the vortices play an essential role for the coupling of the superfluid to the normal component of the star, because the pure superfluid has no viscosity at all and, as such, can circulate without affecting any normal component. But the vortices interact with the normal component. Without entering the details, one can distinguish two types of interaction between the vortices and the normal part of a neutron star:

- *Pinning*
  This is the situation where a vortex, or a segment of it, is anchored to the (inner) crust. The pinning is due to the interaction between the vortex and the nucleus, which can be attractive or repulsive depending on the matter density [16,17].

- *Friction*
  This is the case where the vortex, moving through the normal component, interacts with it in some dissipative processes, which results in an effective friction force (per unit length) exerted on the vortex by the normal component of the form

$$\boldsymbol{F}_{\text{drag}} = \mathcal{C}(\boldsymbol{v}_V - \boldsymbol{v}_c). \tag{5.12}$$

Let us assume here that we are in the friction case. The vortices, interacting with the normal component, are also sensitive to the motion of the superfluid through a Magnus force term,

$$\boldsymbol{F}_{\text{M}} = n_n \boldsymbol{\kappa} \times (\boldsymbol{v}_V - \boldsymbol{v}_n), \tag{5.13}$$

which is a force per unit vortex length. The effective motion of the vortices is obtained by requiring that the two forces (5.12) and (5.13) just cancel. This implies that the vortices must have a radial motion in addition to the angular motion. One finds easily that the angular motion is given by

$$\Omega_V = \frac{c_r^{-1}\Omega_n + c_r\Omega_c}{c_r^{-1} + c_r}, \tag{5.14}$$

which means that the angular velocity of the vortices is simply a weighted average of the superfluid and normal angular velocities, the weight depending on the dimensionless friction coefficient

$$c_r = \frac{\mathcal{C}}{\kappa n_n}. \tag{5.15}$$

The two extreme cases are: the case where the friction is zero, the vortices being then in corotation with the superfluid; the case where the friction is huge, the vortices being then in corotation with the normal component. As for the radial velocity of the vortices, it is given by

$$v_V^r = \frac{\Omega_n - \Omega_c}{c_r^{-1} + c_r} r. \tag{5.16}$$

One can then compute the effective coupling timescale between the two components of the stars, by noting that the evolution of the angular velocity of the superfluid component is directly given by the radial velocity of the vortex array, according to the expression

$$\dot{\Omega}_n = -2\frac{v_V^r}{r}\Omega_n, \qquad (5.17)$$

simply because the angular momentum of the superfluid is proportional to the vortex density. Let us consider now the evolution of the angular velocity lag between the two components,

$$\omega \equiv \Omega_n - \Omega_c, \qquad (5.18)$$

and let us define the coupling timescale as $\tau_d \equiv |\omega/\dot{\omega}|$. Combining the expressions obtained above, one finds

$$\tau_d \simeq \frac{c_r + c_r^{-1}}{2}\left(\frac{I_c}{I_c + I_n}\right)\Omega^{-1}. \qquad (5.19)$$

The above Newtonian analysis has been generalized to a general relativistic context in [18]. The coupling between the crust and the superfluid can also be evaluated when vortices are pinned, in which case vortices can move outwards by vortex creep, the temperature being a crucial parameter [19].

5.4.3 Origin of the Glitches

There is no certainty at present on the physical origin of observed glitches. The scenario which seems to have attracted the greatest attention is the one proposed by Anderson and Itoh [20]. In this model, vortices are supposed to be pinned in the crust due to the nucleus-vortex interaction. Because the normal component is slowing down, a Magnus force, due to the superfluid motion relative to the vortices, progressively builds up until this force reaches a critical amplitude for which the pinning sites simply break. The vortices then can suddenly move outwards, thus transfering angular momentum from the superfluid to the crust, which explains the sudden spin-up of the crust.

Other models have been proposed to explain glitches. For example, a glitch could be induced by a sudden heat wave propagating in the star, which would increase the effective coupling between the superfluid and the crust [21]. Such a heating could be generated for instance by a crustquake. In this respect, one must mention that the differential rotation between the superfluid and the crust generates a centrifugal buoyancy force which might increase the crust stresses and maybe lead to a glitch [22].

5.5 Relativistic Description

Until very recently, only Newtonian theory was used in the studies of superfluidity in neutron stars, whereas, in parallel, general relativistic studies of neutron stars, even numerical, were based on perfect fluid matter. The purpose of the work summarized here is to establish a bridge between these two approaches, taking into account both the highly relativistic nature of neutron stars and their superfluid interior. We shall present a macroscopic formalism [23] allowing for the average effect of vorticity quantisation in a rotating superfluid, which can then be extended [24] to describe superconducting fluids as well , such as protons in the core of neutron stars. The Newtonian version of this formalism can be found in [25]. In the next section, a simplified relativistic two-component model of neutron star, which allows for differential rotation of the superfluid component, will be used to study oscillations of neutron stars.

5.5.1 Perfect Fluid in General Relativity

Let us start by recalling the simplest case, that of the perfect fluid. In general relativity, a perfect fluid is characterized by

- a four-velocity vector u^μ (where $\mu = 0, 1, 2, 3$ is a spacetime index denoting the time coordinate and the three space coordinates), which satisfies the normalization $g_{\rho\sigma} u^\rho u^\sigma = -1$ where $g_{\rho\sigma}$ is the metric tensor describing the geometry of spacetime (and which is used to raise or lower the indices, e.g. $u_\rho \equiv g_{\rho\sigma} u^\sigma$),
- a particle number density scalar field $n(x^\mu)$ giving in each spacetime point the density of particles *as seen by an observer comoving with the fluid*,
- an equation of state of the form $\rho = \rho(n)$ giving the energy density ρ as a function of the particle number density n.

Note that we have chosen here the simpler case of a barotropic perfect fluid where the equation of state depends on only one parameter. Note also that one could characterize the fluid by its energy density instead of its particle number density, and then derive the number density by inverting the equation of state.

The equations of motion for the relativistic fluid will be generalizations of the well-known Newtonian equations of fluid mechanics, namely,

- the matter conservation equation, which can be written in the present context in the very simple form,

$$\nabla_\mu n^\mu = 0, \tag{5.20}$$

where $n^\mu \equiv n u^\mu$ is the *particle current* and ∇_μ stands for the *covariant derivative*,

- the relativistic Euler equation, which can be written in the very condensed form

$$n^\rho \nabla_{[\rho} \mu_{\sigma]} = 0, \tag{5.21}$$

where the brackets represent antisymmetrization on the indices and where one has conveniently introduced the momentum covector $\mu_\rho \equiv \mu u_\rho$, μ being the relativistic *chemical potential* defined by

$$\mu = \frac{d\rho}{dn}. \tag{5.22}$$

It is possible to derive directly the above equations of motion from a variational principle, the Lagrangian density being simply the energy density considered as a function of the particle number current n^μ. Any variation of n^μ is not allowed, but only the *convective* variations, which correspond to variations of the particle flow lines (see [31] for more details).

5.5.2 Relativistic Superfluid

What will characterize a "pure superfluid" (i.e. a superfluid at zero temperature, without gas of excitations), with respect to the more general class of perfect fluids, is the fact that *its motion is locally irrotational*. In the standard language, this is a consequence of the fact that the 'superfluid velocity' can be expressed as the gradient of a quantum scalar phase. Strictly speaking, the standard 'superfluid velocity' is in fact a *momentum* divided by a somewhat arbitrary mass. In the relativistic context, this will be generalized by the expression

$$\mu_\sigma = \hbar \nabla_\sigma \varphi, \tag{5.23}$$

so that the property of irrotational flow can be written as the vanishing of the vorticity tensor

$$w_{\rho\sigma} \equiv 2\nabla_{[\rho} \mu_{\sigma]}. \tag{5.24}$$

The previous condition is valid locally. However, as explained above, a superfluid can exhibit a non-irrotational flow at the price of being threaded by quantized vortices. This means that macroscopically, on distances bigger than the typical intervortex separation, one would like to describe the superfluid as well as the vortices in an average way. This can be done by considering the vorticity tensor $w_{\rho\sigma}$, now non-zero since there are vorticies, as a fundamental quantity. In fact, this tensor contains information about the density of vortices as well as about the local average direction of the vortex array.

The idea then is to construct a generalized Lagrangian density that depends not only on the particle current n^σ but also on the vorticity tensor $w_{\rho\sigma}$. The variations with respect to these fundamental variables will define the canonical momenta,

$$\delta \Lambda = \mu_\sigma \delta n^\sigma - \frac{1}{2} \lambda^{\rho\sigma} \delta w_{\rho\sigma}. \tag{5.25}$$

This can be seen as a generalization of the perfect fluid where $\Lambda = -\rho$ and only the first term is present on the right hand side.

The equations of motion are the matter conservation equation (5.20), as in the perfect fluid case, and a generalized Euler-type equation of the form,

$$w_{\mu\nu}(n^\nu - \nabla_\rho \lambda^{\rho\nu}) = 0. \tag{5.26}$$

Of course, the expression for $\lambda^{\rho\sigma}$, which essentially represents the energy density and momentum of an individual vortex, must be specified by resorting to a 'microphysical' model of the vortices (see [27] and [28] for relativistic descriptions of vortices).

5.5.3 Superfluid-Superconducting Mixtures

It is useful to extend the previous formalism to the case where the superfluid particles (or Cooper pairs) are electrically charged, like electrons in laboratory superconductors, in order to be able to describe the protons in the cores of neutron stars. Let us consider several species labelled by the index X, with respective particle currents n_X^ρ and respective electric charge per particle e^X. The total electric current is then given by

$$j^\rho = \sum_X e^X n_X^\rho. \tag{5.27}$$

Proceeding as before, in order to obtain the equations of motion for the global system of fluids, one uses a variational principle based on a Lagrangian density $\mathcal{L}$ that will be the sum of three contributions:

- a "matter" contribution Λ_M, which depends only on the 'hydrodynamical' part of the system, and which is a function only of the particle currents n_X^ρ, or more exactly of all scalar combinations obtained by their mutual contractions,
- an electromagnetic interaction term depending on the electromagnetic gauge form A_ρ,

$$\Lambda_I = j^\rho A_\rho, \tag{5.28}$$

- an electromagnetic field contribution

$$\Lambda_F = \frac{1}{16\pi} F_{\rho\sigma} F^{\sigma\rho}. \tag{5.29}$$

where the electromagnetic field tensor $F_{\rho\sigma}$ is defined in terms of the gauge form by the usual expression $F_{\rho\sigma} = 2\nabla_{[\rho} A_{\sigma]}$.

Considering the variations of $\mathcal{L} = \Lambda_M + \Lambda_F + \Lambda_I$ with respect to the particle currents n_X^ρ naturally suggests to define the canonical momentum covectors

$$\pi_\rho^X = \mu_\rho^X + e^X A_\rho \tag{5.30}$$

which are the sum of a pure 'hydrodynamical' (or 'chemical') part and of the electromagnetic gauge form weighted by the electric charge of the species. From this, it is convenient to define generalized vorticity tensors defined by

$$w^X_{\rho\sigma} = 2\nabla_{[\rho}\pi^X_{\sigma]} = 2\nabla_{[\rho}\mu^X_{\sigma]} + e^X F_{\rho\sigma}. \tag{5.31}$$

The equations of motion for the system consist of the separate matter conservation equation for each species (one can generalize to allow for chemical reactions between various species [18]),

$$\nabla_\rho n^\rho_X = 0, \tag{5.32}$$

and of Euler-type equations, which can be written in the very compact form

$$n^\sigma_X w^X_{\sigma\rho} = 0. \tag{5.33}$$

Note that in the case of a charged component, this equation contains the electromagnetic field tensor and thus automatically includes for instance the Lorentz force exerted on the fluid.

It is also useful to write for the system under investigation the associated total stress-energy-momentum tensor, which appears on the right hand side of the Einstein's equations,

$$R_{\mu\nu} - \frac{1}{2}Rg_{\mu\nu} = 8\pi G T_{\mu\nu}, \tag{5.34}$$

if one needs to determine the spacetime metric. Once again, the variational principle is very useful because variation of the Lagrangian density with respect to the metric directly yields

$$T^{\rho\sigma} = T_M^{\rho\sigma} + T_F^{\rho\sigma}, \tag{5.35}$$

where the part derived from the material Lagrangian density contribution Λ_M is given by

$$T_M{}^\rho{}_\sigma = \sum_X n^\rho_X \mu^X_\sigma + s\Theta u^\rho u_\sigma + \Psi_M g^\rho_\sigma, \quad \Psi_M = \Lambda_M - \sum_X n^\sigma_X \mu^X_\sigma + s\Theta, \tag{5.36}$$

while the electromagnetic contribution has the usual Maxwellian form

$$T_F{}^\rho{}_\sigma = \frac{1}{4\pi}\left(F^{\nu\rho}F_{\nu\sigma} - \frac{1}{4}F^{\mu\nu}F_{\mu\nu}g^\rho_\sigma\right). \tag{5.37}$$

The above relations are valid for ordinary perfectly conducting fluids. The condition characterizing superconductors, generalizing the superfluid condition (5.23), is that the *generalized momentum* should be the gradient of a quantum phase,

$$\pi^X{}_\rho \equiv \mu^X{}_\rho + e^X_\rho = \hbar\nabla_\rho \varphi^X, \tag{5.38}$$

so that superconductors (including the case of superfluids) are characterized *locally* by a vanishing generalized vorticity tensor. Of course, it is then possible, in analogy with the treatment of the previous subsection, to extend the formalism in order to describe on average a superfluid-superconducting mixture threaded by superfluid vortices as well as magnetic flux tubes.

5.6 Relativistic Neutron Stars

Let us now see how one can construct a concrete model of neutron star in a general relativistic framework. To simplify, we will assume that the neutron star is made of simply two components, each being treated as a perfect fluid. First, we will consider the case of a static, i.e. nonrotating, neutron star. Then we will consider small, i.e. linear, oscillations of the two-component neutron star.

Let us call, loosely, the component corresponding to the free neutron superfluid the "neutron" component and the component corresponding to the normal fluid the "proton" component. Their respective particle currents will be denoted $n^\rho = nu^\rho$ and $p^\rho = pv^\rho$, where u^ρ and v^ρ are unit four-velocities, which are not aligned if the two components are not comoving. Their equations of motion consist of two separate conservation equations of the type (5.20) and of two Euler equations of the type (5.21). Here, the two components are treated as ordinary perfect fluids and one ignores the refinements due to the presence of vortices in the superfluid component.

5.6.1 Static Neutron Star

We first consider the configuration corresponding to a non-rotating neutron star, which is thus *static* and *spherically symmetric* and for which the metric can be written in the form

$$ds^2 = g_{\mu\nu}dx^\mu dx^\nu = -e^{\nu(r)}dt^2 + e^{\lambda(r)}dr^2 + r^2\left(d\theta^2 + \sin^2\theta d\phi^2\right). \quad (5.39)$$

The particle currents, since the two fluids are motionless, are necessarily of the form

$$n^\mu = n\bar{u}^\mu, \quad p^\mu = p\bar{u}^\mu, \quad \bar{u}^\mu = \{u^0, 0, 0, 0\}. \quad (5.40)$$

Writing Einstein's equations with the above ansätze, one obtains the well-known TOV (Tolman–Oppenheimer–Volkov) equations:

$$\frac{dP}{dr} = -G\frac{(\rho(r) + P(r))(m(r) + 4\pi r^3 P(r))}{r^2(1 - 2Gm(r)/r)} = -\frac{1}{2}(\rho(r) + P(r))\frac{d\nu}{dr}, \quad (5.41)$$

and

$$e^{\lambda(r)} = \left(1 - 2\frac{m(r)}{r}\right)^{-1}, \quad (5.42)$$

with $m(r) \equiv 4\pi G \int_0^r dr' \rho(r') r'^2$. The mass of the star is given by $M = m(r)$, where R is the star radius defined by the condition that the pressure vanishes.

5.6.2 Oscillations of Superfluid Neutron Stars

Oscillations of relativistic stars have been studied by many authors in the case of a *single* perfect fluid. Here, we summarize the first investigation for a two-component relativistic star [29], thus generalizing two previous studies of two-component Newtonian stars [30].

In a relativistic approach, one must consider not only the perturbations of the quantities describing the matter, for example the particle density and the velocity, but also the perturbations of the spacetime metric $g_{\mu\nu}$. After decomposition of the perturbations into spherical harmonics, labelled by l and m, it is convenient to distinguish even-parity (or polar) perturbations and odd-parity (or axial) modes depending on their transformation under parity. We will restrict ourselves here to the case of even-parity perturbations, which is the most interesting since scalar quantities, such as the matter densities, give only this type of perturbations. Also, one needs to study only $m = 0$ perturbations, since there is a degeneracy in m for each given l because of the spherical symmetry of the background.

Choosing a specific gauge, one will consider metric perturbations of the form

$$\delta g_{00} = -e^{\nu} r^l H_0 e^{i\omega t} P_l(\theta),\ \delta g_{0r} = \delta g_{r0} = -i\omega r^{l+1} H_1 e^{i\omega t} P_l(\theta),$$
$$\delta g_{rr} = -e^{\lambda} r^l H_2 e^{i\omega t} P_l(\theta),\ \delta g_{\theta\theta} = \delta g_{\phi\phi}/\sin^2\theta = -r^{l+2} K e^{i\omega t} P_l(\theta), \quad (5.43)$$

where P_l stands for the Legendre polynomial of order l. The matter perturbations will be described, in a Lagrangian way, by the matter displacements,

$$\xi_n^r = r^{l-1} e^{-\lambda/2} W_n e^{i\omega t} P_l(\theta), \quad \xi_n^\theta = -r^{l-2} V_n e^{i\omega t} \partial_\theta P_l(\theta),$$
$$\xi_p^r = r^{l-1} e^{-\lambda/2} W_p e^{i\omega t} P_l(\theta), \quad \xi_p^\theta = -r^{l-2} V_p e^{i\omega t} \partial_\theta P_l(\theta), \quad (5.44)$$

from which one can compute the perturbed velocities, by taking the time derivative or the variations δn and δp by using the perturbed conservation equations. Inserting the above expressions for the perturbations into the perturbed Einstein's equations as well as the perturbed Euler equations, one ends up with a system of linear equations, consisting of two constraints (one is $H_2 = H_0$, the other expresses H_0 in terms of the other perturbations) and of first order differential equations of the form

$$\frac{dY}{dr} = Q_{l,\omega} Y, \quad (5.45)$$

where Y is 6-dimensional column vector containing H_1, K, W_n, V_n, W_p and V_p, and $Q_{l,\omega}$ is a 6×6 matrix with r-dependent coefficients that depend only on the background configuration, as well as on l and ω. Considering the boundary conditions at the center and at the surface of the star, this system can be solved, up to a global amplitude, for *any value* of ω. The physically relevant modes, however, also called quasi-normal modes, correspond the specific values of ω for which the metric outside the star represents *only outgoing gravitational waves*.

A numerical investigation for a very crude, and unrealistic, model of two independent polytropes (with different adiabatic indices) has shown that the two-component star will exhibit new modes, *superfluid* modes, which are specific of the existence of two components since the two fluids are counter-moving for these modes.

References

1. F.C. Michel, *Theory of neutron star magnetospheres*, University of Chicago Press, Chicago (1991)
2. S.L. Shapiro, S.A. Teukolsky, *Black holes, white dwarfs and neutron stars: the physics of compact objects*, John Wiley & Sons, New York (1983)
3. C.J. Pethick, D.G. Ravenhall, Annu. Rev. Nucl. Part. Sci. **45**, 429 (1995)
4. J. A. Sauls, p. 457, in *Timing neutron stars*, Eds H. Ögelman, E.P.J. van den Heuvel; Dordrecht, Kluwer (1989)
5. M. Hoffberg, A.E. Glassgold, R.W. Richardson, M. Ruderman, Phys. Rev. Lett. **24**, 775 (1970)
6. T. Takatsuka, Prog. Theor. Phys. **48**, 1517 (1972)
7. N.C. Chao, J.W. Clark, C.H. Yang, Nucl. Phys. PA **179**, 320 (1972)
8. H. Heiselberg, V. Pandharipande, "Recent progress in neutron star theory", astro-ph/0003276
9. N.K. Glendenning, *Compact stars*, Springer–Verlag (1997)
10. J. M. Lattimer, M. Prakash, "Neutron Star Structure and the Equation of State", astro-ph/0002232
11. D. Page, M. Prakash, J.M. Lattimer, A. Steiner, "Prospects of Detecting Baryon and Quark Superfluidity from Cooling Neutron Stars", hep-ph/0005094
12. G. Baym, C.J. Pethick, D. Pines, M. Ruderman, Nature **224**, 872 (1969)
13. B. Carter, R. Prix, D. Langlois, "Energy of flux tubes in rotating superconductor", to appear in *Phys. Rev.* B (2000) [cond-mat/9910240]
14. G.A. Vardanian, D.M. Sedrakian, Zh. Eskp. Teor. Fiz. **81**, 1731 (1981) [English translation: Soviet Phys. JETP **54**, 919 (1981)]
15. M.A. Alpar, S.A. Langer, J.A. Sauls, Astrophys. J. **282**, 533 (1984)
16. R.I. Epstein, G. Baym, ApJ **328**, 680 (1988)
17. P.M. Pizzochero, L. Viverit, R.A. Broglia, Phys. Rev. Lett. **79**, 3347 (1997)
18. D. Langlois, D. M. Sedrakian, B. Carter, *Month. Not. R.A.S.* **297**, 1189 (1998)
19. M.A. Alpar, P.W. Anderson, D. Pines, J. Shaham, ApJ **276**, 325 (1984)
20. P.W. Anderson, N. Itoh, Nature **256**, 25 (1975)
21. B. Link, R.I. Epstein, ApJ **457**, 844 (1996) "Thermally driven neutron star glitches"
22. B. Carter, D. Langlois, D. Sedrakian, "Centrifugal buoyancy as a mechanism for neutron star glitches", to appear in *Astro. Astrophys.* (2000) [astro-ph/0004121]
23. B. Carter, D. Langlois, Nuclear Physics **B 454**, 402 (1995)
24. B. Carter, D. Langlois, *Nuclear Physics* **B 531** , 478 (1998)
25. G. Mendell, L. Lindblom, *Ann. Phys.* **205**, 110 (1991)
26. B. Carter, in *Relativistic Fluid Dynamics*, ed. A. Anile, M. Choquet Bruhat, *Lecture Notes in Mathematics* **1385**, pp 1–64 (Springer–Verlag, Heidelberg, 1989)
27. B. Carter, D. Langlois, Phys. Rev. **D 52**, 4640 (1995)

28. R. Prix, to appear in Phys. Rev. D (2000), gr-qc/0004076
29. G.L. Comer, D. Langlois, Lap Ming Lin, *Phys. Rev.* **D 60**, 104025 (1999)
30. L. Lindblom, G. Mendell, ApJ **421**, 689 (1994); U. Lee, AA **303**, 515 (1995)

6 Superconducting Superfluids in Neutron Stars

Brandon Carter

Summary. For treatment of the layers below the crust of a neutron star it is useful to employ a relativistic model involving three independently moving constituents, representing superfluid neutrons, superfluid protons, and degenerate negatively charged leptons. A Kalb–Ramond type formulation is used here to develop such a model for the specific purpose of application at the semi macroscopic level characterised by lengthscales that are long compared with the separation between the highly localised and densely packed proton vortices of the Abrikosov type lattice that carries the main part of the magnetic flux, but that are short compared with the separation between the neutron vortices.

6.1 Introduction

The purpose of this article is to present a concise overview of a class of macroscopic relativistic superconducting superfluid models developed [12,15] as a generalisation of previous non conducting relativistic superfluid models [13,4,5] with a view to applications concerning the layers below the crust of a neutron star, which are believed to be well described by three constituent superconducting superfluid models of the kind that was introduced (as a charged generalisation of the Andreev–Bashkin model [6]) for a superfluid mixture) by Vardanyan and Sedrakyan [7], and that has more recently been further developed (though still using a non-relativistic treatment) by Mendell and Lindblom [8]. The three basic ingredients in a description of this kind are, firstly, a condensate of superfluid neutrons, secondly an independently moving – effectively superconducting – condensate of superfluid protons, and thirdly a negatively charged degenerate leptonic constituent (consisting mainly of electrons but also including a significant proportion of muons) that is of "normal", i.e. non-superfluid, kind. Such a treatment does not include thermal effects (whose inclusion would involve a fourth constituent representing entropy) but should nevertheless be applicable as a very good first approximation except during a short lived high temperature phase immediately after the birth of the neutron star.

As the relativistic analogue of the kind of phenomenological description introduced in a Newtonian context by Bekarevich and Khalatnikov [9] (as a generalisation of Landau's original two-constituent model) it will first be shown how to set up a general category of three-constituent perfectly conducting perfect fluid models of the type that is needed, as a preliminary for the more specific developments that follow. This category [15] includes, as a specialisation, the case in which the neutronic and the protonic constituents are both characterised by strictly irrotational behaviour of the kind that is relevant in neutron stars on a "mesoscopic" scale, meaning a scale large compared with that of the underlying microscopic particle description, but small compared with the macroscopic scale separation between the vortex defects within which the superfluid comportment is locally violated. Models of this irrotational kind are just a specially simple limit within the more general category that is needed for the purpose of treating the superconducting superfluid on a "macroscopic" scale, meaning a scale that is large compared with the separation between vortices.

The present discussion will be focussed on an intermediate "semi macroscopic" scale meaning a scale that is small compared with the spacing between the superfluid neutron vortices (which will be rather widely separated due to the relatively low angular velocity of the star, though they contain quite a lot of energy due to their "global" nature) but large compared with the Abrikosov lattice spacing between the "local" proton vortices, which are expected to be much more numerous in order to carry the rather large magnetic fluxes that are thought to be present.

6.2 Generic Category of 3-Constituent Superconducting Superfluid Models

In so far as its contribution to the mass density is concerned, the most important of the the three independent constituents under consideration is that of the superfluid neutrons, with number current 4-vector n_n^ρ, say. The second contribution is that of the superconducting protons – which make up a small but significant part of the mass density – with number current four-vector n_p^ρ. The third constituent is that of the degenerate non-superconducting background of negatively charged leptons – consisting mainly of electrons, but including also a certain proportion of muons at the high densities under consideration – with a corresponding lepton number number current vector n_e^ρ say. This negatively charged "normal" (i.e. non superconducting) constituent contributes only a very small fraction of the mass density, but it nevertheless has a crucially important role, not just because the corresponding unit vector u^ρ defined by

$$n_\mathrm{e}^\rho = n_\mathrm{e} u^\rho, \qquad n_\mathrm{e}^2 = -n_\mathrm{e}^\rho n_{\mathrm{e}\rho} \tag{6.1}$$

characterises the natural reference frame of rigid corotation in an equilibrium configuration, but more generally, in so far as electromagnetic effects are concerned, because in terms of the electron charge coupling constant e the corresponding total electric current 4-vector will be given by

$$J^\mu = e\left(n_{\rm p}^\rho - n_{\rm e}^\rho\right). \tag{6.2}$$

It will be convenient to express formulae such as this in a condensed notation system based on the use of the summation convention for "chemical" indices represented by capital Latin letters running over the three relevant values, namely X=n, X=p, X=e. Using this convention, the equation (6.2) for the electric current density can be rewritten in the concise form

$$J^\rho = e^{\rm X} n_{\rm X}^\rho, \tag{6.3}$$

where the the charges per neutron, proton, and electron are given respectively by $e^{\rm n} = 0$, $e^{\rm p} = e$, and $e^{\rm e} = -e$.

Since each of the three independent currents involved is conserved, it will be possible to use a Kalb–Ramond formulation in which, instead of imposing the three corresponding conservation laws

$$\nabla_\rho n_{\rm X}^\rho = 0, \tag{6.4}$$

as dynamical equations, they will be obtained as identities by postulating that the currents should have the form

$$n_{\rm X}^\rho = \nabla_\sigma b_{\rm X}^{\rho\sigma}, \tag{6.5}$$

for corresponding antisymmetric gauge bivector fields $b_{\rm X}^{\rho\sigma}$ which are physically defined only modulo Kalb–Ramond gauge transformations of the form

$$b_{\rm X}^{\rho\sigma} \mapsto b_{\rm X}^{\rho\sigma} + \nabla_\nu \theta_{\rm X}^{\nu\rho\sigma}, \tag{6.6}$$

for arbitrary antisymmetric trivector fields $\theta_{\rm X}^{\nu\rho\sigma}$.

Since our present treatment will be restricted to the conservative limit in which dissipative effects are neglected, the analysis will be assumed to be expressible in terms of a variational principle based on a Lagrangian density, in which as usual the representation of the electromagnetic field requires the introduction of a Maxwellian gauge 1-form A_ρ, in terms of which the gauge invariant electromagnetic field tensor is given by

$$F_{\rho\sigma} = 2\nabla_{[\rho} A_{\sigma]}, \tag{6.7}$$

(using square brackets to indicate index antisymetrisation). The implementation of the Kalb Ramond formulation requires that the set of independent currents $n_{\rm X}^\rho$ be supplemented [15] by the introduction of a corresponding set of vorticity 2-forms $w^{\rm X}{}_{\rho\sigma}$ each of which is characterised both by an algebraic degeneracy condition of the form

$$w_{[\mu\nu} w_{\rho]\sigma} = 0, \tag{6.8}$$

and by a closure condition of the form

$$\nabla_{[\nu} w^{\text{X}}_{\rho\sigma]} = 0, \qquad (6.9)$$

so that each such 2-form $w_{\rho\sigma}$ is interpretable as a pullback of a prescribed area measure on a two-dimensional base space. This means that in terms of suitably chosen local vorticity base space coordinates χ^1, χ^2, the corresponding pair of scalar fields χ^1_{X}, χ^2_{X} induced on the four dimensional spacetime background will specify the vorticities by prescriptions of the form $w_{\rho\sigma} = 2\chi^1_{\text{X},[\rho}\chi^2_{\text{X},\sigma]}$.

In terms of these quantities, a Lagrangian of the appropriate kind will will be expressible in the generic form

$$\mathcal{L} = \Lambda + J^\rho A_\rho + \frac{1}{2} b^{\sigma\rho}_{\text{X}} w^{\text{X}}_{\rho\sigma}, \qquad (6.10)$$

which consists of a pair of gauge dependent coupling terms preceded by a first term Λ that is a function only of the relevant gauge independent field quantities, which in addition, of course to the spacetime metric $g_{\rho\sigma}$ and the electromagnetic field tensor $F_{\rho\sigma}$ consist of the three independent currents n^ρ_{X} and the three corresponding vorticity 2-forms $w^{\text{X}}_{\rho\sigma}$. This means that its most general infinitesimal variation will be given by an expression of the form

$$\delta\Lambda = \mu^{\text{X}}_\rho \delta n^\rho_{\text{X}} + \frac{1}{2}\lambda^{\sigma\rho}_{\text{X}} \delta w^{\text{X}}_{\rho\sigma} + \frac{1}{8\pi}\mathcal{H}^{\sigma\rho}\delta F_{\rho\sigma} + \frac{\partial\Lambda}{\partial g_{\rho\sigma}}\delta g_{\rho\sigma}, \qquad (6.11)$$

where the coefficients of the metric variations are not independent of the others but must satisfy a Noether type identity of the form

$$\frac{\partial\Lambda}{\partial g_{\rho\sigma}} = \frac{\partial\Lambda}{\partial g_{\sigma\rho}} = \frac{1}{2}\mu^{\text{X}}_\nu n^\rho_{\text{X}} g^{\nu\sigma} + \frac{1}{2}\lambda^{\nu\rho}_{\text{X}} w^{\text{X}\sigma}_\nu + \frac{1}{16\pi}\mathcal{H}^{\nu\rho}F_\nu{}^\sigma. \qquad (6.12)$$

The triplet of covectorial quantities μ^{X}_ρ is simply interpretable as representing usual 4-momenta (per particle) of the neutrons, protons, and leptons. The triplet of rather less familiar bivectorial coefficients $\lambda^{\sigma\rho}_{\text{X}} = -\lambda^{\rho\sigma}_{\text{X}}$ in this expansion characterises the macroscopic anisotropy arising respectively from the concentration of energy and tension in mesoscopic vortices of the neutron and proton superfluids, as a consequence of their vorticity quantisation conditions, in the manner discussed in our previous work [5] on the single constituent model. These new four dimensional bivectorial coefficients replace the three dimensional (space) vectorial coefficients introduced for a similar purpose in a more restricted Newtonian framework by Bekarevich and Khalatnikov [9]. Finally the bivectorial coefficient $\mathcal{H}^{\rho\sigma} = -\mathcal{H}^{\rho\sigma}$ will be interpretable as an electromagnetic displacement tensor, in terms of which the total electromagnetic field tensor (6.7) will be given by an expression of the form

$$F^{\rho\sigma} = \mathcal{H}^{\rho\sigma} + 4\pi\mathcal{M}^{\rho\sigma}, \qquad (6.13)$$

in which $\mathcal{M}^{\rho\sigma}$ is what can be interpreted as the magnetic polarisation tensor. In the application that we are considering, this – typically dominant – Abrikosov polarisation contribution $4\pi\mathcal{M}^{\rho\sigma}$ is to be thought of as representing the part of the magnetic field confined in the vortices, while the – typically much smaller – remainder $\mathcal{H}^{\rho\sigma}$ represents the average contribution from the field in between the vortices, which can be expected to vanish by the "Meissner effect" in strictly static configurations, but which can be expected to acquire a non zero value in rotating configurations due to the London mechanism that will be discussed below.

In the application of such a generalised Kalb–Ramond type variation principle, the gauge fields $b_{\text{x}}^{\sigma\rho}$ and A_ρ are to be considered as free variables, but n_{x}^ρ and $w_{\rho\sigma}^{\text{x}}$ are not. Each current n_{x}^ρ is to be considered as fully determined by the corresponding gauge bivector $b_{\text{x}}^{\rho\sigma}$ according to the prescription (6.5) while each vorticity 2 form $w_{\rho\sigma}^{\text{x}}$ is to be considered as being determined by corresponding freely chosen scalar base coordinate pullback fields χ_{x}^1, χ_{x}^2, which means that the variation of any vorticity 2 form $w_{\rho\sigma}$ will be determined by a corresponding freely chosen displacement vector field ξ^ρ according to a prescription [5] of the form $\delta w_{\mu\nu} = -2\nabla_{[\mu}(w_{\nu]\rho}\xi^\rho)$.

Subject to these rules, the "diamond" variational integrand

$$\Diamond\mathcal{L} = \|g\|^{-1/2}\delta\left(\|g\|^{1/2}\mathcal{L}\right) = \delta\mathcal{L} + \tfrac{1}{2}\mathcal{L}g^{\mu\nu}\delta g_{\mu\nu}, \qquad (6.14)$$

needed for the application of the variational principle will be given by

$$\Diamond\mathcal{L} = \left(\nabla_\rho \pi_\sigma^{\text{x}} - \tfrac{1}{2}w_{\rho\sigma}^{\text{x}}\right)\delta b_{\text{x}}^{\rho\sigma} + f_\rho^{\text{x}}\xi_{\text{x}}^\rho$$
$$+\left(J^\rho - \frac{1}{4\pi}\nabla_\sigma\mathcal{H}^{\rho\sigma}\right)\delta A_\rho + \tfrac{1}{2}T^{\mu\nu}\delta g_{\mu\nu} + \nabla_\sigma\mathcal{R}^\sigma, \qquad (6.15)$$

in which it is useful to allow for the possibility of varying the background spacetime metric $g_{\rho\sigma}$, not only for the purpose of dealing with cases in which one may be concerned with General Relativistic gravitational coupling, but even for dealing with cases in which one is concerned only with a flat Minkowski background, since, as will be made explicit below, the effect of virtual virtuations with respect to the relevant curved or flat background can be used for evaluating the relevant "geometric" stress energy momentum density tensor $T^{\rho\sigma}$. The coefficients of the current variations are the usual gauge dependent total momentum covectors given by

$$\pi_\rho^{\text{x}} = \mu_\rho^{\text{x}} + e^{\text{x}} A_\rho. \qquad (6.16)$$

The coefficients f_ρ^{x} of the three independent displacement displacement vector fields ξ_{x}^ρ will be interpretable as the effective force densities acting on the corresponding constituent currents. The generic expression for these force densities can be read out for the neutrons X=n, protons and X=p and leptons X=e respectively as

$$f_\rho^{\text{n}} = \left(n_{\text{n}}^\sigma + \nabla_\nu\lambda_{\text{n}}^{\sigma\nu}\right)w_{\sigma\rho}^{\text{n}}, \qquad (6.17)$$

$$f_\rho^{\rm p} = \left(n_{\rm p}^\sigma + \nabla_\nu \lambda_{\rm p}^{\sigma\nu}\right) w_{\sigma\rho}^{\rm p}, \tag{6.18}$$

$$f_\rho^{\rm e} = \left(n_{\rm e}^\sigma + \nabla_\nu \lambda_{\rm e}^{\sigma\nu}\right) w_{\sigma\rho}^{\rm e}. \tag{6.19}$$

Although it is of no relevance for the application of the variation principle, it can be noted for the record that the current appearing in the final divergence term of (6.15) will be given by

$$\mathcal{R}^\sigma_{\rho} = \pi^{\rm x}_\rho \delta b^{\rho\sigma}_{\rm x} - \left(b^{\rho\sigma}_{\rm x} + \lambda^{\rho\sigma}_{\rm x}\right) w^{\rm x}_{\rho\nu} \xi^\nu_{\rm x}$$
$$+ \frac{1}{2} \pi^{\rm x}_\rho b^{\rho\sigma}_{\rm x} g^{\mu\nu} \delta g_{\mu\nu} + \frac{1}{4\pi} H^{\sigma\nu} \delta A_\nu. \tag{6.20}$$

An entity of much greater practical interest is the corresponding stress momentum energy density tensor, which can be seen to be given by

$$T_\sigma^{\rho} = n^\rho_{\rm x} \mu^{\rm x}_\sigma + \lambda^{\nu\rho}_{\rm x} w^{\rm x}_{\nu\sigma} + \frac{1}{8\pi} \mathcal{H}^{\nu\rho} F_{\nu\sigma} + \Psi g^\rho_{\sigma}, \tag{6.21}$$

where the generalised pressure function is given by

$$\Psi = \Lambda - n^\nu_{\rm x} \mu^{\rm x}_\nu + b^{\rho\sigma}_{\rm x} \left(\nabla_\rho \pi^{\rm x}_\sigma - \frac{1}{2} w^{\rm x}_{\rho\sigma}\right). \tag{6.22}$$

The last term in (6.15) will evidently drop out when we impose the condition of invariance with respect to infinitesimal variations of the bivectorial gauge potentials $b^{\rho\sigma}_{\rm n}$ and $b^{\rho\sigma}_{\rm p}$ is imposed, a requirement which can be seen from (6.15) to give field equations of the form

$$w^{\rm x}_{\rho\sigma} = 2\nabla_{[\rho} \pi^{\rm x}_{\sigma]} = 2\nabla_{[\rho} \mu^{\rm x}_{\sigma]} + e^{\rm x} F_{\rho\sigma}, \tag{6.23}$$

which are evidently equivalent to what in other formulations could be considered just as defining relations for the vorticity two-forms. The remaining field equations obtained from (6.15) will consist of

$$\nabla_\sigma \mathcal{H}^{\rho\sigma} = 4\pi J^\rho, \tag{6.24}$$

together with the condition that the force density coefficients should all vanish, i.e.

$$f^{\rm x}_\rho = 0 \tag{6.25}$$

for each of the three relevant chemical index values X=n, X=p, X=e.

6.3 The Semi-Macroscopic Application

To be more specific, we need to specify the scale of application for which the model is intended. At a mesoscopic level, meaning on scales large compared with the dimensions of individual molecules or Cooper type pairs, but small compared with the intervortex spacing, the appropriately specialised model will be of purely fluid type, meaning that the function Λ should not depend on the vorticity forms $w^x_{\rho\sigma}$ which implies the vanishing of the Bekarevich–Khalatnikov coefficients, i.e. the restriction $\lambda^{\sigma\rho}_x = 0$. A further restriction that applies at this mesoscopic level is that for the superfluid constituents, namely the neutrons and the protons, but not for the degenerate lepton constituent, the corresponding vorticities themselves should be zero, i.e. for X $\neq$ e we should have $w^x_{\rho\sigma} = 0$, which is the integrability condition for the corresponding momenta to have the form $2\pi^n_\rho = \hbar\nabla_\rho \varphi^n$, $2\pi^p_\rho = \hbar\nabla_\rho \varphi^p$ in which the scalars φ^n and φ^p will be interpretable as the phases of underlying bosonic quantum condensates, with periodicity 2π, and in which the preceding factors of 2 have been inserted to allow for the fact that the relevant bosons will consist not of single neutrons and protons but of Cooper type pairs thereof.

At a much larger, fully macroscopic scale, involving averaging over large numbers of the neutron and proton vortices (that arise as topological defects of the mesoscopic phase fields) the neutronic and protonic constituents will be characterised by not just by non vanishing effective large scale vorticities $w^n_{\rho\sigma} \neq 0$ and $w^p_{\rho\sigma} \neq 0$, but also by non vanishing Bekarevich–Khalatnikov coefficients, $\lambda^{\sigma\rho}_n \neq 0$ and $\lambda^{\sigma\rho}_p \neq 0$, so that only the degenerate electrons still behave in an effectively fluid manner, in accordance with the restriction

$$\lambda^{\sigma\rho}_e = 0. \tag{6.26}$$

The purpose of the present article is to focus on an intermediate scale that will be referred to as "semi macroscopic" meaning that it deals with averages over scales that are large compared with the spacing between proton vortices, but small compared with the spacing between the neutron vortices, which are expected to be relatively widely spaced in typical circumstances within neutron stars, whose angular velocities are very low as measured by local physical timescales, whereas their magnetic fields are typically rather large. On such "semi macroscopic" scales the behaviour of the neutrons constituent will not just be of strictly fluid type, meaning that it will be characterised by

$$\lambda^{\sigma\rho}_n = 0, \tag{6.27}$$

but it will also be subject to the mesoscopic superfluidity condition

$$2\pi^n_\rho = \hbar\nabla_\rho \varphi^n, \tag{6.28}$$

and hence

$$w^n_{\rho\sigma} = 0, \tag{6.29}$$

so that the corresponding dynamical equation, i.e. the requirement (6.25) that the effect that the net force density (6.17) should vanish, will be automatically satisfied everywhere outside the microscopic cores of the neutron superfluid vortices (which are of "global" type, meaning that their energy would diverge logarithmically in the absence of the "infra red" cut off imposed by the presence of neighbouring vortices). However since we are considering scales large compared with the separation distance between the much more numerous proton vortices (which are of "local" type, meaning that their energy density falls off exponentially on a microscopic lengthscale ℓ whose evaluation will be discussed below) the proton constituent will be characterised not only by non vanishing averaged vorticity, $w^{\rm P}_{\rho\sigma} \neq 0$ but also by a non vanishing Bekarevich–Khalatnikov coefficient, $\lambda^{\sigma\rho}_{\rm p} \neq 0$. This means that the dynamical requirement that the corresponding net force density (6.18) should vanish will provide a rather complicated dynamical equation, expressible in the form

$$2n^\sigma_{\rm p} \nabla_{[\sigma} \mu^{\rm P}_{\rho]} + en^\sigma_{\rm p} F_{\sigma\rho} + w^{\rm P}_{\sigma\rho} \nabla_\nu \lambda^{\sigma\nu}_{\rm p} = 0, \qquad (6.30)$$

in which the first term is interpretable as the negative of the Joukovski force density due to the "Magnus effect" acting on the proton vortices, the middle term is the Lorentz force density representing the effect of the magnetic field on the passing protons, while the last term (which was absent in the mesoscopic description) represents the extra force density on the fluid due to the effect of the tension of the vortices. As a consequence of its "normality" property (6.26), the leptonic constituent is governed by a dynamical equation of the simpler form

$$2n^\sigma_{\rm e} \nabla_{[\sigma} \mu^{\rm e}_{\rho]} + en^\sigma_{\rm e} F_{\sigma\rho} = 0. \qquad (6.31)$$

In view of the highly localised nature of the proton vortices, it is reasonable [10] to suppose that their action contribution should be fully determined just by the Abrikosov lattice density of such vortices, and that the only independent contribution from the electromagnetic field $F_{\rho\sigma}$ should be the part provided by the residual – weaker but much more widely extended – part of the flux outside the vortex tubes, as given by the field $\mathcal{H}^{\rho\sigma}$ that is given by the relation (6.13), on the understanding that the polarisation contribution $4\pi\mathcal{M}^{\rho\sigma}$ represents the part of the flux attributable to the the vortex tubes. This implies that that the gauge independent term Λ in the action will be decomposable in the form

$$\Lambda = \Lambda_{\rm MV} + \Lambda_{\rm F}, \qquad (6.32)$$

where the macroscopic contribution $\Lambda_{\rm MV}$ is required to be functionally independent of $F_{\mu\nu}$, so that it is determined just by the vorticity $w^{\rm P}_{\rho\sigma}$ and the currents $n^\rho_{\rm X}$, while for consistency with the variational definition (6.11) there will be no loss of generality in taking the remaining, electromagnetic field dependent, contribution to have the simple quadratic form

$$\Lambda_{\rm F} = \frac{1}{16\pi} \mathcal{H}_{\rho\sigma} \mathcal{H}^{\sigma\rho}, \qquad (6.33)$$

which, in the absence of the polarisation contribution $4\pi\mathcal{M}^{\rho\sigma}$ in (6.13), would reduce just to the usual action contribution for an electromagnetic field in vacuum.

Despite of being much more densely packed than the neutron vortices, the fact that (unlike the neutron vortices) the proton vortices are exponentially localised within a microscopic confinement radius means that their mutual interactions (unlike those of the long range interacting neutron vortices) should remain entirely negligible even for extremely high magnetic fields, so that their contribution to the action should be simply proportional to their density. This means that it will be possible to make the further decomposition

$$\Lambda_{\rm MV} = \Lambda_{\rm M} + \Lambda_{\rm V}\,, \tag{6.34}$$

in which $\Lambda_{\rm M}$ is entirely independent of the vorticity $w^{\rm p}_{\rho\sigma}$, so that it depends only on the three independent currents $n^\rho_{\rm x}$, while the remainder $\Lambda_{\rm V}$ is just linearly proportional to the protonic vorticity density, so that it will be expressible in the form

$$\Lambda_{\rm V} = \lambda_{\rm p} w^{\rm p} \tag{6.35}$$

where $w^{\rm p}$ is the protonic vorticity magnitude as defined by

$$w^{\rm p} = \sqrt{w^{{\rm p}\,\rho\sigma} w^{\rm p}{}_{\rho\sigma}/2}\,, \tag{6.36}$$

and where, like $\Lambda_{\rm M}$, the coefficient $\lambda_{\rm p}$ depends only on the currents $n^\rho_{\rm x}$. On the basis of dimensional considerations – which should be valid provided the Ginzburg landau ratio of the London penetration length ℓ that will be discussed below to the relevant Pippard correlation length is not too far from the order of unity value that characterises the Bogomol'nyi limit [12] – it can be anticipated that, as is confirmed by more detailed analysis [13,10] the Bekarevich–Khalatnikov coefficient $\lambda_{\rm p}$ will have an order of magnitude given in terms of the relevant charged particle mass, which in the neutron star case under consideration is the proton mass $m_{\rm p}$ (but which in an ordinary metallic superconductor would be the electron mass $m_{\rm e}$) by by $\lambda \approx \hbar n_{\rm p}/m_{\rm p}$.

The appropriate form for the polarisation tensor $4\pi\mathcal{M}^{\rho\sigma}$ in (6.13), can be seen by decomposing the vector potential A_ρ as the sum of a gauge dependent contribution proportional to the proton momentum covector $\pi^{\rm p}_\rho$ and a gauge independent remainder $\mathcal{A}_\rho$ in the form

$$A_\rho = \frac{1}{e}\pi^{\rm p}_\rho + \mathcal{A}_\rho\,, \tag{6.37}$$

from which, by exterior differentiation, one obtains a corresponding decomposition of the form

$$F_{\rho\sigma} = \frac{1}{e} w^{\rm p}_{\rho\sigma} + \mathcal{H}_{\rho\sigma}\,, \tag{6.38}$$

with

$$\mathcal{H}_{\rho\sigma} = 2\nabla_{[\rho}\mathcal{A}_{\sigma]}\,. \tag{6.39}$$

This decomposition has the required form (6.13) provided one makes the identification

$$\mathcal{M}_{\rho\sigma} = \frac{1}{4\pi e} w^{\mathrm{p}}_{\rho\sigma}, \tag{6.40}$$

for what will be referred to as the Abrikosov polarisation tensor.

6.4 Phenomenological Interpretation

As discussed in more detail in particular cases [15,10] the quantity $4\pi\mathcal{M}_{\rho\sigma}$ defined by (6.40) will be interpretable as representing the part of the magnetic flux confined to the proton vortices, whose action contribution will be included in the term Λ_{V} given by (6.35), while $\mathcal{H}_{\rho\sigma}$ accounts for the remainder of the flux, which will be distributed over the region outside the proton vortices, and whose contribution to the action will be given by (6.33). The covector $\mathcal{A}$ will be given by

$$\mathcal{A}_\rho = -\frac{1}{e}\mu^{\mathrm{p}}_\rho, \tag{6.41}$$

so that it will be obtainable from the equation of state function for μ^{p}_ρ as derived from Λ_{M} as a linear combination of the form

$$\mathcal{A}_\rho = \mathcal{A}^{\mathrm{n}}_\rho + \mathcal{A}^{\mathrm{L}}_\rho + \mathcal{A}^{\mathrm{M}}_\rho, \tag{6.42}$$

in which the terms are proportional respectively to the neutron 4-momentum, the "normal" reference state unit vector u^ρ (as specified according to (6.1) by the leptonic current), and the (semi macroscopic) electric current J^ρ, so that they will be expressible as

$$\mathcal{A}^{\mathrm{n}}_\rho = \frac{1}{e}\alpha^{\mathrm{p}}_{\mathrm{n}} \pi^{\mathrm{n}}_\rho, \qquad \mathcal{A}^{\mathrm{L}}_\rho = -\frac{\mu^{\mathrm{L}}}{e} u_\rho, \qquad \mathcal{A}^{\mathrm{M}}_\rho = -4\pi\ell^2 J_\rho, \tag{6.43}$$

with proportionality factors that depend (just) on the form of the master function specifying Λ_{M} in terms of the relevant currents. In particular the dimensionless parameter α would be zero if there were no entrainment, but in view of the effect originally predicted by Andreev and Bashkin [6] can be expected [11] to be of the order of unity, while the effective London mass parameter μ^{L} will have a magnitude that is the same as that of the relevant charge carriers, in this case protons, to within a factor comparable with unity, from which it differs by an amount that also depends on the entrainment effect. The third parameter ℓ is interpretable as the relevant London penetration lengthscale that characterises the effective thickness of the individual proton vortices of the Abrikosov lattice. If it is assumed, as most authors have done, that the entrainment effect only couples the neutrons and protons but does not significantly involve the leptonic background (so that the a master function Λ_{M} of semi separable form [15] can be used) then it can be estimated that this length scale will be given roughly as a function of the effective London mass $\mu^{\rm L}$ and the lepton number density $n_{\rm e}$ (which must be very close to the proton number density) by

$$\ell^2 \simeq \frac{\mu^{\rm L}}{4\pi e^2 n_{\rm e}} \,. \tag{6.44}$$

It follows from (6.43) that there will be a corresponding decomposition

$$\mathcal{H}_{\rho\sigma} = \mathcal{H}^{\rm n}_{\rho\sigma} + \mathcal{H}^{\rm L}_{\rho\sigma} + \mathcal{H}^{\rm M}_{\rho\sigma}\,, \tag{6.45}$$

with

$$\mathcal{H}^{\rm n}_{\rho\sigma} = 2\nabla_{[\rho}\mathcal{A}^{\rm n}_{\sigma]}\,, \qquad \mathcal{H}^{\rm L}_{\rho\sigma} = 2\nabla_{[\rho}\mathcal{A}^{\rm L}_{\sigma]}\,, \qquad \mathcal{H}^{\rm M}_{\rho\sigma} = 2\nabla_{[\rho}\mathcal{A}^{\rm M}_{\sigma]}\,, \tag{6.46}$$

in which far as the averaged flux is concerned, the main contribution will typically be that of the London field $\mathcal{H}^{\rm L}_{\rho\sigma}$, meaning the part attributable to the rotation of the "normal" (i.e. non-superfluid) negatively charged background. (In an ordinary metallic superconductor the analogous London field contribution arises as a well known consequence of rotation of the positively charged ionic background). In the absence of entrainment, the neutron vortex contribution $\mathcal{H}^{\rm n}_{\rho\sigma}$ would vanish. However the expectation [11] that the entrainment coefficient $\alpha^{\rm p}_{\rm n}$ will actually be of the order order unity implies that although it can be expected to be extremely small outside the immediate neighbourhood of a neutron vortex core (with a confinement radius of the same microscopic order of magnitude ℓ as that of a proton vortex) the integrated flux arising from neutron vortex contribution $\mathcal{H}^{\rm n}_{\rho\sigma}$ can be expected to be comparable with that provided by the more smoothly distributed (unconfined) London contribution $\mathcal{H}^{\rm L}_{\rho\sigma}$.

In normal circumstances, the least important term in the sum (6.45) will be what we shall refer to as the Meissner residue, meaning the residual contribution $\mathcal{H}^{\rm M}_{\rho\sigma}$ arising from the semi-macroscopic current J^ρ if any. On the assumption that the lengthscales characterising variation of the coefficients $\mu^{\rm L}$ and ℓ are very long compared with the London penetration lengthscale ℓ itself, it can be seen that – except within the microscopic defects forming the actual vortex cores where the mesoscopic superfluid description breaks down so that $w^{\rm n}_{\rho\sigma}$ is locally non zero – the dominant contribution in the source equation (6.24) will be the part arising from the semi-macroscopic current J^μ itself, so that, as a very good approximation, the source equation will reduce to the well known London form

$$\nabla^\sigma \nabla_\sigma J^\rho \simeq \frac{1}{\ell^2} J^\mu\,, \tag{6.47}$$

whose homogeneous linear character entails that the only spacially and temporally uniform solution is that for which the current J^μ simply vanishes. What this implies is that after any initial high frequency osillations that may have been present have had time to radiate away or otherwise be dissipated, the medium will tend to settle towards a state in which the current actually is zero outside a very small radius of order ℓ surrounding each individual vortex, so that its average $\langle J^\rho \rangle$ over scales large compared with ℓ will also tend to vanish,

$$\langle J^\rho \rangle \simeq 0 \,. \tag{6.48}$$

The same conclusion therefore applies to the corresponding residual Meissner field contribution, $\mathcal{H}^\mathrm{M}_{\rho\sigma}$, which will end up in a state such that

$$\langle \mathcal{H}^\mathrm{M}_{\rho\sigma} \rangle \simeq 0 \,. \tag{6.49}$$

This last result is interpretable as a generalisation to "type II" (London–Abrikosov) superconductors of the Meissner effect that was originally observed in laboratory examples of "type I", meaning cases in which the Ginzburg Landau ratio of the penetration lengthscale ℓ to the relevant Pippard correlation lenghthscale is so small that, instead of condensing into an Abrikosov vortex lattice, the magnetic flux tends to be entirely expelled into domains where the superconductivity breaks down. In a type I situation the polarisation $\mathcal{M}_{\rho\sigma}$ associated with the Abrikosov lattice will be absent, so there will be no distinction between the total flux $F_{\rho\sigma}$ and the contribution $\mathcal{H}_{\rho\sigma}$ as defined here, which means that in this experimentally more familiar case, expulsion of $\mathcal{H}_{\mu\nu}$ is equivalent to complete expulsion of $F_{\rho\sigma}$. On the other hand in the type II case, although there will be the same tendency to expulsion of $\mathcal{H}_{\rho\sigma}$, this will not entail the complete expulsion of $F_{\rho\sigma}$ because the Abrikosov polarisation contribution $\mathcal{M}_{\mu\nu}$ will still remain.

Both in the type II and – as originally remarked by London – in the type I case, the tendency for $\mathcal{H}_{\mu\nu}$ to vanish will be partially thwarted in a rotating background, for which the London contribution $\mathcal{H}^\mathrm{L}_{\rho\sigma}$ will still remain even after the residual Meissner contribution $\mathcal{H}^\mathrm{M}_{\rho\sigma}$ has been dissipated in accordance with (6.49). In the usual laboratory applications, whether of type I or type II, this London contribution is all that will remain, but in the neutron star case there will also be the neutron vortex contribution $\mathcal{H}^\mathrm{n}_{\rho\sigma}$. If the gradient of the (weakly density dependent) coefficient a^p_n is not entirely negligible, then as well as having a dominant part that is of magnetic character, may also include a small contribution to the electric displacement vector D_ρ as defined with respect to the "normal" leptonic background frame by the decomposition

$$\mathcal{H}_{\rho\sigma} = H_{\rho\sigma} + 2u_{[\rho}D_{\sigma]} \,, \qquad D_\rho = \mathcal{H}_{\rho\sigma}u^\sigma \,. \tag{6.50}$$

Since the alignment of the neutron momentum covector will not on average be very different from that of the background frame, it can be seen that what is to be expected is that under typical equilibrium conditions the macroscopically averaged value of the first contribution in (6.45) will be given by

$$\langle \mathcal{H}^\mathrm{n}_{\rho\sigma} \rangle \simeq 2u_{[\rho}\langle D^\mathrm{n}_{\sigma]}\rangle + \langle H^\mathrm{n}_{\rho\sigma} \rangle \,, \tag{6.51}$$

in which the main contribution is the purely magnetic part given by

$$H^{\mathrm{n}}_{\rho\sigma} \simeq \frac{1}{e} \langle \alpha^{\mathrm{p}}_{\mathrm{n}} w^{\mathrm{n}}_{\rho\sigma} \rangle, \qquad (6.52)$$

while the – typically very small – electric part will be given by

$$\langle D^{\mathrm{n}}_{\rho} \rangle \simeq -\frac{1}{e} \langle \mu^{\mathrm{n}} \nabla_{\rho} \alpha^{\mathrm{p}}_{\mathrm{n}} \rangle, \qquad (6.53)$$

where the relevant neutron Fermi energy parameter is given by $\mu^{\mathrm{n}} = -u^{\sigma} \mu^{\mathrm{n}}_{\sigma}$.

The analogous macroscopic average of the London contribution can be instructively formulated in terms of the "normal" background's acceleration tensor $\dot{u}^{\rho}$ and rotation tensor $\Omega_{\rho\sigma}$ (whose magnitude $\Omega = \sqrt{\Omega_{\rho\sigma} \Omega^{\rho\sigma}/2}$ is the local angular velocity) as defined by

$$\nabla_{[\rho} u_{\sigma]} = \Omega_{\rho\sigma} - u_{[\rho} \dot{u}_{\sigma]}, \qquad \dot{u}^{\rho} = u^{\sigma} \nabla_{\sigma} u^{\rho}. \qquad (6.54)$$

The ensuing result (correcting a misplaced factor of 2 in the preceding presentation [15]) is expressible as

$$\langle \mathcal{H}^{\mathrm{L}}_{\rho\sigma} \rangle \simeq 2 u_{[\rho} \langle D^{\mathrm{L}}_{\sigma]} \rangle + \langle H^{\mathrm{L}}_{\rho\sigma} \rangle, \qquad (6.55)$$

in which the main contribution is the magnetic part which can be seen to be given by

$$\langle H^{\mathrm{L}}_{\rho\sigma} \rangle = -\frac{2}{e} \langle \mu^{\mathrm{L}} \Omega_{\rho\sigma} \rangle. \qquad (6.56)$$

As originally observed by London, this is proportional to the angular velocity Ω, with a proportionality factor μ^{L} that in the neutron star application will be given roughly (but due to the "entrainment" effect not exactly) by the proton mass (whereas in the ordinary metallic superconductors originally envisaged by London it is given by the electron mass). As well as this well known magnetic contribution, there will also be a corresponding, but typically much less important, electric displacement contribution given by

$$\langle D^{\mathrm{L}}_{\rho} \rangle \simeq \frac{1}{e} \langle \mu^{\mathrm{L}} \dot{u}_{\sigma} + \nabla_{\sigma} \mu^{\mathrm{L}} \rangle. \qquad (6.57)$$

(This electric displacement field is usually ignored in discussions of laboratory applications, but even in the ideally simplified case of a motionless incompressible sample that is postulated to be strictly homogeneous so that the gradient term would be absent, a small residual electric field of this type would still be needed to balance the effect on the conducting particles – which in that case would be electrons – of the ordinary terrestrial gravitation field.)

I wish to thank Silvano Bonazzola, David Langlois, and Reinhard Prix for many helpful discussions, and I particularly wish to express my appreciation to Isaac Khalatnikov for instructive conversations on numerous occasions since my interest in this subject was originally inspired by his classic textbook [14].

References

1. B. Carter, "The canonical treatment of heat conduction and superfluidity in relativistic hydrodynamics", in *A random walk in Relativity and Cosmology (Essays in honour of P.C. Vaidya & A.K. Raychaudhuri)*, ed. N. Dadhich, J. Krishna Rao, J.V. Narlikar, C.V. Visveshwara, pp 48–62 (Wiley Eastern, Bombay, 1985).
2. B. Carter, D. Langlois, "Relativistic model for superconducting superfluid mixtures", *Nucl. Phys.* **B531**, pp 478–504 (1998) [gr-qc/9806024].
3. V.V. Lebedev, I.M. Khalatnikov, "Relativistic hydrodynamics of a superfluid", *Sov. Phys. J.E.T.P.* **56**, pp 923–930 (1982).
4. B. Carter, I.M. Khalatnikov, "Momentum, Vorticity, and Helicity in Covariant Superfluid Dynamics", *Ann. Phys.* **219**, pp 243–265 (1992).
5. B. Carter, D. Langlois, "Kalb–Ramond coupled vortex fibration model for relativistic superfluid dynamics", *Nuclear Physics* **B 454**, 402–424 (1995) [hep-th/9611082].
6. A.F. Andreev, E.P. Bashkin, "Three velocity hydrodynamics of superfluid solutions" *Sov. Phys. J.E.T.P.*, **42**, 164–646 (1975).
7. G.A. Vardanyan, D.M. Sedrakyan, "Magnetohydrodynamics of superfluid solutions", *Sov. Phys. J.E.T.P.* **54**, 919–921 (1981).
8. G. Mendell, L. Lindblom, "The coupling of charged superfluid mixtures to the electromagnetic field", *Ann. Phys.* **205**, 110–129 (1991).
9. I.L. Bekarevich and I.M. Khalatnikov, "Phenomenological derivation of the equations of vortex motion in HeII", *Sov. Phys. J.E.T.P.* **13**, 643 (1961).
10. B. Carter, R. Prix, D. Langlois, "Energy of Magnetic Vortices in Rotating Superconductor" *Phys. Rev.* **B62** [cond-mat/9910240].
11. M.A. Alpar, S.A. Langer, J.A. Sauls, "Rapid postglitch spin-up of the superfluid core in pulsars", *Astroph. J.* **282**, 533–541 (1984).
12. B. Carter, R. Prix, D. Langlois, "Bogomoln'yi limit for magnetic vortices in rotating superconductor" *Phys. Rev.* **B62** (2000) [cond-mat/9910263].
13. G. Mendel, "Superfluid hydrodynamics in rotating neutron stars. I Nondissipative equations" *Astroph. J* **380**, pp 515–529 (1991).
14. I. M. Khalatnikov, *Introduction to the Theory of Superfluidity* (Benjamin, New York, 1965).

Part II

Vortex Dynamics, Spectral Flow and Aharonov–Bohm Effect

7 Vortex Dynamics and the Problem of the Transverse Force in Clean Superconductors and Fermi Superfluids

N.B. Kopnin

Summary. We review the basic ideas and results on the vortex dynamics in clean superfluid Fermi systems. The forces acting on moving vortices are discussed including the problem of the transverse force which was a matter of confusion for quite a time. We formulate the equations of the vortex dynamics which include all the forces and the inertial term associated with excitations bound to the moving vortex.

7.1 Introduction

The study of the vortex dynamics opens promising perspectives for understanding the most fundamental properties of condensed matter, especially for superconductors and superfluids. Among them, clean systems, i.e., the systems where the mean free path of quasiparticles is much longer than the characteristic coherence length, offer more intriguing physics than dirty superconductors. For example, one of the fundamental problems can be formulated as follows: Speaking of clean superconductors one can, in particular, think of such a system where no relaxation processes are available, i.e., the mean free path of excitations is infinite. In particular, vortices in such systems should move without dissipation since there is no mechanism to absorb the energy. The vortex velocity should then be parallel to the transport current, which makes the induced electric field perpendicular to the current; the dissipation thus vanishes $\mathbf{j} \cdot \mathbf{E} = 0$. This contrasts with what we know of dirty superconductors: the vortex motion there is dissipative, and each moving vortex experiences a large friction; it generates an electric field parallel to the transport current which produces energy dissipation. Clearly, a crossover should occur from dissipative to non-dissipative vortex motion as the quasiparticle mean free path increases. The question is what is the condition which controls the crossover?

This question is of a fundamental importance for our understanding of the dynamics of superconductors and, in a more broad sense, for the understanding of dynamic properties of quantum condensed matter in general. To illustrate the problem one can consider a simple example as follows. One can argue that a time dependent non-dissipative superconducting state, similarly to any other quantum state, can be described by a Hamiltonian dynamics based on a time dependent Schrödinger equation. Such a description for a weakly interacting Bose gas has been suggested by Pitaevskii and Gross [1,2]. It is widely used for superfluid helium II as well. The Gross–Pitaevskii equation is essentially a nonlinear Schrödinger equation, it has the imaginary factor $i\hbar$ in front of the time derivative of the condensate wave function $\partial \Psi/\partial t$. On the other hand, the time dependent Ginzburg–Landau model which is a particular case of a more general Model F dynamics [41] is believed to describe a relaxation dynamics of superconductors near the transition temperature. In contrast to the Gross–Pitaevskii equation, it has the time derivative $\partial \Psi/\partial t$ with a real factor in front of it. The question which we are interested in can be formulated as follows: What is the condition when the imaginary prefactor transforms into a real one?

It seems that there is no universal answer to this simple question in general. However, the problem of crossover from non-dissipative to dissipative behavior of a condensed matter state can be solved for the particular example of vortex dynamics. It is known that the relaxation constant in the time dependent Ginzburg–Landau model has in fact a small imaginary part [4–6] which results in appearance of a small transverse component of the electric field with respect to the current. We shall see later that the transverse component of the electric field increases at the expense of the longitudinal component as the mean free path of excitations grows. The crossover condition, however, does not coincide simply with the condition that divides superconductors between dirty and clean ones. The criterion rather involves the spectrum of excitations localized in the vortex cores; the distance between their levels takes the part of the energy gap. The condition for a non-dissipative vortex motion requires that the relaxation rate of localized excitations is smaller than the distance between the levels. This implies a much longer mean free path of excitations than the condition for a superconductor to be just in a clean limit.

In this chapter we review the basic ideas and results on the vortex dynamics in clean Fermi systems. We concentrate largely on superconductors, the case of superfluid ^{3}He can be easily incorporated if one takes the limit of zero charge of carriers in the final results (to be discussed it in more detail later). We consider forces acting on moving vortices and clarify the conditions for the crossover from dissipative to non-dissipative vortex dynamics. A great deal of attention is given to the problem of the so called transverse force, i.e., the force perpendicular to the vortex velocity, which was a matter of confusion for quite a time, since its discovery. Finally, we formulate the equations of the vortex dynamics which include all the forces and the inertial term associated with excitations bound to the moving vortex.

All our consideration is restricted to the vortex dynamics in isothermal conditions when the temperature is uniform in space. In presence of a temperature gradient, additional forces appear as explained in Chap. 19 [7]. The next chapter in this book [8] is devoted to the discussion of the transverse force in a superfluid Bose liquid; in particular, it gives a detailed derivation of that part of the force which is called the Iordanskii force. More contributions of smaller magnitudes to the total force can be found if one includes into consideration other subtle effects such as vortex-motion-induced variations of the pairing interaction itself between particles comprising the Cooper pairs. Discussion of these effects can be found in [4–6,9,10]. One of such issues associated with a charge of a vortex is considered in Chap. 17 [47] in this book.

7.2 Boltzmann Kinetic Equation Approach

The forces on a vortex come from several different sources including the hydrodynamic Magnus force, the force produced by excitations scattered from the vortex, and the force associated with the momentum flow from the heat bath to the vortex through the localized excitations. The whole rich and exciting physics involved in the vortex dynamics can be successfully described by the general microscopic time dependent theory. Here we rather discuss a general picture using a simple semi-classical approach based on the Boltzmann kinetic equation. The semi-classical approach assumes that the wavelength of excitations is much shorter than the superfluid coherence length, $p_F \xi \gg 1$ (from now on we put $\hbar = 1$). This condition is quite safely fulfilled in almost all superconductors and in superfluid ^{3}He. Only in some high-temperature superconductors may its accuracy not be very good. For simplicity, we consider s–wave superconductors. We concentrate on isolated vortices such that their cores do not overlap, i.e., on the region of magnetic fields $H \ll H_{c2}$.

One can distinguish two types of excitations: Excitations localized in the vortex core, and those which are not localized but move in the vortex potential under the action of magnetic filed. We start our discussion with the localized excitations.

7.2.1 Localized Excitations

We remind that the profile of the order parameter $\Delta(\mathbf{r})$ near the vortex core produces a potential well where localized states with a discrete spectrum exist. The localized states correspond to energies $|\epsilon| < \Delta_\infty$. The spectrum has the so called anomalous chiral branch with the radial quantum number $n = 0$ whose energy varies from $-\Delta_\infty$ to $+\Delta_\infty$ as the particle impact parameter b changes from $-\infty$ to $+\infty$; the chiral branch crosses $\epsilon = 0$ being an odd function of b. For low $\epsilon \ll \Delta_\infty$, the chiral branch is $\epsilon_0 = -\omega_0 \mu$ where $\mu = -b p_\perp$ is the angular momentum, and $p_\perp$ is the momentum in the plane perpendicular to the vortex axis [2]. In a s–wave superconductor with an axisymmetric vortex, the angular momentum μ is quantized and so is the spectrum; ω_0 is thus the distance between the discrete levels in the vortex core. The spectrum also has branches with $n n_e 0$ which are separated from the one with $n = 0$ by energies of the order of Δ. In fact, these branches are

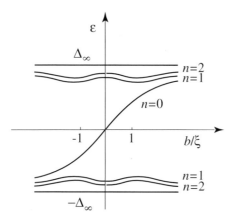

Fig. 7.1. Spectrum of excitations localized in the vortex core.

concentrated very close to the gap edges $\pm\Delta_\infty$ and are almost absorbed by the continuous spectrum [3]. An energy $\epsilon_n(b)$ for $n n_e 0$ does not cross zero of energy and approaches $+\Delta_\infty$ (or $-\Delta_\infty$) for both $b \to +\infty$ and $b \to -\infty$ (see Fig. 7.1; a sketch of the levels is also shown in Fig. 19.4 of Freimuth's chapter [7] in this book). The spectrum has the general symmetry $\epsilon(b) = -\epsilon(-b)$. We denote the separation between the levels with neighboring angular momenta through

$$\omega_n = p_\perp^{-1} \frac{\partial \epsilon_n}{\partial b} = -\frac{\partial \epsilon_n}{\partial \mu}. \tag{7.1}$$

The interlevel spacing $\omega_0(b)$ is an even function of b for $n = 0$. In fact, the spacings ω_n for $n n_e 0$ are much smaller than ω_0.

We choose the direction of the z-axis in such a way that the vortex has a positive circulation. The z-axis is thus parallel to the magnetic field for positive charge of carriers, and it is antiparallel to it for negative charge: $\hat{\mathbf{z}} = \hat{\mathbf{h}} \operatorname{sign}(e)$. Since the particle velocity $\mathbf{v}_\perp$ in the plane perpendicular to the vortex axis makes an angle α with the x-axis, the cylindrical coordinates of the position point (ρ, ϕ) are connected with the impact parameter and the coordinate along the trajectory through $\rho^2 = b^2 + s^2$ where

$$b = \rho \sin(\phi - \alpha) \, ; \quad s = \rho \cos(\phi - \alpha). \tag{7.2}$$

The coordinates are shown in Fig. 7.2.

The first step is as follows. We assume that the quasiclassical spectrum $\epsilon_n(b)$ of a particle plays the role of its effective Hamiltonian. We can thus invoke the Boltzmann equation in the canonical form

$$\frac{\partial f}{\partial t} + \frac{\partial f}{\partial \alpha} \frac{\partial \epsilon_n}{\partial \mu} - \frac{\partial \epsilon_n}{\partial \alpha} \frac{\partial f}{\partial \mu} = \left(\frac{\partial f}{\partial t} \right)_{\text{coll}}, \tag{7.3}$$

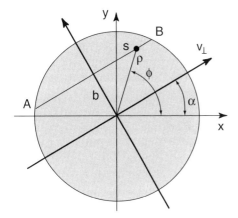

Fig. 7.2. Transformation from cylindrical coordinates to the frame associated with a particle moving in the vortex core. The line AB is the particle trajectory making an angle α with the x-axis and passing through the position point (ρ, ϕ) at a distance b from the vortex axis.

to describe the quasiparticle distribution [14]. Equation (7.3) has been derived in [15] from the set of microscopic kinetic equations.

In the time derivative, the energy ϵ_n contains a time dependence through $\mu(t) = [(\mathbf{r} - \mathbf{v}_\mathrm{L} t) \times \mathbf{p}] \cdot \hat{\mathbf{z}}$ such that

$$\frac{\partial f^{(0)}}{\partial t} = \frac{\partial f^{(0)}}{\partial \epsilon} \frac{\partial \epsilon_n}{\partial \mu} \frac{\partial \mu}{\partial t} = \frac{\partial f^{(0)}}{\partial \epsilon} \frac{\partial \epsilon_n}{\partial \mu} \left([\mathbf{p}_\perp \times \mathbf{v}_\mathrm{L}] \cdot \hat{\mathbf{z}} \right). \tag{7.4}$$

Kinetic equation takes the form

$$\frac{\partial f_1}{\partial t} + \frac{\partial f^{(0)}}{\partial \epsilon} \frac{\partial \epsilon_n}{\partial \mu} \left([\mathbf{p}_\perp \times \mathbf{v}_\mathrm{L}] \cdot \hat{\mathbf{z}} \right) + \frac{\partial f_1}{\partial \alpha} \frac{\partial \epsilon_n}{\partial \mu} - \frac{\partial \epsilon_n}{\partial \alpha} \frac{\partial f_1}{\partial \mu} = \left(\frac{\partial f}{\partial t} \right)_\mathrm{coll}, \tag{7.5}$$

where the distribution function is separated into an equilibrium and a non-equilibrium parts, $f = f^{(0)} + f_1$, respectively. Here $f^{(0)} = 1 - 2n_\epsilon = \tanh(\epsilon/2T)$ with n_ϵ being the Fermi function.

We shall simplify the collision integral in (7.3) using the relaxation–time approximation

$$\left(\frac{\partial f}{\partial t} \right)_\mathrm{coll} = -\frac{f_1}{\tau_n} \tag{7.6}$$

where $\tau_n \sim \tau$. With this approximation, the mean free time can be of any origin. To be definite, we assume that the most effective relaxation is brought about by impurities, as is the case in almost all practical superconducting compounds. However, the relaxation time τ can be due to the electron–phonon interaction, as well, or due to quasiparticle–quasiparticle scattering. The latter is, in fact, the only relaxation mechanism available in ^{3}He [16].

Therefore, the term "clean" does not necessarily mean a low concentration of impurities but simply refers to a situation when the mean free time is long.

Equation (7.3) with the collision integral in the form of (7.6) is easy to solve. For an axi-symmetric s-wave vortex the energies ϵ_n do not depend on α and the term $\partial \epsilon_n / \partial \alpha$ vanishes. Let us take the distribution function in the form

$$f_1 = -\frac{\partial f^{(0)}}{\partial \epsilon} [\gamma_O ([\mathbf{v}_L \times \mathbf{p}_\perp] \cdot \hat{\mathbf{z}}) + \gamma_H (\mathbf{v}_L \cdot \mathbf{p}_\perp)] \tag{7.7}$$

where the factors $\gamma_{O,H}$ are to be found from the Boltzmann equation (7.3). The result for a steady vortex motion is

$$\gamma_O = \frac{\omega_n \tau_n}{\omega_n^2 \tau_n^2 + 1}, \qquad \gamma_H = \frac{\omega_n^2 \tau_n^2}{\omega_n^2 \tau_n^2 + 1}. \tag{7.8}$$

This generalizes the result first obtained in [17].

7.2.2 Delocalized Excitations

A delocalized particle moves mostly far from the vortex core where the order parameter is constant and the superfluid velocity potential $\mathbf{pv}_s$ is small compared to Δ. The kinetic equation for delocalized excitations can thus be written as for a particle in a magnetic field with a semi-classical spectrum

$$\epsilon_p = \sqrt{\xi_p^2 + \Delta_\infty^2} \tag{7.9}$$

where $\xi_p = p^2/2m - E_F$. As shown in [18] the kinetic equation for a particle moving in a vortex array has the conventional form

$$\frac{\partial f}{\partial t} + \frac{\partial f}{\partial \mathbf{p}} \cdot \mathbf{f} + \mathbf{v}_g \cdot \frac{\partial f}{\partial \mathbf{r}} = \left(\frac{\partial f}{\partial t}\right)_{coll}.$$

The force is the elementary Lorentz force

$$\mathbf{f} = \frac{d\mathbf{p}}{dt} = \frac{e}{c} [\mathbf{v}_g \times \mathbf{H}] = \frac{\omega_c}{g} [\mathbf{p}_F \times \hat{\mathbf{z}}] \tag{7.10}$$

where $\omega_c = |e| H/mc$ is the cyclotron frequency,

$$\mathbf{v}_g = \frac{\partial \epsilon_p}{\partial \mathbf{p}} = \frac{\mathbf{v}_F}{g} \tag{7.11}$$

is the group velocity, and

$$g = \frac{\epsilon}{\sqrt{\epsilon^2 - \Delta_\infty^2}}.$$

The driving term can be written as $\partial f/\partial t = \left(\partial f^{(0)}/\partial \epsilon\right) (\partial \epsilon / \partial t)$ where

$$\frac{\partial \epsilon}{\partial t} = \mathbf{f} \cdot \mathbf{v}_L = \frac{\omega_c}{g} \mathbf{p}_\perp \cdot [\hat{\mathbf{z}} \times \mathbf{v}_L].$$

The kinetic equation becomes

$$\frac{\partial f_1}{\partial t} + \frac{\partial f^{(0)}}{\partial \epsilon}\frac{\omega_c}{g}\mathbf{p}_\perp \cdot [\hat{\mathbf{z}} \times \mathbf{v}_L] + \frac{\partial f_1}{\partial \mathbf{p}} \cdot \frac{\omega_c}{g}[\mathbf{p}_F \times \hat{\mathbf{z}}] = \left(\frac{\partial f}{\partial t}\right)_{coll}. \quad (7.12)$$

We omit the spatial derivative of the distribution function since it is constant in space.

For the energy spectrum of (7.9) the collision integral is [19]

$$\left(\frac{\partial f}{\partial t}\right)_{coll} = -\frac{f_1}{g\tau}.$$

Kinetic equation (7.12) takes the final form

$$\mathbf{p}_\perp \cdot [\hat{\mathbf{z}} \times \mathbf{v}_L]\frac{\partial f^{(0)}}{\partial \epsilon} - \frac{\partial f_1}{\partial \alpha} = -\frac{1}{\omega_c \tau}f_1.$$

Its solution is (7.7) with

$$\gamma'_O = \frac{\omega_c \tau}{\omega_c^2 \tau^2 + 1}, \qquad \gamma'_H = \frac{\omega_c^2 \tau^2}{\omega_c^2 \tau^2 + 1}. \quad (7.13)$$

7.3 Forces

At the second step we calculate the force acting on a vortex from the environment. This force is exerted via excitations which travel near the vortex through transfer of their momenta to the vortex. Consider first localized excitations. The transferred momentum is

$$\mathbf{F}^{(loc)}_{env} = \frac{1}{2}\sum_n \int \frac{dp_z}{2\pi}\frac{d\alpha d\mu}{2\pi}\frac{\partial \mathbf{p}_n}{\partial t}f_1$$

$$= -\frac{1}{2}\sum_n \int \frac{dp_z}{2\pi}\frac{d\alpha d\mu}{2\pi}\frac{\partial \epsilon_n}{\partial b}[\hat{\mathbf{z}} \times \hat{\mathbf{v}}_\perp]f_1. \quad (7.14)$$

Here we make use of the Hamilton equation

$$\frac{\partial \mathbf{p}_n}{\partial t} = -\nabla \epsilon_n = -\frac{\partial \epsilon_n}{\partial b}[\hat{\mathbf{z}} \times \hat{\mathbf{v}}_\perp]. \quad (7.15)$$

The second equality follows from the fact that, in the coordinate frame (s, b) of Fig. 7.2, the energy only depends on the particle impact parameter b.

The transfer of the momentum from delocalized excitations in the semi-classical approximation has the form

$$\mathbf{F}^{(del)}_{env} = \frac{1}{2}\sum_n \int \frac{dp_z}{2\pi}\int \frac{d\alpha}{2\pi}\frac{\partial \mathbf{p}_n}{\partial t}f_1$$

where $\partial \mathbf{p}_n \partial t$ is given by (7.10). The force becomes

$$\mathbf{F}^{(del)}_{env} = \frac{1}{2}\sum_n \int \frac{dp_z}{2\pi}\int \frac{d\alpha}{2\pi}[\mathbf{p}_F \times \hat{\mathbf{z}}]\frac{\omega_c}{g}f_1.$$

The sum is over quantized states of a semi-classical quasiparticle in a magnetic field. The energy spectrum (7.9) becomes quantized [18] in a magnetic field and turns into a modified Landau spectrum ϵ_n such that

$$\sqrt{\epsilon_n^2 - \Delta_\infty^2} = \omega_c n - \mathbf{p}_\perp^2/2m.$$

The quantity

$$\frac{g}{\omega_c} = \frac{\partial n}{\partial \epsilon}$$

is thus the density of states. As a result,

$$\sum_n \to \int \frac{\partial n}{\partial \epsilon} d\epsilon$$

and the force becomes [20]

$$\mathbf{F}_{\text{env}}^{(\text{del})} = \int_{\epsilon > \Delta_\infty} d\epsilon \int \frac{dp_z}{2\pi} \int \frac{d\alpha}{2\pi} \left[\mathbf{p}_\perp \times \hat{\mathbf{z}} \right] f_1. \tag{7.16}$$

The total force is

$$\mathbf{F}_{\text{env}} = \mathbf{F}_{\text{env}}^{(\text{loc})} + \mathbf{F}_{\text{env}}^{(\text{del})}. \tag{7.17}$$

With the Ansatz (7.7), the total force (7.17) splits into two terms $\mathbf{F}_{\text{env}} = \mathbf{F}_\parallel + \mathbf{F}_\perp$, with the friction $\mathbf{F}_\parallel$ and transverse $\mathbf{F}_\perp$ forces given by

$$\mathbf{F}_\parallel = -\pi N \left[\left\langle\!\!\left\langle \sum_n \int \omega_n \gamma_O \frac{\partial f^{(0)}}{\partial \epsilon} \frac{d\mu}{2} \right\rangle\!\!\right\rangle \right.$$
$$\left. + \left(1 - \tanh \frac{\Delta_\infty}{2T} \right) \gamma_O' \right] \mathbf{v}_L, \tag{7.18}$$

$$\mathbf{F}_\perp = \pi N \left[\left\langle\!\!\left\langle \sum_n \int \omega_n \gamma_H \frac{\partial f^{(0)}}{\partial \epsilon} \frac{d\mu}{2} \right\rangle\!\!\right\rangle \right.$$
$$\left. + \left(1 - \tanh \frac{\Delta_\infty}{2T} \right) \gamma_H' \right] [\hat{\mathbf{z}} \times \mathbf{v}_L] \tag{7.19}$$

where N is the quasiparticle (electron) density, $\langle\!\langle \ldots \rangle\!\rangle$ is the average over the Fermi surface with the weight $\pi p_\perp^2$,

$$\langle\!\langle \ldots \rangle\!\rangle = V_F^{-1} \int \pi p_\perp^2 \, dp_z \, (\ldots), \tag{7.20}$$

and V_F is the volume encompassed by the Fermi surface.

We emphasize that the force $\mathbf{F}_{\text{env}}$ is defined as the response of the whole environment to the vortex displacement. It is therefore the *total* force acting on the vortex from the ambient system, including all partial forces such as the longitudinal friction force and the non-dissipative transverse force. The transverse force, in turn, includes various parts which can be identified historically [21] as the Iordanskii force [22], the spectral flow force [23–25], and the Magnus force. We shall discuss this later in more detail in Sect. 7.4.

7.3.1 Flux-Flow Conductivity

The force from the environment is balanced by the Lorentz force:

$$\mathbf{F}_\mathrm{L} = \frac{\Phi_0}{c}[\mathbf{j}_\mathrm{tr} \times \hat{\mathbf{z}}]\mathrm{sign}(e), \qquad (7.21)$$

with the flux quantum $\Phi_0 = \pi c/|e|$. The force balance equation

$$\mathbf{F}_\mathrm{L} + \mathbf{F}_\mathrm{env} = 0 \qquad (7.22)$$

determines the transport current in terms of the vortex velocity and thus allows us to find the flux-flow conductivity tensor. The *longitudinal force* $\mathbf{F}_\parallel$ defines the friction coefficient in the vortex equation of motion and determines the Ohmic component of the conductivity σ_O. Expressing the vortex velocity $\mathbf{v}_\mathrm{L}$ through the average electric field $\mathbf{E}$, as $\mathbf{v}_\mathrm{L} = c\,[\mathbf{E} \times \hat{\mathbf{z}}]/B\,\mathrm{sign}(e)$, we find

$$\sigma_\mathrm{O} = \frac{N|e|c}{B}\left[\left\langle\!\!\left\langle \sum_n \int \frac{\omega_n \tau_n}{\omega_n^2 \tau_n^2 + 1}\frac{\partial f^{(0)}}{\partial \epsilon}\frac{\omega_n d\mu}{2}\right\rangle\!\!\right\rangle\right.$$
$$\left. + \left(1 - \tanh\frac{\Delta_\infty}{2T}\right)\frac{\omega_c \tau}{\omega_c^2 \tau^2 + 1}\right]. \qquad (7.23)$$

The *transverse force* determines the Hall conductivity

$$\sigma_\mathrm{H} = \frac{Nec}{B}\left[\left\langle\!\!\left\langle \sum_n \int \frac{\omega_n^2 \tau_n^2}{\omega_n^2 \tau_n^2 + 1}\frac{\partial f^{(0)}}{\partial \epsilon}\frac{\omega_n d\mu}{2}\right\rangle\!\!\right\rangle\right.$$
$$\left. + \left(1 - \tanh\frac{\Delta_\infty}{2T}\right)\frac{\omega_c^2 \tau^2}{\omega_c^2 \tau^2 + 1}\right]. \qquad (7.24)$$

The contributions from the localized states are determined mainly by the chiral branch $n = 0$ because $\omega_n \ll \omega_0$ for $n \ne 0$ as explained earlier.

The main conclusion is that the Ohmic and Hall conductivities depend on the purity of the sample through the parameters $\omega_0 \tau$ and $\omega_c \tau$. Note that $\omega_c \sim (H/H_{c2})\omega_0$ thus $\omega_c \ll \omega_0$. One can distinguish two regimes: moderately clean $\omega_0 \tau \ll 1$ and superclean $\omega_0 \tau \gg 1$. Note that the moderately clean regime still requires that the superconductor is clean in the usual sense $\Delta_\infty \tau \gg 1$.

In the moderately clean limit $\omega_0 \tau \ll 1$, the conductivity roughly follows the Bardeen and Stephen expression [26] at low temperatures though it exhibits an extra temperature-dependent factor Δ_∞/T_c on approaching T_c [20]

$$\sigma_\mathrm{O} \sim \sigma_n \frac{H_{c2}}{H}\frac{\Delta_\infty}{T_c}.$$

The factor Δ_∞/T_c appears because the number of delocalized quasiparticles contributing to the vortex dynamics decreases near T_c. This extra factor has been recently identified experimentally [27]. The Hall conductivity and the Hall angle are small $\sigma_\mathrm{H} \sim (\omega_0 \tau)\sigma_\mathrm{O}$ and $\tan\Theta_\mathrm{H} \sim (\omega_0 \tau)$, respectively. The contribution from delocalized states are not important since $\omega_c \ll \omega_0$.

In the superclean limit $\omega_0 \tau \gg 1$, on the contrary, the Ohmic conductivity is small. The vortex dynamics becomes non-dissipative. In particular, if $\omega_c \tau \ll 1$, the Hall conductivity reduces to a very simple expression

$$\sigma_H = \frac{Nec}{B} \tanh\left(\frac{\Delta_\infty}{2T}\right).$$

If $\omega_c \tau \gg 1$, the hyperbolic tangent should be replaced with unity, as explained below.

7.4 Transverse Force

Let us now discuss the forces acting on a moving vortex in more detail. The friction force is determined by the (7.18). It is proportional to the mean free path of excitations in the moderately clean regime and vanishes in the superclean limit. The transverse force (7.19) deserves a more careful discussion because it has been a matter of controversy for a long time since first calculated for vortices in helium II by Lifshitz and Pitaevskii [28] and then by Iordanskii [22]. The review [21] tells about old disputes (see also [29] and Chap. 8 [8] in this book). Recently, the presence of the transverse force has been questioned in [30,31].

As we see, however, the microscopic picture gives a finite transverse force in a full accordance with the symmetry arguments which allow a transverse force in a chiral system such as a moving vortex in presence of a superflow. In general, we can identify several contributions to the transverse force. One can present the full transverse force (7.19) in the form

$$\mathbf{F}_\perp = -\pi N_s [\mathbf{v}_L \times \hat{\mathbf{z}}] - \pi N_n [\mathbf{v}_L \times \hat{\mathbf{z}}] + \mathbf{F}_{sf}. \tag{7.25}$$

To derive this, we write $\gamma_H = 1 - [\omega_n^2 \tau_n^2 + 1]^{-1}$ in the first term in square brackets of (7.19) and separate the sum of integrals of the full derivative of the distribution function which gives

$$\sum_n \int \frac{\partial f^{(0)}}{\partial \epsilon} \frac{d\epsilon_n}{2} = \tanh\left(\frac{\Delta_\infty}{2T}\right). \tag{7.26}$$

Indeed, the terms with $n n_e 0$ in the sum vanish because $\epsilon_n(\mu)$ returns to the same energy $+\Delta_\infty$ (or $-\Delta_\infty$) for $\mu \to \pm\infty$. The chiral branch with $n = 0$ does not disappear because $\epsilon_0(\mu)$ varies from $+\Delta_\infty$ to $-\Delta_\infty$ as μ runs from $-\infty$ to $+\infty$. Doing the same with the term γ'_H we obtain $1 - \tanh(\Delta_\infty/2T)$ for the corresponding full derivative. Together with (7.26) this gives $\pi N = \pi N_s + \pi N_n$ in (7.19) that leads to the first two terms in (7.25).

The force

$$\mathbf{F}_{sf} = \pi N \left\langle\!\!\left\langle \sum_n \int \frac{d\mu}{2} \frac{\partial f^{(0)}}{\partial \epsilon} \frac{\omega_n}{\omega_n^2 \tau_n^2 + 1} \right\rangle\!\!\right\rangle [\mathbf{v}_L \times \hat{\mathbf{z}}]$$

$$+ \pi N \left(1 - \tanh\frac{\Delta_\infty}{2T}\right) \frac{1}{\omega_c^2 \tau^2 + 1} [\mathbf{v}_L \times \hat{\mathbf{z}}] \tag{7.27}$$

is called the spectral flow force [23]. The spectral flow force $\mathbf{F}_{\text{sf}}$ is due to the momentum flow from normal excitations to the vortex along the spectral branches $\epsilon_n(\mu)$. Due to the time dependent angular momentum of the excitations $\mu(t) = [(\mathbf{r} - \mathbf{v}_\text{L} t) \times \mathbf{p}] \cdot \hat{\mathbf{z}}$, there appears a flow (with the velocity $\partial \mu/\partial t$) of spectral levels characterized by the angular momentum μ. Each particle on a level carries a momentum p_F which is transferred to the vortex if the quasiparticle relaxation occurs quickly. The factor $(\omega_n^2 \tau_n^2 + 1)^{-1}$ in the first line of (7.27) accounts for the relaxation on localized levels: the relaxation and hence the momentum transfer is complete for $\omega_0 \tau \ll 1$, and vanishes in the opposite limit. The largest contribution is due to the momentum flow from the Fermi sea of normal excitations to the moving vortex via the gapless spectral branch going through the vortex core from negative to positive energies [24,25] because $\omega_0 \gg \omega_n$ for $n n_e 0$. The first term in (7.27) thus describes a disorder-mediated momentum flow along the anomalous chiral branch $\epsilon_0(\mu)$ for energies below the gap. The second line in (7.27) accounts for the spectral flow for energies above the gap. The corresponding factor that takes into account the relaxation rate is $(\omega_c^2 \tau^2 + 1)^{-1}$.

Note that $\mathbf{F}_{\text{env}}$ is calculated in the reference frame where the normal component is at rest $\mathbf{v}_n = 0$ for stationary vortices. In superconductors, the normal velocity is always zero in equilibrium, the excitations being at rest in the reference frame associated with the crystal lattice. In superfluid ^{3}He, the situation is similar: the normal component has a rather large viscosity so that it is normally at rest in the container frame of reference.

Let us now turn to the force balance equation (7.22). In presence of an electric field, the transport current is not entirely due to a supercurrent, a part of it being carried by delocalized quasiparticles. Far from the vortex core, the quasiparticle current is

$$\mathbf{j}^{(\text{qp})} = -2\nu(0) e \int_{\epsilon > \Delta_\infty} \mathbf{v}_\perp g f_1 \, d\epsilon \frac{d\Omega_\mathbf{p}}{4\pi} \qquad (7.28)$$

where $\nu(0)$ is the single–spin normal density of states, and $d\Omega_\mathbf{p}$ is the elementary solid angle in the direction of the momentum $\mathbf{p}$. Using (7.7) and (7.13) we find

$$\mathbf{j}^{(\text{qp})} = N_n e \gamma'_\text{H} \mathbf{v}_\text{L} + N_n e \gamma'_\text{O} [\mathbf{z} \times \mathbf{v}_\text{L}] \qquad (7.29)$$

where the density of normal quasiparticles is

$$N_n = N \int_{\epsilon > \Delta_\infty} g \frac{\partial f^{(0)}}{\partial \epsilon} \, d\epsilon. \qquad (7.30)$$

Writing the transport current as $\mathbf{j}_{\text{tr}} = N_s e \mathbf{v}_s + \mathbf{j}^{(\text{qp})}$ we get the force balance (7.22) in the form

$$\mathbf{F}_\text{M} + \mathbf{F}_\text{L}^{(\text{qp})} + \mathbf{F}_\text{I} + \mathbf{F}_{\text{sf}} + \mathbf{F}_\| = 0. \qquad (7.31)$$

Here

$$\mathbf{F}_M = \pi N_s \left[(\mathbf{v}_s - \mathbf{v}_L) \times \hat{\mathbf{z}}\right]$$

is the Magnus force;

$$\mathbf{F}_L^{(qp)} = \frac{\Phi_0}{c} [\mathbf{j}^{(qp)} \times \hat{\mathbf{z}}] \mathrm{sign}\,(e) \qquad (7.32)$$

is the Lorentz force from the quasiparticle current (7.29), and

$$\mathbf{F}_I = \pi N_n \left[(\mathbf{v}_n - \mathbf{v}_L) \times \hat{\mathbf{z}}\right]$$

is called the Iordanskii force [22]. The Iordanskii force is the counterpart of the Magnus force for normal excitations. We have included the (zero) normal velocity $\mathbf{v}_n$ to make this similarity more transparent.

The spectral flow force vanishes for an ideal superconductor $\tau \to \infty$ when both $\omega_0 \tau \gg 1$ and $\omega_c \tau \gg 1$. The transverse force is

$$\mathbf{F}_\perp = \pi N [\hat{\mathbf{z}} \times \mathbf{v}_L]. \qquad (7.33)$$

The balance (7.22) results in the transport current in the form

$$\mathbf{j}_{tr} = N e \mathbf{v}_L. \qquad (7.34)$$

In this limit $\gamma'_O = 0$ while $\gamma'_H = 1$. The quasiparticle current is $\mathbf{j}^{(qp)} = N_n e \mathbf{v}_L$ so that the force from the quasiparticle current compensates the Iordanskii force. The force balance (7.31) reduces to $\mathbf{F}_M = 0$. The vortex thus moves with the superfluid velocity $\mathbf{v}_L = \mathbf{v}_s$. This is consistent with the Helmholtz theorem of conservation of circulation in an ideal fluid: vortices move together with the flow. The correction to the distribution function of excitations is simply

$$f_1 = -\frac{\partial f^{(0)}}{\partial \epsilon}(\mathbf{v}_L \cdot \mathbf{p}_\perp) \qquad (7.35)$$

such that the full distribution function is $f = f^{(0)}(\epsilon - \mathbf{v}_L \cdot \mathbf{p})$: excitations move together with the vortex being in equilibrium with the vortex array. Therefore, the entire systems moves as a whole with the velocity $\mathbf{v}_L$ which amounts to the total current as in (7.34).

To understand this better let us consider first the mean free path of delocalized excitations with respect to their collisions with vortices. If the vortex cross section is σ_v, the mean free path is $\ell_v = 1/\sigma_v n_v$ where $n_v = B/\Phi_0$ is the density of vortices. We shall see in a moment that the vortex cross section is $\sigma_v \sim p_F^{-1}$ so that

$$\ell_v \sim p_F/n_v \sim v_F/\omega_c, \qquad (7.36)$$

i.e., $\ell \sim r_L$ where r_L is the Larmor radius. In the limit $\omega_c \tau \gg 1$, the vortex mean free path ℓ_v becomes shorter than the impurity mean free path $\ell_{imp} =$

$v_F\tau$ so that the delocalized excitations scatter on vortices more frequently than on impurities and thus come to equilibrium with moving vortices. A similar consideration also applies to localized excitations: In the limit $\omega_0\tau \gg 1$ interaction with a vortex is more effective than relaxation on impurities, and the excitations come to equilibrium with the moving vortex.

In the intermediate case when $\omega_c\tau \to 0$ but $\omega_0\tau \sim 1$, only the localized excitations are out of equilibrium. The excitations with energies above the energy gap Δ_∞ have $\ell_v \gg \ell_{\text{imp}}$ and are now in equilibrium with the heat bath: both $\gamma'_O = 0$, and $\gamma'_H = 0$. As a result these excitations do not influence considerably the vortex motion. Delocalized excitations, however, do affect the vortex motion if $\omega_c\tau$ is comparable with unity.

Finally, in the moderately clean limit $\omega_0\tau, \omega_c\tau \ll 1$ the spectral flow force has its maximum value. Equation (7.27) gives in this limit

$$\mathbf{F}_{\text{sf}} = \pi N[\mathbf{v}_L \times \hat{\mathbf{z}}].$$

This completely compensates the first two terms in (7.25), i.e, the Iordanskii force and the part of the Magnus force that contains the vortex velocity; the transverse force vanishes. The quasiparticle current vanishes even faster because $\omega_c \ll \omega_0$. The Lorentz force is balanced only by a friction force $\mathbf{F}_\parallel$. As a result, the dissipative dynamics is restored.

7.4.1 Low-Field Limit and Superfluid ^{3}He

The low-field limit when $\omega_c\tau \ll 1$ is most practical for superconductors. Moreover, this regime is realized in electrically neutral superfluids such as ^{3}He. At the first glance, it is simply because ω_c vanishes together with the charge of carriers. However, this is not completely correct. In fact, to estimate a deviation from equilibrium of delocalized excitations in this case one has again to compare the mean free path of excitations with their mean free path with respect to scattering by vortices. Keeping in mind that the vortex density is $n_v = 2\Omega/\kappa$ where Ω is an angular velocity of a rotating container and $\kappa = \pi/m$ is the circulation quantum, (7.36) gives $\ell_v \sim v_F/\Omega$. We observe that the cyclotron frequency is replaced with the rotation velocity in a full compliance with the Larmor theorem. The ratio of the particle–particle mean free path ℓ to the vortex mean free path is $\ell/\ell_v \sim \Omega\tau$. With the practical rotation velocity Ω of a few radians per second, one always has ℓ_v exceedingly larger than ℓ. Delocalized excitations are thus at rest in the container frame.

Consider this regime in more detail. Since delocalized excitations are in equilibrium the quasiparticle current vanishes. The force balance becomes

$$\mathbf{F}_M + \mathbf{F}_I + \mathbf{F}_{\text{sf}} + \mathbf{F}_\parallel = 0 \qquad (7.37)$$

where the spectral flow force is

$$\mathbf{F}_{\text{sf}} = \pi N \left\langle\!\!\!\left\langle \sum_n \int \frac{d\mu}{2} \frac{\partial f^{(0)}}{\partial \epsilon} \frac{\omega_n}{\omega_n^2 \tau_n^2 + 1} \right\rangle\!\!\!\right\rangle [\mathbf{v}_L \times \hat{\mathbf{z}}]$$

$$+\pi N\left(1 - \tanh\frac{\Delta_\infty}{2T}\right)[\mathbf{v}_L \times \hat{\mathbf{z}}]. \tag{7.38}$$

Note that the spectral flow force does not appear in Bose superfluids. The corresponding force balance equation can be found in the following chapter [8] in this book.

It is interesting to observe that the spectral flow force from delocalized states in this case is related to the anomalous contribution to the transverse vortex cross section for scattering of delocalized quasiparticles. The vortex cross sections were calculated in [32,33]. The transverse cross section is

$$\sigma_\perp = \frac{\pi}{p_\perp}\left[\frac{\epsilon}{\sqrt{\epsilon^2 - \Delta_\infty^2}} - 1\right] \tag{7.39}$$

per unit vortex length. Note that the transport cross section which is responsible for the scattering contribution to the longitudinal force is vanishingly small in the semi-classical limit. Inserting (7.39) into the expression for the force exerted on the vortex by scattered excitations,

$$\mathbf{F}_\perp^{(\mathrm{sc})} = \int_{\epsilon > \Delta_\infty} \frac{d\epsilon}{2}\frac{\partial f^{(0)}}{\partial \epsilon}\int\frac{dp_z}{2\pi}p_\perp^3 \sigma_\perp[\hat{\mathbf{z}} \times \mathbf{v}_L], \tag{7.40}$$

we recover those contributions to (7.25) which are due to the normal excitations, namely the term $-\pi N_n[\mathbf{v}_L \times \hat{\mathbf{z}}]$ and the last term in the spectral flow force (7.38).

The first term in (7.39) corresponds to the cross section of a vortex in a Bose superfluid [21]

$$\sigma'_\perp = 2\pi/m_B v_g \tag{7.41}$$

where v_g is the group velocity which, in our case, is defined by (7.11), and m_B is the mass of a Bosonic atom. Note that, in our case, $m_B = 2m$. The corresponding part of the transverse cross section can be easily obtained from the semi-classical description. Indeed, the Doppler energy due to the vortex velocity is $\mathbf{p}\cdot\mathbf{v}_s = \mathbf{p}\cdot\nabla\chi/m_B$. Its contribution to the quasiparticle action is

$$A = -\frac{1}{m_B}\int \mathbf{p}\cdot\nabla\chi\, dt = -\frac{p}{m_B v_g}\int\frac{\partial\chi}{\partial s}ds = -\frac{p}{m_B v_g}\delta\chi.$$

Here s is the coordinate along the particle trajectory as in Fig. 7.2, and $\delta\chi$ is the variation of the order parameter phase along the trajectory. The change in the transverse momentum of the particle is $\delta p_\perp = \partial A/\partial b$, hence the transverse cross section becomes

$$\sigma'_\perp = \int\frac{\delta p_\perp}{p}db = \frac{A_+ - A_-}{p}$$

where $A_\pm$ is the action along the trajectory passing on the left (right) side of the vortex. Since $\delta\chi_+ - \delta\chi_- = -2\pi$ we recover (7.41). With this expression for the cross section, (7.40) gives the Iordanskii force $\mathbf{F}_I = -\pi N_n [\mathbf{v}_L \times \hat{\mathbf{z}}]$. The derivation of the Iordanskii force for phonons in Helium II can be found in the following chapter [8].

The second term in (7.39) originates [32] from the fact that here, as distinct from the situation in a Bose superfluid, the phase of the single-particle wave function changes by π upon encircling the vortex, while it is the order parameter phase which changes by 2π. It is this singularity produced by the vortex in the single-particle wave function which results in the anomalous contribution to the cross section in (7.39). Inserted into (7.40), it exactly reproduces the second term in (7.38). We see that the spectral flow force is related to a single-particle anomaly associated with the vortex.

7.5 Vortex Momentum

The vortex mass in superfluids and superconductors has been a long standing problem in vortex physics and remains to be an issue of controversies. There are different approaches to its definition. In early works on this subject, the vortex mass was determined through an increase in the free energy of a superconductor calculated as an expansion in slow time derivatives of the order parameter. The quasiparticle distribution was assumed to be essentially as in equilibrium. First used by Suhl [34] (see also [35]) this approach yields the mass of the order of *one quasiparticle mass* (electron, in case of superconductor) per atomic layer. Another approach consists in calculating an electromagnetic energy $E^2/8\pi$ which is proportional to the square of the vortex velocity. This gives rise to the so called electromagnetic mass [36] which, in good metals, is of the same order of magnitude (see [37] for a review).

A crucial disadvantage of the above definitions of the vortex mass is that they do not take into account the kinetics of excitations disturbed by a moving vortex. We shall see that the inertia of excitations contributes much more to the vortex mass than what the old calculations predict. The kinetic equation approach described here is able to incorporate this effect. To implement this method we find the force necessary to support an unsteady vortex motion. Identifying then the contribution to the force proportional to the vortex acceleration, one defines the vortex mass as a coefficient of proportionality. This method was first applied for vortices in superclean superconductors in [38] and then was used by other authors (see for example, [39,40]). The resulting mass is of the order of the *total mass of all electrons* within the area occupied by the vortex core. We will refer to this mass as to the *dynamic mass*. Since the dynamic mass originates from the inertia of excitations localized in the vortex core it can also be calculated through the *momentum carried by localized excitations* [41]. We shall see that dynamic mass displays a nontrivial feature: it is a tensor whose components depend on the quasiparticle mean free time. In *s*-wave superconductors, this tensor is diagonal in the superclean limit. The diagonal mass decreases rapidly as a function of the mean free time, and the off-diagonal components dominate in the moderately clean regime. These results were obtained in [42]. Our results agree with the previous work [38,39,41] in the limit $\tau \to \infty$.

7.5.1 Equation of Vortex Dynamics

To introduce the vortex momentum we consider a non-steady motion of a vortex such that its acceleration is small. We again start with the delocalized excitations. Multiplying (7.5) by $\mathbf{p}_\perp/2$ and summing up over all the quantum numbers, we obtain

$$\mathbf{F}_{\text{env}}^{(\text{loc})} = \mathbf{F}_{\text{coll}}^{(\text{loc})} - \frac{\partial \mathbf{P}^{(\text{loc})}}{\partial t} - \pi N \tanh\left(\frac{\Delta_\infty}{2T}\right)[\mathbf{v}_L \times \hat{\mathbf{z}}] \qquad (7.42)$$

where the l.h.s. of (7.42) is the force from the environment on a moving vortex (7.14).

The first term in the r.h.s. of (7.42) is the force exerted on the vortex by the heat bath via excitations localized in the vortex core:

$$\mathbf{F}_{\text{coll}}^{(\text{loc})} = -\frac{1}{2}\sum_n \int \mathbf{p}_\perp \left(\frac{\partial f}{\partial t}\right)_{\text{coll}} \frac{dp_z}{2\pi} \frac{d\alpha\, d\mu}{2\pi}. \qquad (7.43)$$

The second term in the r.h.s. of (7.42) is the change in the vortex momentum

$$\mathbf{P}^{(\text{loc})} = -\frac{1}{2}\sum_n \int \mathbf{p}_\perp f_1 \frac{dp_z}{2\pi} \frac{d\alpha\, d\mu}{2\pi}. \qquad (7.44)$$

We turn now to delocalized excitations. Multiplying (7.12) by $\mathbf{p}_\perp/2$ again and summing over all the states we find

$$\mathbf{F}_{\text{env}}^{(\text{del})} = \mathbf{F}_{\text{coll}}^{(\text{del})} - \frac{\partial \mathbf{P}^{(\text{del})}}{\partial t} + \frac{1}{2}\sum_n \int \frac{dp_z}{2\pi}\frac{d\alpha}{2\pi}\frac{\partial f^{(0)}}{\partial \epsilon}\frac{\omega_c}{g}\mathbf{p}_\perp\left(\mathbf{p}_\perp \cdot [\hat{\mathbf{z}} \times \mathbf{v}_L]\right)$$

$$= \mathbf{F}_{\text{coll}}^{(\text{del})} - \frac{\partial \mathbf{P}^{(\text{del})}}{\partial t} - \pi N\left[1 - \tanh\left(\frac{\Delta_\infty}{2T}\right)\right][\mathbf{v}_L \times \hat{\mathbf{z}}]. \qquad (7.45)$$

Here

$$\mathbf{F}_{\text{coll}}^{(\text{del})} = -\frac{1}{2}\sum_n \int \mathbf{p}_\perp\left(\frac{\partial f}{\partial t}\right)_{\text{coll}} \frac{dp_z}{2\pi}\frac{d\alpha}{2\pi} = \int_{\epsilon>\Delta_\infty}\mathbf{p}_\perp \frac{f_1}{\omega_c\tau}\frac{dp_z}{2\pi}\frac{d\alpha}{2\pi}d\epsilon$$

is the force from the heat bath, and the corresponding contribution to the vortex momentum is

$$\mathbf{P}^{(\text{del})} = -\frac{1}{2}\sum_n \int \mathbf{p}_\perp f_1 \frac{dp_z}{2\pi}\frac{d\alpha}{2\pi} = -\int_{\epsilon>\Delta_\infty}\mathbf{p}_\perp\frac{g}{\omega_c}f_1\frac{dp_z}{2\pi}\frac{d\alpha}{2\pi}d\epsilon. \qquad (7.46)$$

The total momentum is $\mathbf{P} = \mathbf{P}^{(\text{loc})} + \mathbf{P}^{(\text{del})}$.

After a little of algebra, the total force from the heat bath $\mathbf{F}_{\text{coll}} = \mathbf{F}_{\text{coll}}^{(\text{loc})} + \mathbf{F}_{\text{coll}}^{(\text{del})}$ can be written in the form

$$\mathbf{F}_{\text{coll}} = \mathbf{F}_{\text{sf}} + \mathbf{F}_{\|}$$

where the friction and spectral flow forces are determined by (7.18) and (7.27), respectively. The total force from environment (7.42) and (7.45) takes the form

$$\mathbf{F}_{\text{env}} = \mathbf{F}_{\text{sf}} + \mathbf{F}_{\|} - \frac{\partial \mathbf{P}}{\partial t} - \pi N[\mathbf{v}_{\text{L}} \times \hat{\mathbf{z}}]. \tag{7.47}$$

This equation agrees with (7.25) for a steady motion of vortices.

The equation of vortex dynamics is obtained by variation of the superconducting free energy plus the external field energy with respect to the vortex displacement. The variation of the superfluid free energy gives the force from the environment, $\mathbf{F}_{\text{env}}$, while the variation of the external field energy produces the external Lorentz force. In the absence of pinning, the total energy is translationally invariant. Therefore, the requirement of zero variation of the free energy gives again the condition $\mathbf{F}_{\text{L}} + \mathbf{F}_{\text{env}} = 0$ in the form the force balance. Using our expression for $\mathbf{F}_{\text{env}}$, the force balance can now be written in the form similar to (7.31)

$$\mathbf{F}_{\text{M}} + \mathbf{F}_{\text{L}}^{(\text{qp})} + \mathbf{F}_{\text{I}} + \mathbf{F}_{\text{sf}} + \mathbf{F}_{\|} = \frac{\partial \mathbf{P}}{\partial t} \tag{7.48}$$

where the r.h.s. contains the time derivative of the vortex momentum due to a non-steady vortex motion. For a steady vortex motion, (7.48) reduces to (7.31). The physical meaning of the (7.48) is simple. The l.h.s. of this equation accounts for all the forces acting on a moving straight vortex line. The r.h.s. of (7.48) comes from the inertia of excitations and is identified as the change in the vortex momentum. The definition of (7.44) is similar to that used in [14,41]. Note that the delocalized quasiparticles do not contribute to the vortex momentum in the limit $\omega_c \tau \ll 1$ because they are in equilibrium with the heat bath.

7.5.2 Vortex Mass

Having defined the vortex momentum, we calculate the vortex mass. The distribution function is given by (7.7). The vortex momentum becomes $P_i = M_{ik} v_{Lk}$; it has both longitudinal and transverse components with respect to the vortex velocity. For a vortex with the symmetry not less than the fourfold, the effective mass tensor per unit length is $M_{ik} = M_{\|} \delta_{ik} - M_{\perp} e_{ikj} \hat{z}_j$ where $M_{\|} = M_{\|\,e} + M_{\|\,h}$ and $M_{\perp} = M_{\perp\,e} - M_{\perp\,h}$. Each component contains contributions from localized and delocalized states such that $M_{\|,\perp} = M_{\|,\perp}^{(\text{loc})} +$

$M_{\|,\perp}^{(\text{del})}$ where

$$M_{\|\,e,h}^{(\text{loc})} = \frac{1}{4}\sum_n \int p_\perp^2 \gamma_H \frac{df^{(0)}}{d\epsilon}\frac{dp_z}{2\pi}\frac{d\mu\,d\alpha}{2\pi}$$

$$= \pi N^{(e,h)} \left\langle\!\!\left\langle \sum_n \int \gamma_H \frac{df^{(0)}}{d\epsilon}\frac{d\mu}{2}\right\rangle\!\!\right\rangle \tag{7.49}$$

and

$$M_{\|\,e,h}^{(\text{del})} = \frac{1}{2}\int_{\epsilon>\Delta_\infty} p_\perp^2 \gamma_H' \frac{g}{\omega_c}\frac{df^{(0)}}{d\epsilon}\frac{dp_z}{2\pi}\frac{d\alpha}{2\pi}\,d\epsilon. \tag{7.50}$$

The same expression holds for $M_{\perp\,e,h}$ where γ_H is replaced with γ_O. The indexes e,h indicate the corresponding momentum integrations over the electron and hole parts of the Fermi surface, respectively.

If the vortex acceleration is slow, one can use expressions (7.8), (7.13) for a steadily moving vortex to calculate the vortex inertia. Consider first the contribution of the states with $|\epsilon| > \Delta_\infty$. Since $\gamma_{H,O}'$ do not depend on energy and momentum, (7.50) gives

$$M_{\|,\perp}^{(\text{del})} = \frac{\pi N_n^{(e,h)}}{\omega_c}\gamma_{H,O}' = mN_n^{(e,h)} S_0 \gamma_{H,O}'.$$

The contribution of the delocalized states decreases as $\omega_c\tau$ gets smaller. In the limit of vanishing $\omega_c\tau$ the vortex mass is determined by localized excitations. The localized excitations dominate also at low temperatures $T \ll \Delta_\infty$. One has in this case

$$M_{\|\,e,h} = \pi N^{(e,h)} \left\langle\!\!\left\langle \frac{\gamma_H}{\omega_0}\right\rangle\!\!\right\rangle,\quad M_{\perp\,e,h} = \pi N^{(e,h)} \left\langle\!\!\left\langle \frac{\gamma_O}{\omega_0}\right\rangle\!\!\right\rangle.$$

For $\omega_c\tau \gg 1$ the mass tensor for delocalized states is diagonal $M_{ik} = M_\| \delta_{ik}$ where $M_\|^{(\text{del})} = mN_n S_0$; it is equal to the mass of normal particles in the area occupied by the vortex. The mass tensor for localized states becomes diagonal in the superclean limit where $T_c^2\tau/E_F \gg 1$ with $M_\|^{(\text{loc})} \sim \pi N \langle\!\langle \omega_0^{-1}\rangle\!\rangle \sim \pi \xi_0^2 mN$. This is the mass of all electrons in the area occupied by the vortex core. The mass decreases with τ. In the moderately clean regime $T_c^2\tau/E_F \ll 1$ where $\omega_n\tau \ll 1$, the diagonal component vanishes as τ^2, and the mass tensor is dominated by the off-diagonal part.

We should emphasize an important point that, in contrast to a conventional physical body, the mass of a vortex is not a constant quantity for a given system: it may depend on the frequency ω of the external drive. Indeed, for a nonzero ω we find from (7.5)

$$\gamma_H = \frac{\omega_n^2\tau_n^2}{\omega_n^2\tau_n^2 + (1-i\omega\tau_n)^2};\quad \gamma_O = \frac{\omega_n\tau_n(1-i\omega\tau_n)}{\omega_n^2\tau_n^2 + (1-i\omega\tau_n)^2}.$$

As a result, all the dynamic characteristics of vortices, including the conductivity and the effective mass, acquire a frequency dispersion. In the limit $\tau \to \infty$, poles in $\gamma_{O,H}$ appear at a frequency equal to the energy spacing between the quasiparticle states in the vortex core which gives rise to resonances in absorption of an external electromagnetic field [38,43].

7.6 Conclusions

All the dynamic characteristics of vortices such as vortex friction, the flux-flow conductivity, the Hall effect, and the vortex mass are determined by the mean free path of excitations which interact with vortices. The key parameter is $\omega_0 \tau$ where ω_0 is the interlevel spacing for the quasiparticle states in the vortex core. For small $\omega_0 \tau$, vortices experience viscous flow; the Lorentz force is opposed by a friction force while the transverse force vanishes. On the contrary, in a superclean regime when $\omega_0 \tau \gg 1$, vortices move with a superflow as in an ideal fluid; the Hall angle is $\pi/2$, the friction force is zero, while the transverse force reaches its maximum value. The vortex mass is a tensor. The longitudinal component dominates in the superclean regime: it is the mass of all excitations in the vortex core. On the contrary, the transverse component is the largest one in the limit $\omega_0 \tau \ll 1$.

I am pleased to acknowledge helpful discussions with E. Sonin and G. Volovik. This work was supported by the Russian Foundation for Basic Research (Grant No. 99-02-16043).

References

1. L. P. Pitaevskii, Sov. Phys. JETP **13**, 451 (1961).
2. E. P. Gross, Il Nouvo Cimento **20**, 454 (1961); J. Math. Phys. **4**, 195 (1963).
3. P. C. Hohenberg and B. I. Halperin, Rev. Mod. Phys. **49**, 435 (1977).
4. H. Fukuyama, H. Ebisawa, and T. Tsuzuki, Progr. Theor. Phys. **46**, 1028 (1971).
5. A.T. Dorsey, Phys. Rev. B **46**, 8376 (1992).
6. N. B. Kopnin, B. I. Ivlev, and V. A. Kalatsky, Pis'ma Zh. Eksp. Teor. Fiz. **55**, 717 (1992) [JETP Lett. **55**, 750 (1992)].
7. A. Freimuth, Chap. 19 of this book.
8. E. B. Sonin, Chap. 8 of this book.
9. A. van Otterlo, M. Feigel'man, V. Geshkenbein, and G. Blatter, Phys. Rev. Lett. **75**, 3736 (1995).
10. N. B. Kopnin, Phys. Rev. B **54**, 9475 (1996).
11. Yu. Matsuda and K. Kumagai, Chap. 17 of this book.
12. C. Caroli, P. G. de Gennes, and J. Matricon, Phys. Lett. **9**, 307 (1964).
13. F. Gygi and M. Schlüter, Phys. Rev. B **43**, 7609 (1991).
14. M. Stone, Phys. Rev. B **54**, 13 222 (1996).
15. G. Blatter, V. Geshkenbein, and N. B. Kopnin, Phys. Rev. B **59**, 14 663 (1999).

16. In a system with the Galilean invariance such as ^{3}He, quasiparticle relaxation in the vortex core occurs through their interaction with delocalized excitations which themselves relax at the container walls. This introduces two effective relaxation times in the vortex dynamics. Qualitatively, however, the physics is very similar to the picture described here. For more detail, see N. B. Kopnin and A. V. Lopatin, Phys. Rev. B **56**, 766 (1997).
17. N. B. Kopnin and V. E. Kravtsov, Pis'ma Zh. Eksp. Teor. Fiz. **23**, 631 (1976) [JETP Lett. **23**, 578 (1976)].
18. N. B. Kopnin and V.M. Vinokur, Phys. Rev. B **62**, 9770 (2000).
19. A. G. Aronov, Yu. M. Galperin, V. L. Gurevich, and V. I. Kozub, Adv. in Phys. **30**, 539 (1981).
20. N. B. Kopnin and A. V. Lopatin, Phys. Rev. B **51**, 15 291 (1995).
21. E. B. Sonin, Rev. Mod. Phys. **59**, 87 (1987).
22. S. V. Iordanskii, Ann. Phys. **29**, 335 (1964).
23. N. B. Kopnin, G. E. Volovik, and Ü. Parts, Europhys. Lett. **32**, 651 (1995).
24. G. E. Volovik, Pis'ma Zh. Eksp. Teor. Fiz. **43**, 428 (1986). [JETP Lett. **43**, 551 (1986)].
25. M. Stone and F. Gaitan, Ann. Phys. **178**, 89 (1987).
26. J. Bardeen and M. J. Stephen, Phys. Rev. **140** A, 1197 (1965).
27. S. Kambe, A. D. Huxley, P. Rodiere, and J. Flouquet, Phys. Rev. Lett. **83**, 1842 (1999).
28. E. M. Lifshitz and L. P. Pitaevskii, Zh. Eksp. Teor. Fiz. **33**, 535 (1957) [Sov. Phys. JETP **6**, 418 (1957)].
29. E. B. Sonin, Phys. Rev. B **55**, 485 (1997).
30. P. Ao and D. J. Thouless, Phys. Rev. Lett. **70**, 2158 (1993); D. J. Thouless, P. Ao, and Q. Niu, Phys. Rev. Lett. **76**, 3758 (1996).
31. P. Ao and X.-M. Zhu, Physica (Amsterdam) **282C–287C**, 367 (1997).
32. N. B. Kopnin and V. E. Kravtsov, Zh. Eksp.Teor. Fiz. **71**, 1644 (1976) [Sov. Phys. JETP **44**, 861 (1976)].
33. Yu. M. Galperin and E. B. Sonin, Fiz. Tverd. Tela **18**, 3034 (1976) [Sov. Phys. Solid State **18**, 1768 (1976)].
34. H. Suhl, Phys. Rev. Lett. **14**, 226 (1965).
35. J.-M. Duan and A. J. Leggett, Phys. Rev. Lett. **68**, 1216 (1992).
36. M. Coffey and Z. Hao, Phys. Rev. B **44**, 5230 (1991).
37. E. B. Sonin, V. B. Geshkenbein, A. van Otterlo, and G. Blatter, Phys. Rev. B **57**, 575 (1998).
38. N. B. Kopnin, Pis'ma Zh. Eksp. Teor. Fiz. **27**, 417 (1978) [JETP Lett. **27**, 390 (1978)].
39. N. B. Kopnin and M. M. Salomaa, Phys. Rev. B **44**, 9667 (1991).
40. E. Šimánek, Journ. Low Temp. Phys. **100**, 1 (1995).
41. G. E. Volovik, Pis'ma Zh. Eksp. Teor. Fiz. **65**, 201 (1997) [JETP Lett. **65**, 217 (1997)].
42. N. B. Kopnin and V. M. Vinokur, Phys. Rev. Lett. **81**, 3952 (1998).
43. N. B. Kopnin, Phys. Rev. **57**, 11 775 (1998).

8 Magnus Force and Aharonov–Bohm Effect in Superfluids

Edouard Sonin

Summary. This chapter addresses the problem of the transverse force (Magnus force) on a vortex in a Galilean invariant quantum Bose liquid. Interaction of quasiparticles (phonons) with a vortex produces an additional transverse force (Iordanskii force). The Iordanskii force is related to the acoustic Aharonov–Bohm effect. The connection between the effective Magnus force and the Berry phase is also discussed.

8.1 Introduction

In classical hydrodynamics it has long been known that if the vortex moves with respect to a liquid there is a force on the vortex normal to the vortex velocity [1]. This is the Magnus force, which is a particular case of a force on a body immersed into a liquid with a flow circulation around it (the Kutta–Joukowski theorem). An example of this is the lift force on an airplane wing.

The key role of the Magnus force in vortex dynamics became clear from the very beginning of studying superfluid hydrodynamics. In the pioneer article on the subject Hall and Vinen [2] defined the superfluid Magnus force as a force between a vortex and a superfluid. Therefore it was proportional to the superfluid density ρ_s. But in the two-fluid hydrodynamics the superfluid Magnus force is not the only force on the vortex transverse to its velocity: there was also a transverse force between the vortex and quasiparticles moving with respect to the vortex. The transverse force from rotons was found by Lifshitz and Pitaevskii [3] from the quasiclassical scattering theory. Later Iordanskii [4] revealed the transverse force from phonons. The analysis done in [5] (see also [6,7]) demonstrated that the Lifshitz–Pitaevskii force for rotons and the Iordanskii force for phonons originate from interference between quasiparticles which move past the vortex on the left and on the right sides with different phase shifts, like in the Aharonov–Bohm effect [8]. Since the phase shift depends on the circulation which is a topological charge for a vortex, this is a clear indication of a connection between the transverse quasiparticle force and topology.

Later on the analogy between wave scattering by vortex and electron scattering by the magnetic-flux tube (the Aharonov–Bohm effect) was studied in classical hydrodynamics for water surface waves (the acoustic Aharonov–Bohm effect [9,10]). Scattering of the light by a vortex also results in the

Aharonov–Bohm interference (the optical Aharonov–Bohm effect [11]). As follows from [5], the Aharonov–Bohm interference always produces a transverse force on the vortex, or the fluxon. For the original Aharonov–Bohm effect this force is discussed by Shelankov [12].

The Magnus force on a vortex in a superconductor was introduced by Nozières and Vinen [13]. The total transverse force on a vortex is responsible for the Hall effect in the mixed state. In superconductors not only quasiparticles, but also impurities produce an additional transverse force on the vortex. The reader can find a discussion of this problem in the review by Kopnin [14].

Despite a lot of work done to understand and calculate the Magnus force, it remained to be a controversial issue. Ao and Thouless [15] pointed out a connection between the Magnus force and the Berry phase [16] which is the phase variation of the quantum-mechanical wave function resulting from transport of the vortex round a close loop. From the Berry-phase analysis Ao and Thouless concluded that the effective Magnus force is proportional to the superfluid density, and there is no transverse force on the vortex from quasiparticles and impurities [15,17,18]. This conclusion disagreed with the previous calculations of the Magnus force in superfluids and superconductors, and therefore generated a vivid discussion [19–21].

The present chapter addresses this problem [22] which is very important for many issues in modern condensed matter physics, field theory, and cosmology. The analysis is based on studying momentum balance in the area around a moving vortex without using any preliminary concept of a force. The word *force* is a label to describe a transfer of momentum between two objects. A careful approach is to give these labels to the various terms in the momentum balance equation only *after* derivation of this equation. I restrict myself with the problem of the Galilean invariant quantum Bose liquid. For a weakly nonideal Bose gas one can use the Gross–Pitaevskii theory [23]. In this theory the liquid is described by the nonlinear Schrödinger equation. At large scales the nonlinear Schrödinger equation yields the hydrodynamics of an ideal inviscous liquid. In presence of an ensemble of sound waves (phonons) with the Planck distribution, which is characterized by a locally defined normal velocity v_n (the drift velocity of the Planck distribution), one obtains two-fluid hydrodynamics. Eventually the problem of the vortex motion in the presence of the phonon normal fluid is a problem of hydrodynamics.

The chapter starts from discussion of the Magnus force in classical hydrodynamics (Sect. 8.2). In Sect. 8.3 I define the superfluid Magnus force and the force from quasiparticles scattered by a vortex. Connection between the Gross–Pitaevskii theory and the two–fluid hydrodynamics is discussed in Sect. 8.4. In Sect. 8.5 I discuss scattering of a sound wave by a vortex and show that the standard scattering theory fails to give a conclusive result on the transverse force because of small-scattering-angle divergence of the scattering amplitude. The next Sect. 8.6 gives a solution of the sound equation around the vortex which is free from the small-angle divergence.

Using this solution in the momentum balance one obtains the equation of vortex motion, which contains the Iordanskii force. The momentum transfer responsible for the Iordanskii force occurs at small scattering angles where a phenomenon analogous to the Aharonov–Bohm effect is important: an interference between the waves on the left and right of the vortex with different phase shifts after interaction with the vortex. Section 8.7 shows how to derive the transverse force from the exact partial-wave solution of the Aharonov–Bohm problem for electrons. In our analysis the force on the vortex originates from scattering of noninteracting quasiparticles. This is a valid assumption since normally the phonon mean-free path essentially exceeds the scale where the force arises (the phonon wavelength) [24]. But in order to know the effect of this force on the whole superfluid, it is important to investigate how the force is transmitted to distances much larger than the mean-free path where the two-fluid hydrodynamics becomes valid. This is done in Sect. 8.8. The effect of the force at very large distances is also important for discussion of the Berry phase in Sect. 8.9. The transverse force creates a circulation of the normal velocity at very large distances from an isolated vortex. Taking into account this circulation, the Berry phase yields a correct value of the transverse force, which agrees with the result derived from the momentum balance.

This chapter does not discuss experimental aspects of this problem which are addressed in other reviews devoted to rotating ^{4}He and ^{3}He [6,26,27]).

8.2 The Magnus Force in Classical Hydrodynamics

It is worth recalling how the Magnus force appears in classical hydrodynamics. Let us consider a cylinder immersed in an incompressible inviscid liquid. There is a potential circular flow around the cylinder with the velocity

$$\boldsymbol{v}_\mathrm{v}(\boldsymbol{r}) = \frac{\boldsymbol{\kappa} \times \boldsymbol{r}}{2\pi r^2} \,. \tag{8.1}$$

Here $\boldsymbol{r}$ is the position vector in the plane xy, the axis z is the axis of the cylinder, and $\boldsymbol{\kappa}$ is the circulation vector along the axis z. In classical hydrodynamics the velocity circulation $\kappa = \oint \boldsymbol{v}_\mathrm{v} \cdot d\boldsymbol{l}$ may have arbitrary values. There is also a fluid current past the cylinder with a transport velocity $\boldsymbol{v}_\mathrm{tr}$ and the net velocity is

$$\boldsymbol{v}(\boldsymbol{r}) = \boldsymbol{v}_\mathrm{v}(\boldsymbol{r}) + \boldsymbol{v}_\mathrm{tr} \,. \tag{8.2}$$

This expression is valid at distances r much larger than the cylinder radius. At smaller distances one should take into account that the flow with the velocity $\boldsymbol{v}_\mathrm{tr}$ cannot penetrate into the cylinder, but velocity corrections due to this effect decrease as $1/r^2$ and are not essential for the further analysis.

The Euler equation for the liquid is

$$\frac{\partial \boldsymbol{v}}{\partial t} + (\boldsymbol{v} \cdot \boldsymbol{\nabla})\boldsymbol{v} = -\frac{1}{\rho}\boldsymbol{\nabla} P \,. \tag{8.3}$$

Here ρ is the liquid density and P is the pressure.

Assuming that the cylinder moves with the constant velocity v_L, i.e., replacing the position vector r by $r - v_L t$, one obtains that

$$\frac{\partial v}{\partial t} = -(v_L \cdot \nabla) v . \tag{8.4}$$

Then the Euler equation (8.3) yields the Bernoulli law for the pressure:

$$P = P_0 - \frac{1}{2}\rho[v(r) - v_L]^2 = P'_0 - \frac{1}{2}\rho v_v(r)^2 - \rho v_v(r) \cdot (v_{tr} - v_L) . \tag{8.5}$$

Here P_0 and $P'_0 = P_0 - \frac{1}{2}\rho(v_{tr} - v_L)^2$ are constants, which are of no importance for the following derivation. Figure 9.1 shows that due to the superposition of two fluid motions given by (8.2) the velocity above the cylinder is higher than below the cylinder. According to the Bernoulli law, the pressure is higher in the area where the velocity is lower. As a result of this, a liquid produces a force on the cylinder normal to the relative velocity of the liquid with respect to the cylinder. This is a lift force, or the Magnus force.

In order to find the whole force, we must consider the momentum balance for a cylindrical region of a radius r_0 around the cylinder (see Fig. 9.1). The momentum conservation law requires that the external force F on the cylinder is equal to the momentum flux through the entire cylindrical boundary in the reference frame moving with the vortex velocity v_L. The momentum-flux tensor is

$$\Pi_{ij} = P\delta_{ij} + \rho v_i(r) v_j(r) , \tag{8.6}$$

or in the reference frame moving with the vortex velocity v_L:

$$\Pi'_{ij} = P\delta_{ij} + \rho(v_i - v_{Li})(v_j - v_{Lj}) . \tag{8.7}$$

The momentum flux through the cylindrical surface of radius r_0 is given by the integral $\int dS_j \Pi'_{ij}$ where dS_j are the components of the vector dS directed along the outer normal to the boundary of the cylindrical region and equal to the elementary area of the boundary in magnitude. Then using (8.1), (8.5), and (8.7), the momentum balance yields the following relation:

$$\rho[(v_L - v_{tr}) \times \kappa] = F . \tag{8.8}$$

On the left-hand side of this equation one can see the Magnus force as it occurs in classical hydrodynamics. The Magnus force balances the resultant external force F applied to the cylinder. In the absence of external forces the cylinder moves with the transport velocity of the liquid: $v_L = v_{tr}$ (Helmholtz's theorem).

In the derivation we used the hydrodynamic equations only at large distance from the vortex line, and the radius of the cylinder does not appear in the final result. Therefore (8.8) is valid even if the circular flow with circulation κ occurs without any cylinder at all. In hydrodynamics such a flow

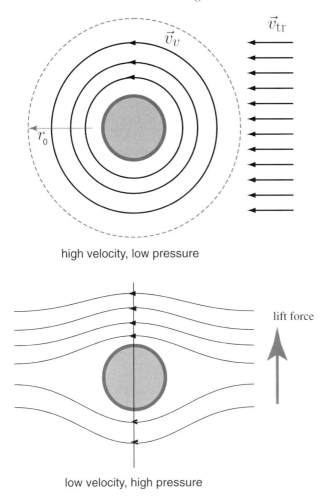

Fig. 8.1. Magnus (lift) force. It is derived from the momentum balance inside a cylinder of radius r_0.

pattern is called a vortex tube, a vortex line, or simply a vortex. Hydrodynamics is invalid at small distance from the vortex line. This area is called the vortex core. But this does not invalidate the derivation of the Magnus force for a Galilean invariant liquid, in which the momentum is a well-defined *conserved* quantity even inside the vortex core where the hydrodynamic theory does not hold.

In the momentum balance in a cylinder around a vortex, half of the Magnus force is due to the Bernoulli contribution to the pressure, equation (8.5); another half is due to the convection term $\propto v_i v_j$ in the momentum flux. However, such decomposition of the resultant momentum flux onto the Bernoulli and the convection parts is not universal and depends on the shape of the area for which we consider the momentum balance. We may consider the momentum balance in a strip, which contains the vortex inside and is oriented normally to the transport flow (see Fig. 8.2). This yields again (8.8), but now the pressure (the Bernoulli term) does not contribute to the transverse force at all, and only convection is responsible for the Magnus force. Then the physical origin of the Magnus force is the following. The liquid enters the strip, which contains a vortex, with one value of the transverse velocity (equal to zero in Fig. 8.2) and exits from the strip with another value of it (Δv in Fig. 8.2). The transverse force is the total variation of the transverse (with respect to the incident velocity v) liquid momentum per unit time. The latter is equal to a product of the current circulation $\rho \oint \Delta v \cdot dl$ and the velocity v.

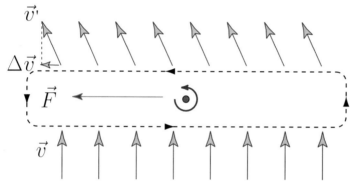

Fig. 8.2. Momentum balance in a strip area. The transverse force is determined by the current circulation $\rho \oint \Delta v \cdot dl$ around the strip.

8.3 The Magnus Force in a Superfluid

In the superfluid state liquid motion is described by two-fluid hydrodynamics: the liquid consists of the superfluid and the normal component with the superfluid and the normal density ρ_s and ρ_n, and the superfluid and the normal velocity v_s and v_n respectively. The circular motion around the vortex line is related to the superfluid motion. Therefore Hall and Vinen [2] suggested that the Magnus force is entirely connected with the superfluid density ρ_s and the superfluid velocity $v_s(r) = v_v(r) + v_{s\,\mathrm{tr}}$. Then instead of (8.8) one has:

$$\rho_s[(v_L - v_s) \times \kappa] = F \ . \tag{8.9}$$

Here and later on we omit the subscript "tr" replacing $v_{s\,\mathrm{tr}}$ by v_s. But one should remember that the superfluid velocity v_s in the expression for the Magnus force is the superfluid velocity far from the vortex line.

The "external" force $\boldsymbol{F}$ is in fact not external for the whole liquid, but only for its superfluid part. The force appears due to interaction with quasiparticles which constitute the normal part of the liquid, and therefore is proportional to the relative velocity $\boldsymbol{v}_{\mathrm{L}} - \boldsymbol{v}_{\mathrm{n}}$ [2]. For a Galilean invariant liquid with axial symmetry the most general expression for the force $\boldsymbol{F}$ is

$$\boldsymbol{F} = -D(\boldsymbol{v}_{\mathrm{L}} - \boldsymbol{v}_{\mathrm{n}}) - D'[\hat{z} \times (\boldsymbol{v}_{\mathrm{L}} - \boldsymbol{v}_{\mathrm{n}})] . \tag{8.10}$$

The force component $\propto D'$ transverse to the velocity $\boldsymbol{v}_{\mathrm{L}} - \boldsymbol{v}_{\mathrm{n}}$ is possible because of broken time invariance in the presence of a vortex and the resulting asymmetry of quasiparticle scattering by a vortex.

Inserting the force $\boldsymbol{F}$ into (8.9), one can rewrite the equation of vortex motion collecting together the terms proportional to the velocity $\boldsymbol{v}_{\mathrm{L}}$:

$$\rho_M[\boldsymbol{v}_{\mathrm{L}} \times \boldsymbol{\kappa}] + D\boldsymbol{v}_{\mathrm{L}} = \rho_{\mathrm{s}}[\boldsymbol{v}_{\mathrm{s}} \times \boldsymbol{\kappa}] + D\boldsymbol{v}_{\mathrm{n}} + D'[\hat{z} \times \boldsymbol{v}_{\mathrm{n}}] . \tag{8.11}$$

The forces on the left-hand side of the equation are the *effective Magnus force* $\propto \rho_M = \rho_{\mathrm{s}} - D'/\kappa$ and the longitudinal friction force $\propto D$. The forces on the right-hand side are driving forces produced by the superfluid and normal flows. In the theory of superconductivity the force $\rho_{\mathrm{s}}[\boldsymbol{v}_{\mathrm{s}} \times \boldsymbol{\kappa}]$, proportional to the superfluid velocity $\boldsymbol{v}_{\mathrm{s}}$, is called the Lorentz force. The left-hand side of (8.11) presents the response of the vortex to these driving forces. The factor ρ_M, which determines the amplitude of the effective Magnus force on the vortex, is not equal to the superfluid density ρ_{s} in general. In the next sections we shall consider the contribution to D' from phonon scattering by a vortex (the Iordanskii force). The contribution to D' from the bound states in the vortex core is discussed by Kopnin [14].

8.4 Nonlinear Schrödinger Equation and Two–Fluid Hydrodynamics

In the Gross–Pitaevskii theory [23] the ground state and weakly excited states of a Bose gas are described by the condensate wave function $\psi = a \exp(i\phi)$ which is a solution of the nonlinear Schrödinger equation

$$i\hbar \frac{\partial \psi}{\partial t} = -\frac{\hbar^2}{2m}\nabla^2 \psi + V|\psi|^2 \psi . \tag{8.12}$$

Here V is the amplitude of two-particle interaction. The nonlinear Schrödinger equation is the Euler–Lagrange equation for the Lagrangian

$$L = \frac{i\hbar}{2}\left(\psi^* \frac{\partial \psi}{\partial t} - \psi \frac{\partial \psi^*}{\partial t}\right) - \frac{\hbar^2}{2m}|\nabla \psi|^2 - \frac{V}{2}|\psi|^4 . \tag{8.13}$$

Noether's theorem yields the momentum conservation law

$$\frac{\partial j_i}{\partial t} + \nabla_j \Pi_{ij} \tag{8.14}$$

where

$$\boldsymbol{j} = -\frac{\partial L}{\partial \dot{\psi}}\boldsymbol{\nabla}\psi - \frac{\partial L}{\partial \dot{\psi}^*}\dot{\psi}^* = -\frac{i\hbar}{2}\left(\psi^*\boldsymbol{\nabla}\psi - \psi\boldsymbol{\nabla}\psi^*\right) \tag{8.15}$$

is the mass current and

$$\begin{aligned}\Pi_{ij} &= -\frac{\partial L}{\partial \nabla_j \psi}\nabla_i\psi - \frac{\partial L}{\partial \nabla_j \psi^*}\nabla_i\psi^* + L\delta_{ij} \\ &= \frac{\hbar^2}{2m}\left(\nabla_i\psi\nabla_j\psi^* + \nabla_i\psi^*\nabla_j\psi\right) + \delta_{ij}P\end{aligned} \tag{8.16}$$

is the momentum–flux tensor. The pressure P in this expression corresponds to the general thermodynamic definition of the pressure via a functional derivation of the energy with respect to the particle density $n = |\psi|^2$:

$$P = L = n\frac{\delta E}{\delta n} - E = n\left[\frac{\partial E}{\partial n} - \boldsymbol{\nabla}\left(\frac{\partial E}{\partial \boldsymbol{\nabla} n}\right)\right] - E = \frac{V}{2}|\psi|^4 - \frac{\hbar^2}{4m}\nabla^2|\psi|^2, \tag{8.17}$$

where

$$E = \frac{\partial L}{\partial \dot{\psi}}\dot{\psi} + \frac{\partial L}{\partial \dot{\psi}^*}\dot{\psi}^* - L = \frac{\hbar^2}{2m}|\boldsymbol{\nabla}\psi|^2 + \frac{V}{2}|\psi|^4 \tag{8.18}$$

is the energy density. But in the hydrodynamic limit (see below) the dependence of the energy on the density gradient is usually neglected.

Using the Madelung transformation [28], the nonlinear Schrödinger equation (8.12) for a complex function may be transformed into two real equations for the liquid density $\rho = ma^2$ and the liquid velocity $\boldsymbol{v} = (\kappa/2\pi)\boldsymbol{\nabla}\phi$ where $\kappa = h/m$ is the circulation quantum. Far from the vortex line these equations are hydrodynamic equations for an ideal inviscous liquid:

$$\frac{\partial \rho}{\partial t} + \boldsymbol{\nabla}(\rho\boldsymbol{v}) = 0 , \tag{8.19}$$

$$\frac{\partial \boldsymbol{v}}{\partial t} + (\boldsymbol{v}\cdot\boldsymbol{\nabla})\boldsymbol{v} = -\boldsymbol{\nabla}\mu . \tag{8.20}$$

Here $\mu = Va^2/m$ is the chemical potential. Equation (8.16) becomes the hydrodynamic momentum-flux tensor $\Pi_{ij} = P\delta_{ij} + \rho v_i(\boldsymbol{r})v_j(\boldsymbol{r})$. Thus the hydrodynamics of an ideal liquid directly follows from the nonlinear Schrödinger equation.

Suppose that a plane sound wave propagates in the liquid generating the phase variation $\phi(\boldsymbol{r}, t) = \phi_0 \exp(i\boldsymbol{k}\cdot\boldsymbol{r} - i\omega t)$. Then the liquid density and velocity are functions of the time t and the position vector $\boldsymbol{r}$ in the plane xy:

$$\rho(\boldsymbol{r}, t) = \rho_0 + \rho_{(1)}(\boldsymbol{r}, t) , \quad \boldsymbol{v}(\boldsymbol{r}, t) = \boldsymbol{v}_0 + \boldsymbol{v}_{(1)}(\boldsymbol{r}, t) , \tag{8.21}$$

where ρ_0 and $\boldsymbol{v}_0$ are the average density and the average velocity in the liquid, $\rho_{(1)}(\boldsymbol{r},t)$ and $\boldsymbol{v}_{(1)}(\boldsymbol{r},t) = \frac{\kappa}{2\pi}\boldsymbol{\nabla}\phi$ are periodical variations of the density and the velocity due to the sound wave ($\langle\rho_{(1)}\rangle = 0$, $\langle\boldsymbol{v}_{(1)}\rangle = 0$). They should be determined from Eqs. (8.19) and (8.20) after their linearization. In particular, (8.20) gives the relation between the density variation and the phase ϕ:

$$\rho_{(1)} = \frac{\rho_0}{c_s^2}\mu_{(1)} = -\frac{\rho_0}{c_s^2}\frac{\kappa}{2\pi}\left\{\frac{\partial\phi}{\partial t} + \boldsymbol{v}_0\cdot\boldsymbol{\nabla}\phi(\boldsymbol{r})\right\}, \tag{8.22}$$

where $c_s = \sqrt{Va^2/m}$ is the sound velocity. Substitution of this expression into (8.19) yields the sound equation for a moving liquid with the wave spectrum $\omega = c_s k + \boldsymbol{k}\cdot\boldsymbol{v}_0$.

The sound propagation is accompanied with the transport of mass. This is an effect of the second order with respect to the wave amplitude. The total mass current expanded with respect to the wave amplitude and averaged over time is

$$\boldsymbol{j} = \rho_0\boldsymbol{v}_0 + \langle\rho_{(1)}\boldsymbol{v}_{(1)}\rangle + \langle\rho_{(2)}\rangle\boldsymbol{v}_0 + \rho_0\langle\boldsymbol{v}_{(2)}\rangle, \tag{8.23}$$

where

$$\langle\rho_{(1)}\boldsymbol{v}_{(1)}\rangle = \boldsymbol{j}^{\text{ph}}(\boldsymbol{p}) = \rho_0\phi_0^2\frac{\kappa^2 k}{8\pi^2 c_s}\boldsymbol{k} = n(\boldsymbol{p})\boldsymbol{p} \tag{8.24}$$

is the mass current and $n(\boldsymbol{p})$ is the number of phonons with the momentum $\boldsymbol{p} = \hbar\boldsymbol{k}$ and the energy $E = \varepsilon(\boldsymbol{p}) + \boldsymbol{p}\cdot\boldsymbol{v}_0$. The phonon mass current is the phonon momentum density in the reference frame moving with the average liquid velocity $\boldsymbol{v}_0$, and $\varepsilon(\boldsymbol{p}) = c_s p$ is the energy in the same reference frame.

Mathematically the second-order corrections to the mass density, $\langle\rho_{(2)}\rangle$, and the average velocity, $\langle\boldsymbol{v}_{(2)}\rangle$, remain undefined, but there are physical constrains to specify them. First of all, we assume that phonon excitations do not change the average mass density, and $\langle\rho_{(2)}\rangle$ must vanish. As for the second-order correction $\langle\boldsymbol{v}_{(2)}\rangle$ to the average velocity, it should produce a second-order correction $\langle\phi_{(2)}\rangle$ to the phase which is impossible in quantum hydrodynamics. The simplest way to see it is to consider the propagation of phonons in an annular channel with the periodic boundary conditions. The phase variation over the channel length is a topological invariant, and weak excitations (phonons) cannot change it. Therefore $\langle\boldsymbol{v}_{(2)}\rangle$ must vanish. More complicated arguments must be given for an open geometry, but intuitively it is clear that this basic physical constrain should not depend on the boundary condition.

It is important to emphasize a difference between sound waves in a liquid and in an elastic solid. The sound wave in the elastic solid is not accompanied by real mass transport: all atoms oscillate near their equilibrium positions in the crystal lattice, but they cannot move in average if the crystal is fixed at a laboratory table. Within our present formalism this means that the second-order contribution $\langle\boldsymbol{v}_{(2)}\rangle$ to the average velocity must not vanish, but compensate the second-order contribution $\langle \rho_{(1)} \boldsymbol{v}_{(1)} \rangle$ to the mass current. Finally a phonon in a solid cannot have a real momentum but only a quasimomentum. The problem of the phonon momentum in liquids and solids has already been discussed for a long time [29] (see also the recent paper by Stone [30]), and they have noticed that the sound wave may have a different momentum using Euler or Lagrange variables. In fact, the momentum should not depend on a choice of variables, but only on physical conditions. However, at various physical conditions a proper choice of variables can make an analysis more straightforward. In solids the Lagrange variables are preferable since in this case the velocity is related to a given particle and coincides with the *center-of-mass* velocity which must not change in average and therefore has no second-order corrections. Using Euler variables in liquids the *average* velocity $\boldsymbol{v}_0$ relates to a given point in the space and has no second-order corrections due to phonons.

Expanding the momentum–flux tensor up to the terms of the second order with respect to the sound wave amplitude one obtains:

$$\Pi_{ij} = P_0 \delta_{ij} + \rho_0 v_{0i} v_{0j} + \Pi_{ij}^{\text{ph}} , \tag{8.25}$$

where the second-order phonon contribution is

$$\Pi_{ij}^{\text{ph}} = \langle P_{(2)} \rangle \delta_{ij} + \langle \rho_{(1)} (v_{(1)})_i \rangle v_{0j} + \langle \rho_{(1)} (v_{(1)})_j \rangle v_{0i} + \rho_0 \langle (v_{(1)})_i (v_{(1)})_j \rangle . \tag{8.26}$$

The second-order contribution $P_{(2)}$ to the pressure can be obtained from the Gibbs-Duhem relation $\delta P = \rho \delta \mu$ at $T = 0$ using expansions $\rho = \rho_0 + \rho_{(1)}$ and $\mu = \mu_0 + \mu_{(1)} + \mu_{(2)}$, where μ_0 is the chemical potential without the sound wave. This yields $P_{(2)} = \rho_0 \mu_{(2)} + \frac{\partial \rho}{\partial \mu} \frac{\rho_{(1)}^2}{2}$, where $\frac{\partial \rho}{\partial \mu} = \rho_0 / c_s^2$. According to the Euler equation (8.3) the second-order contribution to the chemical potential is $\mu_{(2)} = -\frac{v_{(1)}^2}{2}$. Then

$$\langle P_{(2)} \rangle = \frac{c_s^2}{\rho_0} \frac{\langle \rho_{(1)}^2 \rangle}{2} - \rho_0 \frac{\langle v_{(1)}^2 \rangle}{2} . \tag{8.27}$$

In the presence of the phonon distribution the total mass current is

$$\boldsymbol{j} = \rho_0 \boldsymbol{v}_0 + \frac{1}{h^3} \int \boldsymbol{j}^{\text{ph}}(\boldsymbol{p}) \, d_3 \boldsymbol{p} = \rho_0 \boldsymbol{v}_0 + \frac{1}{h^3} \int n(\boldsymbol{p}) \boldsymbol{p} \, d_3 \boldsymbol{p} , \tag{8.28}$$

In the thermal equilibrium at $T > 0$, the phonon numbers are given by the Planck distribution $n(\boldsymbol{p}) = n_0(E, \boldsymbol{v}_n)$ with the drift velocity $\boldsymbol{v}_n$ of quasiparticles:

$$n_0(E, \boldsymbol{v}_n) = \frac{1}{\exp \frac{E(\boldsymbol{p}) - \boldsymbol{p} \cdot \boldsymbol{v}_n}{T} - 1} = \frac{1}{\exp \frac{\varepsilon(\boldsymbol{p}) + \boldsymbol{p} \cdot (\boldsymbol{v}_0 - \boldsymbol{v}_n)}{T} - 1} . \tag{8.29}$$

Linearizing (8.29) with respect to the relative velocity $\boldsymbol{v}_0 - \boldsymbol{v}_n$, one obtains from (8.28) that

$$\boldsymbol{j} = \rho_0 \boldsymbol{v}_0 + \rho_n (\boldsymbol{v}_n - \boldsymbol{v}_0) . \tag{8.30}$$

This expression is equivalent to the two–fluid expression $\boldsymbol{j} = \rho \boldsymbol{v}_\mathrm{s} + \rho_\mathrm{n}(\boldsymbol{v}_\mathrm{n} - \boldsymbol{v}_\mathrm{s}) = \rho_\mathrm{s} \boldsymbol{v}_\mathrm{s} + \rho_\mathrm{n} \boldsymbol{v}_\mathrm{n}$ assuming that $\rho = \rho_0 = \rho_\mathrm{s} + \rho_\mathrm{n}$, $\boldsymbol{v}_0 = \boldsymbol{v}_\mathrm{s}$, and the normal density is given by the usual two-fluid-hydrodynamics expression:

$$\rho_\mathrm{n} = -\frac{1}{3h^3} \int \frac{\partial n_0(\varepsilon,0)}{\partial E} p^2 \, d_3\boldsymbol{p} \,. \tag{8.31}$$

In the same manner one can derive the two–fluid momentum–flux tensor [7]:

$$\Pi_{ij} = P\delta_{ij} + \rho_\mathrm{s} v_{\mathrm{s}i} v_{\mathrm{s}j} + \rho_\mathrm{n} v_{\mathrm{n}i} v_{\mathrm{n}j} \,. \tag{8.32}$$

This analysis demonstrates that two-fluid-hydrodynamics relations can be derived from the hydrodynamics of an ideal liquid in the presence of thermally excited sound waves, as was shown by Putterman and Roberts [31]. In order to obtain a complete system of equations of the two–fluid theory, one should take into account phonon–phonon interaction, which is essential for the phonon distribution function being close to the equilibrium Planck distribution. In the two–fluid theory the locally defined superfluid and normal velocities $\boldsymbol{v}_\mathrm{s}$ and $\boldsymbol{v}_\mathrm{n}$ correspond to the average velocity of a liquid in a fixed point of the space and to the drift velocity of the phonon gas respectively. The two-fluid hydrodynamics is valid only at scales exceeding the phonon mean-free path l_ph.

8.5 Scattering of Phonons by the Vortex in Hydrodynamics

Phonon scattering by a vortex line was initially studied in the work of Pitaevskii [32] and Fetter [33]. We consider a sound wave propagating in the plane xy normal to a vortex line (the axis z). In the linearized hydrodynamic equations of the previous section the fluid velocity $\boldsymbol{v}_0$ should be replaced by the velocity $\boldsymbol{v}_\mathrm{v}(\boldsymbol{r})$ around the vortex line:

$$\frac{\partial \rho_{(1)}}{\partial t} + \rho_0 \nabla \cdot \boldsymbol{v}_{(1)} = -\boldsymbol{v}_\mathrm{v} \cdot \nabla \rho_{(1)} \,, \tag{8.33}$$

$$\frac{\partial \boldsymbol{v}_{(1)}}{\partial t} + \frac{c_\mathrm{s}^2}{\rho_0} \nabla \rho_{(1)} = -\left[(\boldsymbol{v}_\mathrm{v} \cdot \nabla)\boldsymbol{v}_{(1)} + (\boldsymbol{v}_{(1)} \cdot \nabla)\boldsymbol{v}_\mathrm{v}\right] \,. \tag{8.34}$$

Using the vector identity

$$(\boldsymbol{v} \cdot \nabla)\boldsymbol{v} = \nabla \frac{v^2}{2} - \boldsymbol{v} \times [\nabla \times \boldsymbol{v}] \tag{8.35}$$

for the velocity $\boldsymbol{v} = \boldsymbol{v}_\mathrm{v} + \boldsymbol{v}_{(1)}$, equation (8.34) can be rewritten as

$$\frac{\partial \boldsymbol{v}_{(1)}}{\partial t} + \frac{c_\mathrm{s}^2}{\rho_0} \nabla \rho_{(1)} = -\nabla(\boldsymbol{v}_\mathrm{v} \cdot \boldsymbol{v}_{(1)}) + [\boldsymbol{v}_{(1)} \times \boldsymbol{\kappa}]\delta_2(\boldsymbol{r}) \,. \tag{8.36}$$

The perturbation from the vortex (the right-hand side) contains a δ-function because the vortex line is not at rest when the sound wave propagates past the vortex. In order to weaken the singularity one can introduce the time-dependent vortex velocity $\bm{v}_\text{v}(\bm{r},t)$ as a zero-order approximation for the velocity field [5]. Then $\bm{r}$ in (8.1) must be replaced by $\bm{r} - \bm{v}_\text{L} t$ and $\partial \bm{v}_\text{v}/\partial t = -(\bm{v}_\text{L} \cdot \bm{\nabla})\bm{v}_\text{v} = -\bm{\nabla}(\bm{v}_\text{L} \cdot \bm{v}_\text{v}) + [\bm{v}_\text{L} \times \bm{\kappa}]\delta_2(\bm{r})$. Since there is no external force on the liquid, the vortex moves with the velocity in the sound wave: $\bm{v}_\text{L} = \bm{v}_{(1)}(0,t)$. Now in the linearization procedure the fluid acceleration in (8.20) must be presented as $\partial \bm{v}/\partial t = \partial \bm{v}_\text{v}/\partial t + \partial \bm{v}_{(1)}/\partial t$. As a result (8.36) is replaced by:

$$\frac{\partial \bm{v}_{(1)}}{\partial t} + \frac{c_\text{s}^2}{\rho_0}\bm{\nabla}\rho_{(1)} = \bm{\nabla}[\bm{v}_\text{v} \cdot \bm{v}_{(1)}(\bm{r})] - \bm{\nabla}[\bm{v}_v \cdot \bm{v}_{(1)}(0)]. \tag{8.37}$$

Equation (8.37) yields:

$$\rho_{(1)} = -\frac{\rho_0}{c_\text{s}^2}\frac{\kappa}{2\pi}\left\{\frac{\partial\phi}{\partial t} + \bm{v}_\text{v} \cdot [\bm{\nabla}\phi(\bm{r}) - \bm{\nabla}\phi(0)]\right\}. \tag{8.38}$$

Substitution of $\rho_{(1)}$ in (19.27) yields the linear sound equation for the phase:

$$\frac{\partial^2\phi}{\partial t^2} - c_\text{s}^2\bm{\nabla}^2\phi = -2\bm{v}_\text{v}(\bm{r}) \cdot \bm{\nabla}\frac{\partial}{\partial t}\left[\phi(\bm{r}) - \frac{1}{2}\phi(0)\right]. \tag{8.39}$$

In the long-wavelength limit $k \to 0$ one may use the Born approximation. The Born perturbation parameter $\kappa k/c_\text{s}$ is on the order of the ratio of the vortex core radius $r_\text{c} \sim \kappa/c_\text{s}$ to the wavelength $2\pi/k$. Then after substituting the plane wave into the right-hand side of (8.39) the solution of this equation is

$$\phi = \phi_0 \exp(-\text{i}\omega t)\left\{\exp(\text{i}\bm{k}\cdot\bm{r}) - \frac{\text{i}k}{4c_\text{s}}\int d_2\bm{r}_1 H_0^{(1)}(k|\bm{r}-\bm{r}_1|)\bm{k}\cdot\bm{v}_\text{v}(\bm{r}_1)[2\exp(\text{i}\bm{k}\cdot\bm{r}_1)-1]\right\}. \tag{8.40}$$

Here $H_0^{(1)}(z)$ is the zero-order Hankel function of the first kind, and $\frac{\text{i}}{4}H_0^{(1)}(k|\bm{r}-\bm{r}_1|)$ is the Green function for the 2D wave equation, i.e., satisfies the equation

$$-(k^2 + \bm{\nabla}^2)\phi(\bm{r}) = \delta_2(\bm{r}-\bm{r}_1). \tag{8.41}$$

In the standard scattering theory they use the asymptotic expression for the Hankel function at large values of the argument:

$$\lim_{z\to\infty} H_0^{(1)}(z) = \sqrt{\frac{2}{\pi z}}e^{\text{i}(z-\pi/4)}. \tag{8.42}$$

If the perturbation is confined to a vicinity of the line, then $r_1 \ll r$ and

$$|\bm{r}-\bm{r}_1| \approx r - \frac{(\bm{r}_1\cdot\bm{r})}{r}. \tag{8.43}$$

After integration in (8.40) the wave at $kr \gg 1$ becomes a superposition of the incident plane wave $\propto \exp(i\mathbf{k} \cdot \mathbf{r})$ and the scattered wave $\propto \exp(ikr)$:

$$\phi = \phi_0 \exp(-i\omega t) \left[\exp(i\mathbf{k} \cdot \mathbf{r}) + \frac{ia(\varphi)}{\sqrt{r}} \exp(ikr) \right] . \tag{8.44}$$

Here $a(\varphi)$ is the scattering amplitude which is a function of the angle φ between the initial wave vector $\mathbf{k}$ and the wave vector $\mathbf{k}' = k\mathbf{r}/r$ after scattering (see Fig. 8.3). In the Born approximation (see the paper [7] and references therein for more details)

$$\begin{aligned}
a(\varphi) &= \sqrt{\frac{k}{2\pi}} \frac{1}{c_s} e^{i\frac{\pi}{4}} [\hat{\kappa} \times \mathbf{k}'] \cdot \mathbf{k} \frac{1}{q^2} \left(1 - \frac{q^2}{2k^2}\right) \\
&= -\frac{1}{2} \sqrt{\frac{k}{2\pi}} \frac{\kappa}{c_s} e^{i\frac{\pi}{4}} \frac{\sin\varphi \cos\varphi}{1 - \cos\varphi} ,
\end{aligned} \tag{8.45}$$

where $\mathbf{q} = \mathbf{k} - \mathbf{k}'$ is the momentum transferred by the scattered phonon to the vortex, and $q^2 = 2k^2(1 - \cos\varphi)$.

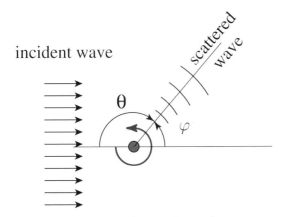

Fig. 8.3. Scattering of a sound wave by a vortex.

Thus the vortex is a line defect which scatters a sound wave. Scattering produces a force on the defect (vortex). If the perturbation by the line defect is confined to a finite vicinity of the line, the force

$$\mathbf{F}^{\text{ph}} = \sigma_\| c_s \mathbf{j}^{\text{ph}} - \sigma_\perp c_s [\hat{z} \times \mathbf{j}^{\text{ph}}] \tag{8.46}$$

is determined by two effective cross-sections [5,7]: the transport cross-section for the dissipative force,

$$\sigma_\| = \int \sigma(\varphi)(1 - \cos\varphi) d\varphi , \tag{8.47}$$

and the transverse cross-section for the transverse force,

$$\sigma_\perp = \int \sigma(\varphi) \sin\varphi \, d\varphi \,. \tag{8.48}$$

The differential cross-section $\sigma(\varphi) = |a(\varphi)|^2$ with $a(\varphi)$ from (8.45) is quadratic in the circulation κ and even in φ. Therefore in the Born approximation the transverse cross-section $\sigma_\perp$ vanishes.

However, due to a very slow decrease of the velocity $v_v \propto 1/r$ far from the vortex, the scattering amplitude is divergent at small scattering angles $\varphi \to 0$:

$$\lim_{\varphi \to 0} a(\varphi) = -\sqrt{\frac{k}{2\pi}} \frac{\kappa}{c_s} e^{i\frac{\pi}{4}} \frac{1}{\varphi} \,. \tag{8.49}$$

This divergence is integrable in the integral for the transport cross-section, (8.47), and its calculation is reliable. In contrast, the integrand in (8.48) for the transverse cross-section has a pole at $\varphi = 0$. The principal value of the integral vanishes, but there is no justification for the choice of the principal value, and the contribution of the small angles requires an additional analysis.

At small scattering angles $\varphi \ll 1/\sqrt{kr}$ the asymptotic expansion (8.44) is invalid, and one cannot use the differential cross-section or the scattering amplitude to describe the small-angle scattering [4,5,9,34]. Meanwhile, the small-angle behavior is crucial for the transverse force as demonstrated below. The accurate calculation of the integral in (8.40) for small angles (see [5] and Appendix B in [7]) yields that at $\varphi \ll 1$

$$\phi = \phi_0 \exp(-i\omega t + i\mathbf{k} \cdot \mathbf{r}) \left[1 + \frac{i\kappa k}{2c_s} \Phi\left(\varphi\sqrt{\frac{kr}{2i}}\right)\right] \,. \tag{8.50}$$

Using an asymptotic expression for the error integral

$$\Phi(z) = \frac{2}{\sqrt{\pi}} \int_0^z e^{-t^2} dt \longrightarrow \frac{z}{|z|} - \frac{1}{\sqrt{\pi}z} \exp(-z^2) \tag{8.51}$$

at $|z| \to \infty$, one obtains for angles $1 \gg \varphi \gg 1/\sqrt{kr}$:

$$\phi = \phi_0 \exp(-i\omega t) \left[\exp(i\mathbf{k} \cdot \mathbf{r}) \left(1 + \frac{i\kappa k}{2c_s} \frac{\varphi}{|\varphi|}\right) \right.$$

$$\left. - \frac{i\kappa}{c_s} \sqrt{\frac{k}{2\pi r}} \frac{1}{\varphi} \exp\left(ikr + i\frac{\pi}{4}\right)\right] \,. \tag{8.52}$$

The second term in square brackets coincides with scattering wave at small angles $\varphi \ll 1$ with the amplitude given by (8.49). But now one can see that the standard scattering theory misses to reveal an important non-analytical correction to the incident plane wave, which changes a sign when the scattering angle φ crosses zero. Its physical meaning is discussed in the next section.

8.6 The Iordanskii Force and the Aharonov–Bohm Effect

The analogy between the phonon scattering by a vortex and the Aharonov–Bohm effect for electrons scattered by a magnetic-flux tube becomes evident if one rewrites the sound equation (8.39) in presence of the vortex as

$$k^2 \phi - \left(-i\boldsymbol{\nabla} + \frac{k}{c_s}\boldsymbol{v}_v\right)^2 \phi = 0 . \tag{8.53}$$

It differs from the sound equation (8.39) by the term of the second order in $v_v \propto \kappa$ and by absence of the contribution from the vortex-line motion [the term $\propto \phi(0)$ on the right-hand side of (8.39)]. This difference is unimportant for the calculation of the transverse force, which is linear in κ. On the other hand, the stationary Schrödinger equation for an electron in the presence of the magnetic flux Φ confined to a thin tube (the Aharonov–Bohm effect [8]) is:

$$E\psi(\boldsymbol{r}) = \frac{1}{2m}\left(-i\hbar\boldsymbol{\nabla} - \frac{e}{c}\boldsymbol{A}\right)^2 \psi(\boldsymbol{r}) . \tag{8.54}$$

Here ψ is the electron wave function with energy E and the electromagnetic vector potential is connected with the magnetic flux Φ by a relation similar to that for the velocity $\boldsymbol{v}_v$ around the vortex line [see (8.1)]:

$$\boldsymbol{A} = \Phi \frac{[\hat{z} \times \boldsymbol{r}]}{2\pi r^2} . \tag{8.55}$$

Let us consider the quasiclassical solution of the sound equation:

$$\phi = \phi_0 \exp\left(-i\omega t + i\boldsymbol{k}\cdot\boldsymbol{r} + \frac{i\delta S}{\hbar}\right)$$
$$= \phi_0 \exp(-i\omega t + i\boldsymbol{k}\cdot\boldsymbol{r})\left(1 + \frac{i\kappa k}{2\pi c_s}\theta\right) , \tag{8.56}$$

where $\delta S = -(\hbar k/c_s)\int^{\boldsymbol{r}} \boldsymbol{v}_v \cdot d\boldsymbol{l} = \hbar\theta\kappa k/2\pi c_s$ is the variation of the action due to interaction with the circular velocity from the vortex along quasiclassical trajectories. The angle θ is an azimuth angle for the position vector $\boldsymbol{r}$ measured from the direction opposite to the wave vector $\boldsymbol{k}$ (see Fig. 8.3). This choice means that the quasiclassical correction vanishes for the incident wave far from the vortex. One can check directly that (8.56) satisfies the sound equation (8.39) in the first order of the parameter $\kappa k/c_s$. For the Aharonov–Bohm effect the phase $\delta S/\hbar$ arises from the electromagnetic vector potential: $\delta S = -(e/c)\int^{\boldsymbol{r}} \boldsymbol{A}\cdot d\boldsymbol{l}$.

The velocity generated by the sound wave around the vortex is

$$\boldsymbol{v}_{(1)} = \frac{\kappa}{2\pi}\boldsymbol{\nabla}\phi = \frac{\kappa}{2\pi}\phi_0 \exp(-i\omega t + i\boldsymbol{k}\cdot\boldsymbol{r})\left(i\boldsymbol{k} - \frac{i k}{c_s}\boldsymbol{v}_v\right) . \tag{8.57}$$

From (8.38) and (8.57) we can obtain the phonon mass current with first-order corrections in v_v.

$$j_{\text{ph}} = \langle \rho_{(1)} v_{(1)} \rangle$$
$$= \frac{1}{8\pi^3} \int d\mathbf{k}\, n(\hbar \mathbf{k}) \hbar \left[\mathbf{k} - \frac{k}{c_s} v_v(r) - (\mathbf{k} \cdot v_v(r)) \frac{\mathbf{k}}{c_s k} \right]. \quad (8.58)$$

Due to the last term in this expression, the phonon mass current is not curl-free.

According to (8.56) the phase ϕ is multivalued, and one must choose a cut for an angle θ at the direction $\mathbf{k}$, where $\theta = \pm\pi$. The jump of the phase on the cut line behind the vortex is a manifestation of the Aharonov–Bohm effect [8]: the sound wave after its interaction with the vortex has a different phase on the left and on the right of the vortex line. This results in an interference [5,7].

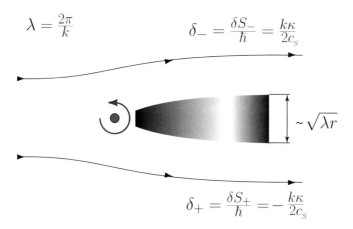

Fig. 8.4. Aharonov–Bohm interference.

In the interference region the quasiclassical solution is invalid and must be replaced by (8.50). The width of the interference region is $d_{\text{int}} \sim \sqrt{r/k}$. Here r is the distance from the vortex line. The interference region corresponds to very small scattering angles $\sim d_{\text{int}}/r = 1/\sqrt{kr}$.

Now we are ready to consider the momentum balance using the condition that $\int dS_j \Pi_{\perp j} = 0$ where subscript $\perp$ is for a component normal to the wave vector $\mathbf{k}$ of the incident wave. The total momentum-flux tensor can be obtained from (8.25) and (8.26) assuming $v_0(r) = v_v(r) + v_s$ and neglecting some unimportant terms:

$$\Pi_{ij} = -\rho_0(v_s - v_L) \cdot v_v \delta_{ij} + \rho_0 v_{0i} v_{0j}$$
$$+ \langle \rho_{(1)}(v_{(1)})_i \rangle v_{vj} + \langle \rho_{(1)}(v_{(1)})_j \rangle v_{vi} + \rho_0 \langle (v_{(1)})_i (v_{(1)})_j \rangle. \quad (8.59)$$

The first two terms in this expression yield the momentum flux without phonons, which gives the Magnus force for a liquid with the density ρ_0 and the velocity $\boldsymbol{v}_{\rm s}$. The term $\propto v_{{\rm v}j}$ does not contribute to the momentum flux through a cylindrical surface around the vortex, since the velocity $\boldsymbol{v}_{\rm v}$ is tangent to this surface. The term $v_{{\rm v}i}\langle \rho_{(1)}(v_{(1)})_j\rangle$, in which the mass current $\langle \rho_{(1)}(v_{(1)})_j\rangle$ is given (8.24) for the plane wave in absence of the vortex, gives a contribution to the momentum flux, which exactly cancels the contribution of the term $\rho_0\langle(v_{(1)})_i(v_{(1)})_j\rangle$ outside the interference region, where the velocity is given by (8.57). Finally only the interference region contributes the momentum flux $\int dS_j \Pi_{\perp j}$.

In the interference region the velocity is obtained by taking the gradient of the phase given by (8.50). Its component normal to $\boldsymbol{k}$,

$$v_{(1)\perp} = \frac{\kappa}{2\pi r}\frac{\partial \phi}{\partial \varphi} = \phi_0 \exp(-i\omega t + i\boldsymbol{k}\cdot\boldsymbol{r})\frac{i\kappa^2 k}{4\pi c_{\rm s} r}\frac{\partial \Phi\left(\varphi\sqrt{kr/2i}\right)}{\partial \varphi}, \quad (8.60)$$

determines the interference contribution to the transverse force:

$$\int dS_j \rho_0 \langle(v_{(1)})_\perp (v_{(1)})_j\rangle = \int \rho_0 \langle(v_{(1)})_\perp (v_{(1)})_r\rangle r d\varphi$$

$$= \frac{1}{8\pi^2}\rho_0 \phi_0^2 \frac{\kappa^2 k}{\hbar}[\delta S_- - \delta S_+]$$

$$= \frac{1}{8\pi^2}\rho_0 \phi_0^2 \frac{\kappa^3 k^2}{c_{\rm s}} = \kappa j^{\rm ph}, \quad (8.61)$$

where $\delta S_\pm$ are the action variations at $\theta \to m_{\rm p}\pi$. Thus the interference region, which corresponds to an infinitesimally small angle interval, yields a finite contribution to the transverse force, which one could not obtain from the standard scattering theory using the differential cross-section. In fact the details of the solution in the interference region are not essential: only a jump of the phase across the interference region is of importance.

Finally the momentum balance condition $\int dS_j \Pi_{\perp j} = 0$ yields the relation

$$\rho_0[(\boldsymbol{v}_{\rm L} - \boldsymbol{v}_{\rm s})\times \boldsymbol{\kappa}] - [\boldsymbol{j}^{\rm ph}(\boldsymbol{p})\times\boldsymbol{\kappa}] = 0. \quad (8.62)$$

The second vector product on the left-hand side is a transverse force which corresponds to the transverse cross-section [see (8.48)]

$$\sigma_\perp = \frac{\delta S_- - \delta S_+}{\hbar k} \quad (8.63)$$

equal to $\kappa/c_{\rm s}$. For the Planck distribution of phonons, $\boldsymbol{j}^{\rm ph}$ must be replaced by $\rho_{\rm n}(\boldsymbol{v}_{\rm n} - \boldsymbol{v}_{\rm s})$:

$$\rho_{\rm s}[(\boldsymbol{v}_{\rm L} - \boldsymbol{v}_{\rm s})\times \boldsymbol{\kappa}] + \rho_{\rm n}[(\boldsymbol{v}_{\rm L} - \boldsymbol{v}_{\rm n})\times\boldsymbol{\kappa}] = 0. \quad (8.64)$$

The force $\propto (\boldsymbol{v}_{\rm L} - \boldsymbol{v}_{\rm n})$ is the Iordanskii force which corresponds to $D' = -\kappa\rho_{\rm n}$ in (8.10). The longitudinal force $\propto D$ is not present in (8.64) since we ignored terms of the second order in κ in order to simplify our derivation. According to (8.64) the vortex moves with the center-of-mass velocity $\boldsymbol{v} = \frac{\rho_\mathrm{s}}{\rho} \boldsymbol{v}_\mathrm{s} + \frac{\rho_\mathrm{n}}{\rho} \boldsymbol{v}_\mathrm{n}$.

Our scattering analysis was done in the coordinate frame where the vortex is at rest and we neglected the relative superfluid velocity $\boldsymbol{v}_\mathrm{s} - \boldsymbol{v}_\mathrm{L}$ with respect to the vortex. If the velocity $\boldsymbol{v}_\mathrm{s} - \boldsymbol{v}_\mathrm{L}$ is high it can affect the value of the effective cross-section. But since the phonon momentum is linear with respect to $\boldsymbol{v}_\mathrm{n} - \boldsymbol{v}_\mathrm{s}$, the dependence of the cross-section on $\boldsymbol{v}_\mathrm{s} - \boldsymbol{v}_\mathrm{L}$ is a nonlinear effect. Thus our momentum balance took into account all effects linear in the superfluid velocity $\boldsymbol{v}_\mathrm{s}$. This is confirmed by a more elaborate analysis of Stone [35].

8.7 Partial-Wave Analysis and the Aharonov–Bohm Effect

Studying interaction of phonons with a vortex we solved the sound equation using the perturbation theory. It is completely justified because the parameter of the perturbation theory $\kappa k/c_\mathrm{s}$ is the ratio between the vortex core and the phonon wavelength, which is always small for phonons. But in the Aharonov–Bohm problem for electrons the corresponding parameter is $\gamma = \Phi/\Phi_1$, where $\Phi_1 = hc/e$ is the magnetic-flux quantum for one electron (two times larger than the magnetic-flux quantum $\Phi_0 = hc/2e$ for a Cooper pair). This parameter can be arbitrary large, and the perturbation theory is not enough to describe an expected periodic dependence on γ. On the other hand, there is an exact solution of the Aharonov–Bohm problem for electrons obtained by the partial-wave expansion [8], and it will be shown now how to derive the transverse force from this solution. Another derivation of the force on the Aharonov–Bohm flux tube using the wave-packet presentation is discussed by Shelankov [12].

We look for a solution of (8.54) as a superposition of the partial cylindrical waves $\psi = \sum_l \psi_l(r) \exp(il\varphi)$. Partial-wave amplitudes ψ_l should satisfy equations in the cylindrical system of coordinates (r, φ):

$$\frac{d^2\psi_l}{dr^2} + \frac{1}{r}\frac{d\psi_l}{dr} - \frac{(l-\gamma)^2}{r^2}\psi_l + k^2\psi_l = 0 \ . \tag{8.65}$$

Here k is the wave number of the electron far from the vortex so that $E = \hbar^2 k^2/2m$. We need a solution of (8.65), which at large distances has an asymptotic behavior given by (8.44):

$$\psi_l = \sqrt{\frac{2n}{\pi k r}} \exp\left[i\frac{\pi}{2}l + i\delta_l\right] \cos\left(kr - \frac{\pi}{2}l + \delta_l - \frac{\pi}{4}\right) , \tag{8.66}$$

where n is the particle density and the partial-wave phase shifts are

$$\delta_l = (l - |l - \gamma|)\pi/2 \ . \tag{8.67}$$

For $\delta_l = 0$, equation (8.66) yields the partial-wave amplitudes of the incident plane wave at large distances r. But for nonzero δ_l there is also the scattered wave in (8.44) with the scattering amplitude

$$a(\varphi) = \sqrt{\frac{1}{2\pi k}} \exp\left(i\frac{\pi}{4}\right) \sum_l [1 - \exp(2i\delta_l)] \exp(il\varphi), \tag{8.68}$$

The transverse force is determined by the transverse cross-section

$$\sigma_\perp = \int |a(\varphi)|^2 \sin\varphi \, d\varphi$$

$$= \frac{1}{2ik} \left\{ \sum_l e^{i2\delta_{l-1}} - \sum_l e^{i2\delta_{l+1}} \right.$$

$$\left. + \sum_l e^{i2(\delta_{l+1} - \delta_l)} - \sum_l e^{i2(\delta_{l-1} - \delta_l)} \right\}. \tag{8.69}$$

Shifting the number l by 2 in the first sum and by one in the fourth sum one obtains the expression for the transverse cross-section in the partial-wave method derived long ago by Cleary [36]:

$$\sigma_\perp = \int |a(\varphi)|^2 \sin\varphi \, d\varphi = \frac{1}{k} \sum_l \sin 2(\delta_l - \delta_{l+1}). \tag{8.70}$$

Using the phase shift values for the Aharonov–Bohm solution, (8.67), the transverse cross-section is

$$\sigma_\perp = -\frac{1}{k} \sin 2\pi\gamma. \tag{8.71}$$

However, the shift of l in the first sum of (8.69) is not an innocent operation because of the divergence of the first and second sum at $l \to \pm\infty$. The derivation of Clearly's formula (8.70) assumes that the first and second divergent sums in (8.69) should exactly cancel after the shift of l. But if one does not shift l, the difference of the first and second sum is finite. Moreover, this difference cancels the contribution of the third and fourth sum, and $\sigma_\perp$ vanishes in the first order with respect to the phase shifts δ_l.

Ambiguity in the calculation of the partial-wave sum for the transverse cross-section is another manifestation of the small-scattering-angle problem in the configurational space: the standard scattering theory does not provide a recipe for treating a singularity at zero scattering angle. The zero-angle singularity is responsible for a divergent partial-wave series. The way to avoid the ambiguity is similar to that in the configurational space: one should not use the concept of the scattering amplitude for calculation of the transverse force.

We must analyze the momentum balance. The momentum-flux tensor for the electron Schrödinger equation (8.54) is

$$\Pi_{ij} = \frac{1}{2m}\text{Re}\left\{\left(-i\hbar\nabla_i - \frac{e}{c}A_i\right)\psi\left(i\hbar\nabla_i - \frac{e}{c}A_i\right)\psi^* \right.$$
$$\left. -\psi^*\left(-i\hbar\nabla_i - \frac{e}{c}A_i\right)\left(-i\hbar\nabla_j - \frac{e}{c}A_j\right)\nabla_j\psi\right\}. \quad (8.72)$$

The transverse force is determined by Π_{yr} if the axis x is directed along the wave vector $\mathbf{k}$ of the incident wave, and $\partial/\partial y = \sin\phi\, \partial/\partial r$. The terms $\propto 1/r$ (A and $\partial/\partial\phi$) are not important. Then the momentum-flux tensor is

$$\Pi_{yr} = \frac{\hbar^2}{2m}\sum_{l'}\sum_l \text{Re}\left\{\sin\phi\left[\frac{\partial\psi_{l'}^*}{\partial r}\frac{\partial\psi_l}{\partial r} - \psi_l^*\frac{\partial^2\psi_l^*}{\partial r^2}\right]e^{i(l-l')\phi}\right\}, \quad (8.73)$$

where ψ_l are given by (8.66). Finally the transverse force is

$$F_\perp = \oint \Pi_{yr}r\,d\phi = \pi r\frac{\hbar^2}{m}\text{Im}\sum_l\left\{\frac{\partial\psi_{l+1}^*}{\partial r}\frac{\partial\psi_l}{\partial r} - \psi_l^*\frac{\partial^2\psi_{l+1}^*}{\partial r^2}\right\}. \quad (8.74)$$

Here we also made a shift of l in some sums, but it is not dangerous because the terms of these sums decrease in the limits $l \to \pm\infty$.

Inserting the partial-wave amplitudes and their radial derivatives into the expression for the force we obtain:

$$F_\perp = -kn\frac{\hbar^2}{m}\sum_l \sin 2(\delta_{l+1} - \delta_l) = -jv\sigma_\perp, \quad (8.75)$$

where $j = \hbar k n$ and $v = \hbar k/m$ are the momentum density and the velocity in the incident plane wave, and $\sigma_\perp$ is the effective cross-section (8.70) derived by Cleary [36]. Thus there is a well-defined effective transverse cross-section, although we cannot directly use its expression (8.48) via the differential cross-section. But formally we can "repair" this expression by a recipe for treating the singularity at small angles: we should add to the differential cross-section a singular term proportional to the derivative of $\delta(\varphi)$, and take the principle value of the integral over the rest of the differential cross-section.

One can obtain the transverse cross-section for phonons directly from (8.70) assuming that $\sin 2\pi\gamma \approx 2\pi\gamma = -\kappa k/c_s$. However, it is useful to follow the connection between Cleary's formula (8.70) and (8.63) obtained from the quasiclassical solution. In the classical limit the partial wave l corresponds to the quasiclassical trajectory with the impact parameter $b = l/k = \hbar l/p$ and the small scattering angle is $\varphi = -d\delta_l/dl = -(1/2\hbar k)d\delta S(b)/db$, where $\delta S(b)$ is the action variation along the trajectory with impact parameter b. Finally replacing the sum by an integral, equation (8.70) can be rewritten as

$$\sigma_\perp = -\frac{2}{k}\int_{-\infty}^\infty dl\frac{d\delta_l}{dl} = -\frac{1}{\hbar k}\int_{-\infty}^\infty db\frac{d\delta S(b)}{db}, \quad (8.76)$$

which yields (8.63) since $\delta S_\pm = \delta S(\pm\infty)$. Strictly speaking (8.76) is valid if $\delta S(b)$ is a continuous function. This is the case for rotons which are scattered quasiclassically [5–7]. But it yields a correct answer even for phonons, despite the fact that, $\delta S(b)$ has a jump at $b \sim 0$. Thus even though phonon scattering cannot be described in the quasiclassical approximation (it yields the scattering angle $\varphi = -(1/\hbar k)d\delta S(b)/db \approx 0$ at $b \neq 0$), the quasiclassical expression (8.76) gives a correct phonon transverse cross-section.

8.8 Momentum Balance in Two-Fluid Hydrodynamics

Up to now we have analyzed spatial scales much less than the mean-free-path l_{ph} of phonons (ballistic region). Sound waves (phonons) interacted with the velocity field generated by a vortex, but phonon–phonon interaction was neglected. Now we shall see what is going on at scales much larger than l_{ph}, where the two-fluid hydrodynamics is valid.

Scattering of phonons by the vortex in the ballistic region produced a force $\boldsymbol{F}$ on the vortex, and according to the third Newton law a force $-\boldsymbol{F}$ on the phonons (the normal fluid) should also arise. A momentum transfer from a phonon to a vortex takes place at distances of the order of the phonon wavelength $\lambda = 2\pi/k$ which is much less than the mean-free path l_{ph}. This means that at hydrodynamic scales the force on the normal fluid is a δ-function force concentrated along the vortex line. The response of the normal fluid to this force is described by the Navier–Stokes equation with the dynamic viscosity η_{n}:

$$\frac{\partial \boldsymbol{v}_{\mathrm{n}}}{\partial t} + (\boldsymbol{v}_{\mathrm{n}} \cdot \boldsymbol{\nabla})\boldsymbol{v}_{\mathrm{n}} = \nu_{\mathrm{n}}\Delta\boldsymbol{v}_{\mathrm{n}} - \frac{\boldsymbol{\nabla}P}{\rho} - \frac{\rho_{\mathrm{s}}S}{\rho_{\mathrm{n}}\rho}\boldsymbol{\nabla}T \, , \tag{8.77}$$

where $\nu_{\mathrm{n}} = \eta_{\mathrm{n}}/\rho_{\mathrm{n}}$ is the kinematic viscosity, S is the entropy per unit volume and T is the temperature. Our problem is similar to the Stokes problem of a cylinder moving through a viscous liquid [1]. Neglecting the nonlinear inertial (convection) term $(\boldsymbol{v}_{\mathrm{n}} \cdot \boldsymbol{\nabla})\boldsymbol{v}_{\mathrm{n}}$ in the Navier–Stokes equation, the δ-function force produces a divergent logarithmic velocity field (the Stokes paradox [1]):

$$\boldsymbol{v}_{\mathrm{n}}(\boldsymbol{r}) = \boldsymbol{v}_{\mathrm{n}} + \frac{\boldsymbol{F}}{4\pi\eta_{\mathrm{n}}}\ln\frac{r}{l_{\mathrm{ph}}} \, , \tag{8.78}$$

where $\boldsymbol{v}_{\mathrm{n}}$ is the normal velocity at a distance of order $r \sim l_{\mathrm{ph}}$, which separates the ballistic and the hydrodynamic regions. In this expression the mean-free path l_{ph} replaces the cylinder radius of the classical Stokes problem in the argument of the logarithm. Such a choice of the lower cut-off assumes that the quasiparticle flux on the vortex is entirely determined by the equilibrium Planck distribution at the border between the ballistic and the two-fluid-hydrodynamics region at $r \sim l_{\mathrm{ph}}$ [37].

However small the relative normal velocity $\boldsymbol{v}_{\mathrm{n}} - \boldsymbol{v}_{\mathrm{L}}$ could be, at distance of the order or larger than $r_{\mathrm{m}} \sim \nu_{\mathrm{n}}/|\boldsymbol{v}_{\mathrm{n}} - \boldsymbol{v}_{\mathrm{L}}|$ the nonlinear convection term becomes important and stops a logarithmic growth of the normal velocity. Due to the force $\boldsymbol{F}$ the normal velocities $\boldsymbol{v}_{n\infty}$ and $\boldsymbol{v}_n$ at large $(r \sim r_m)$ and small $(r \sim l_{ph})$ distances from the vortex line are different (the viscous drag [2]):

$$\boldsymbol{F} = \frac{4\pi \eta_n}{\ln(r_m/l_{ph})} (\boldsymbol{v}_{n\infty} - \boldsymbol{v}_n) \ . \tag{8.79}$$

At very large distances $r \gg r_m$ the nonlinear convection term is more important than the viscous term. Thus the scale r_m, which we shall call Oseen's length, divides the hydrodynamic region onto the viscous and convection subregions [38]. All relevant scales are shown in Fig 8.5. For a longitudinal force the solution of the Navier–Stokes equation, valid for both the viscous and convection subregions, was obtained by Oseen long ago (see [1]).

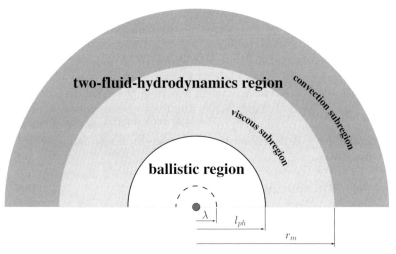

Fig. 8.5. Relevant scales: the phonon wavelength λ, the phonon mean free path l_{ph}, and Oseen's length $r_m \sim \nu_n/|\boldsymbol{v}_n - \boldsymbol{v}_L|$.

The force $-\boldsymbol{F}$ is transmitted to large distances by a constant momentum flux. In the viscous subregion momentum transport in the normal fluid occurs via viscosity: $F_i = -\oint \tau_{ij} dS_j$, where $\tau_{ij} = -\eta_n (\nabla_i v_{nj} + \nabla_j v_{ni})$ is the viscous stress tensor. On the other hand, the total momentum flux for the whole liquid should vanish: $\oint \Pi_{ij} dS_j + \oint \tau_{ij} dS_j = 0$, where Π_{ij} is given by (8.32). The normal velocity field does not contain the circular velocity $\boldsymbol{v}_v$, and there is no normal circulation in the viscous subregion. Therefore the momentum flux $\oint \Pi_{ij} dS_j$ yields the superfluid Magnus force, i.e., the momentum balance in the viscous subregion confirms that the force $\boldsymbol{F}$ satisfies (8.9).

In the convection region the viscosity becomes ineffective and the momentum conservation gives again $\oint \Pi_{ij} dS_j = 0$, like in the ballistic region. The superfluid part of the momentum flux is related to the superfluid Magnus force $\propto \rho_s$ and the normal part of the flux should transmit the same force $\boldsymbol{F}$. The relation between the force $\boldsymbol{F}$ and the relative normal velocity $\boldsymbol{v}_{n\infty} - \boldsymbol{v}_L$ in the convection subregion can be derived from (8.10) and (8.79):

$$\boldsymbol{F} = -\tilde{D}(\boldsymbol{v}_L - \boldsymbol{v}_{n\infty}) - \tilde{D}'[\hat{z} \times (\boldsymbol{v}_L - \boldsymbol{v}_{n\infty})] \,, \tag{8.80}$$

where the parameters $\tilde{D}$ and $\tilde{D}'$ are connected with D and D' in (8.10) by a complex relation

$$\frac{1}{\tilde{D} + i\tilde{D}'} = \frac{\ln(r_m/l_{\text{ph}})}{4\pi\eta_n} + \frac{1}{D + iD'} \,. \tag{8.81}$$

Since in the convection region the viscosity is ineffective and the normal fluid behaves as an ideal one, the only way to transmit the transverse component of the force $\boldsymbol{F}$ to infinity is to create a circulation of the normal velocity [39]. But because of the viscous drag, separation into the longitudinal and the transverse force should be done relative to the normal velocity $\boldsymbol{v}_{n\infty} - \boldsymbol{v}_L$, and not $\boldsymbol{v}_n - \boldsymbol{v}_L$. Thus the normal circulation is determined by $\tilde{D}'$, and not D':

$$\kappa_n = \oint d\boldsymbol{l} \cdot \boldsymbol{v}_n = -\frac{\tilde{D}'}{\rho_n} = \frac{\kappa}{1 + \left[\frac{\kappa \rho_n \ln(r_m/l_{\text{ph}})}{4\pi\eta_n}\right]^2} \,, \tag{8.82}$$

where we neglected the longitudinal component $\propto D$ in the ballistic region and used the relation $D' = -\kappa\rho_n$ for the Iordanskii force.

In the convection subregion it is more problematic to transmit not the transverse, but the longitudinal component of the force. It is impossible to do if the viscosity is neglected completely: a body moving through an ideal liquid does not produce any dissipative force (d'Alembert's paradox [40]). The paradox is resolved by finding that a *laminar wake* should arise behind a moving body. Within the wake one cannot neglect viscosity even deeply in the convection area when $r \gg r_m$. The width of the laminar wake is growing as $\sim \sqrt{rr_m}$ far from the moving body [1]. Solving the Navier–Stokes equation by Oseen's method one can find out how the laminar wake and the normal circulation are formed during the crossover from the viscous to the convection subregion. One can find a detailed analysis of this crossover in the presence of the longitudinal and the transverse force in [41].

The force $\boldsymbol{F}$ is the mutual friction force introduced by Hall and Vinen [2] for the analysis of propagation of the second sound in rotating superfluid. With the help of (8.9), (8.10), and (8.79) one obtains a linear relation between the mutual friction force and the counterflow velocity $\boldsymbol{v}_s - \boldsymbol{v}_{n\infty}$ bearing in mind that the average normal velocity practically coincides with the velocity $\boldsymbol{v}_{n\infty}$:

$$\boldsymbol{F} = \frac{\kappa \rho_s \rho_n}{2\rho} B(\boldsymbol{v}_s - \boldsymbol{v}_{n\infty}) - \frac{\kappa \rho_s \rho_n}{2\rho} B'[\hat{z} \times (\boldsymbol{v}_s - \boldsymbol{v}_{n\infty})] \,, \tag{8.83}$$

The Hall-Vinen parameters B and B' are given by a complex relation [6]

$$\frac{2\rho}{\kappa\rho_s\rho_n}\frac{1}{B-iB'} = -\frac{1}{i\rho_s\kappa} + \frac{1}{\tilde{D}+i\tilde{D}'}$$
$$= -\frac{1}{i\rho_s\kappa} + \frac{\ln(r_m/l_{ph})}{4\pi\eta_n} + \frac{1}{D+iD'} . \quad (8.84)$$

According to this relation a strong viscous drag (large logarithm $\ln(r_m/l)$, or small viscosity η_n) suppresses the effect of the transverse force ($\propto D'$) on the mutual friction [41]. But the effect of the superfluid Magnus force and the longitudinal force $\propto D$ is also suppressed in this limit. In fact, this is a limit when the force from quasiparticle scattering (transverse, or longitudinal) is so strong that the normal fluid sticks to the vortex, and $v_L = v_n$. Then the resultant force F depends only on viscosity, as in the classical Stokes problem.

8.9 Magnus Force and the Berry Phase

Now let us consider a connection between the transverse force and the Berry phase. We shall use the hydrodynamic description with the Lagrangian obtained from (8.13) after the Madelung transformation:

$$L = \frac{\kappa\rho}{2\pi}\frac{\partial\phi}{\partial t} - \frac{\kappa^2\rho}{8\pi}\nabla\phi^2 - \frac{V}{2}\rho^2 . \quad (8.85)$$

The first term with the first time derivative of the phase ϕ (Wess–Zumino term) is responsible for the Berry phase $\Theta = \Delta S_B/\hbar$, which is the variation of the phase of the quantum-mechanical wave function for an adiabatic motion of the vortex around a closed loop [16]. Here

$$\Delta S_B = \int d\mathbf{r}\, dt\, \frac{\kappa\rho}{2\pi}\frac{\partial\phi}{\partial t} = -\int d\mathbf{r}\, dt\, \frac{\kappa\rho}{2\pi}(\mathbf{v}_L \cdot \boldsymbol{\nabla}_L)\phi . \quad (8.86)$$

is the classical action variation around the loop and $\boldsymbol{\nabla}_L\phi$ is the gradient of the phase $\phi[\mathbf{r} - \mathbf{r}_L(t)]$ with respect to the vortex position vector $\mathbf{r}_L(t)$. However, $\boldsymbol{\nabla}_L\phi = -\boldsymbol{\nabla}\phi$, where $\boldsymbol{\nabla}\phi$ is the gradient with respect to $\mathbf{r}$. Then the loop integral $\oint d\mathbf{l}$ yields the circulation of total current $\mathbf{j} = (\kappa/2\pi)\langle\rho\boldsymbol{\nabla}\phi\rangle$ for points inside the loop, but vanishes for points outside. As a result, the Berry-phase action is given by [18]

$$\Delta S_B = V\frac{\kappa}{2\pi}\oint (d\mathbf{l} \cdot \mathbf{j}) . \quad (8.87)$$

where V is the volume inside the loop (a product of the loop area and the liquid height along a vortex). Contrary to (8.86), the loop integral in (8.87) is related to the variation of the position vector $\mathbf{r}$, the vortex position vector $\mathbf{r}_L$ being fixed.

Since the Berry phase is proportional to the current circulation which determines the transverse force, there is a direct connection between the Berry phase and the amplitude of the transverse force on a vortex, as was shown, e.g., in [17]. Thus the problem is reduced to a calculation of the current circulation. If the circulation of the normal velocity at large distances vanished (as assumed in [15,17,18]), the current circulation would be $\oint (d\boldsymbol{l} \cdot \boldsymbol{j}) = \rho_s \kappa$, and the Berry phase (as well as the effective Magnus force) would be proportional to ρ_s. However, according to Sect. 8.8 and [41], at very large distances (in the convection subregion) the normal circulation does not vanish and is proportional to the transverse force. Using a proper value of the asymptotic normal circulation given by (8.82) the Berry phase yields the same transverse force as determined from the momentum balance.

In order to obtain a correct value of the transverse force from the Berry phase, one should choose a loop radius much larger than Oseen's length r_m. If the loop radius is chosen in the viscous subregion $l_{ph} < r < r_m$, the total-current circulation is proportional ρ_s, but the Berry phase does not yield the total transverse force, since a part of it, namely, the Iordanskii force, is presented by the viscous momentum flux, which cannot be obtained in the Lagrange formalism. And if the loop radius is chosen in the ballistic region, the total-current circulation is not defined at all and depends on a shape of the loop, since the phonon mass current is not curl-free, as pointed out after (8.58).

8.10 Discussion and Conclusions

The momentum-balance analysis definitely confirms an existence of the transverse force on a vortex from phonon scattering (the Iordanskii force). This agrees with the results of the latest analysis of Thouless et al. [41].

The Berry phase yields the same value of the transverse force as the momentum balance, if a proper value of the normal circulation at large distances from the vortex is used for the calculation of the Berry phase. However, the Berry-phase analysis itself cannot provide the normal-circulation value, since the latter is determined by the processes at small distances from the vortex, which are beyond of the Berry-phase analysis. The small-distance processes determine the force between the superfluid and the normal component, which is present in the small-distance boundary condition for the Navier-Stokes equation in the two-fluid-hydrodynamics region. The force on the normal component is transmitted to infinite distances by the constant momentum flux, which requires a normal circulation far from the vortex. This circulation should be used for determination of the Berry phase.

Ambiguity in the Berry–phase analysis of the transverse force originates from ambiguity of the transverse force in the Lagrange formalism, which was discussed in the end of Sect. V in [7]). Adding a constant C to the total density in the Wess–Zumino term, i.e., replacing ρ by $\rho + C$, one obtains different amplitudes of the Berry phase and the transverse force without any effect on the field equations for the condensate wave function (see also discussion of this constant in Fermi liquids by Volovik [42]). This arbitrariness has a profound physical meaning. Derivation of the Magnus force in the Lagrange formalism dealt only with large distances much exceeding the vortex core size. However, the processes inside the core affect the total Magnus force in general. Deriving the Magnus force from the momentum balance we use the condition that the total momentum flux $\int \Pi_{ij} dS_j$ through a surface of large radius around the vortex line vanishes. This is true only for a Galilean invariant liquid satisfying the momentum conservation law. If the liquid in the vortex core interacts with the external world (e.g., with crystal impurities in superconductors), the momentum balance condition must be $\oint \Pi_{ij} dS_j = f_i$ where $\bm{f}$ is the force on the vortex core which could also have a transverse component (the Kopnin–Kravtsov force [14]). One can calculate such a force only from the vortex-core analysis. The results of this analysis must be used for determining an unknown constant in the Wess–Zumino term.

Acknowledgements. I thank N. Kopnin, L. Pitaevskii, A. Shelankov, M. Stone, D. Thouless, and G. Volovik for interesting discussions. This work was supported by a grant from the Israel Academy of Sciences and Humanities.

References

1. H. Lamb, *Hydrodynamics* (Cambridge University Press, New York 1975).
2. H.E. Hall and W.F. Vinen, Proc. Roy. Soc. **A238**, 204 (1956).
3. E.M. Lifshitz and L.P. Pitaevskii, Zh. Eksp. Teor. Fiz. **33**, 535 (1957) [Sov. Phys.-JETP **6**, 418 (1958)].
4. S.V. Iordanskii, Zh. Eksp. Teor. Fiz. **49**, 225 (1965) [Sov. Phys.-JETP **22**, 160 (1966)].
5. E.B. Sonin, Zh. Eksp. Teor. Fiz. **69**, 921 (1975) [Sov. Phys.-JETP **42**, 469 (1976)].
6. E.B. Sonin, Rev. Mod. Phys. **59**, 87 (1987).
7. E.B. Sonin, Phys. Rev. B **55**, 485 (1997).
8. Y. Aharonov and D. Bohm, Phys. Rev. **115**, 485 (1959).
9. M.V. Berry, R.G. Chambers, M.D. Large, C. Upstill, and J.C. Walmslay, Eur. J. Phys. **1**, 154 (1980).
10. P. Roux, J. de Rosny, M. Tanter, and M. Fink, Phys. Rev. Lett. **79**, 3170 (1997).
11. U. Leonhardt and P. Piwnicki, Phys. Rev. Lett. **84**, 822 (2000).
12. A.L. Shelankov, Chap. 9 of this book.
13. P. Nozières and W.F. Vinen, Phil. Mag. **14**, 667 (1966).
14. N.B. Kopnin, Chap. 7 of this book.
15. P. Ao and D.J. Thouless, Phys. Rev. Lett. **70**, 2158 (1993).
16. M.V. Berry, Proc. R. Soc. London A **392**, 45 (1984).
17. D.J. Thouless, P. Ao, and Q. Niu, Phys. Rev. Lett. **76**, 3758 (1996).
18. M.R. Geller, C. Wexler, and D.J. Thouless, Phys. Rev. B **57**, R8119 (1998).
19. H.E. Hall and J.R. Hook, Phys. Rev. Lett., **80**, 4356 (1998); C. Wexler, D.J. Thouless, P. Ao, and Q. Niu, *ibid.*, **80**, 4357 (1998).

20. E.B. Sonin, Phys. Rev. Lett. **81**, 4276 (1998).
21. C. Wexler, D.J. Thouless, P. Ao, and Q. Niu, Phys. Rev. Lett. **81**, 4277 (1998).
22. A short version of this paper is to be published in *Proceedings of the Isaac Newton Institute Workshop on Quantized Vortex Dynamics and Superfluid Turbulence, August 2000*.
23. E.P. Gross, Nuovo Cimento **20**, 454 (1961); L.P. Pitaevskii, Zh. Eksp. Teor. Fiz. **40**, 646 (1961) [Sov. Phys.-JETP **13**, 451 (1961)].
24. But this assumption becomes invalid for the roton normal fluid at higher temperatures, especially close to the critical temperature. In this case other methods based on the modified Ginzburg–Pitaevskii theory should be used [6,25].
25. E.B. Sonin, J. Low Temp. Phys. **42**, 417 (1981).
26. N.B. Kopnin, Physica B **210**, 267 (1995).
27. T.D.C. Bevan et al, J. Low Temp. Phys. **109**, 423 (1997).
28. R.J. Donnelly, *Quantized vortices in helium II* (Cambridge University Press, Cambridge, 1991), Sect. 2.8.3.
29. R. Peierls, *Surprises in Theoretical Physics*, (Princeton, 1979).
30. M.Stone, Phys. Rev. E **62**, 1341 (2000).
31. S.J. Putterman and P.H. Roberts, Physica **117** A, 369 (1983).
32. L.P. Pitaevskii Zh. Eksp. Teor. Fiz. **35**, 1271 (1958) [Sov. Phys.-JETP **8**, 888 (1959)].
33. A. Fetter, Phys. Rev. **136A**, 1488 (1964).
34. A.L. Shelankov, Europhys. Lett., **43**, 623 (1998)
35. M. Stone, Phys. Rev. B **61**, 11 780 (2000).
36. R.M. Cleary, Phys. Rev. **175**, 587 (1968).
37. However, scattered quasiparticles colliding with other quasiparticles immediately after scattering can return and modify the quasiparticle flux on the vortex. As a result of this, the lower cut-off could be a combination of the mean-free path l_{ph} and some effective cross-section of the vortex. This logarithmically weak effect was briefly discussed in the end of Sect. 1 in [5].
38. Note that the choice of Oseen's length as an upper cut-off of the logarithmic factor is valid for an isolated vortex under a stationary fluid flow. In general the upper cut-off should be the smallest of the three scales: Oseen's length, the frequency-dependent viscous depth, and the intervortex spacing [6].
39. Existence of the normal circulation at large distances in the presence of the transverse force on the vortex was pointed out by Pitaevskii (unpublished).
40. L.D. Landau and E.M. Lifshitz, *Fluid Mechanics*, (Pergamon Press, Oxford 1982).
41. D.J. Thouless, M.R. Geller, W.F. Vinen, J-Y. Fortin, and S.W. Rhee, cond-mat/0101297.
42. G.E. Volovik, Pis'ma v ZhETF, **65**, 641 (1997) [JETP Letters, **65**, 676 (1997)].

9 The Lorentz Force Exerted by the Aharonov–Bohm Flux Line

Andrei Shelankov and A.F. Ioffe

Vortices in superfluids share certain similarities with the magnetic flux lines introduced to quantum mechanics by Aharonov and Bohm [1–3]. The Aharonov–Bohm (AB) line is a source of the vector potential A defined via the circulation $\oint dl \cdot A = \Phi$, the same for any contour around the line. The magnetic field $B = \operatorname{rot} A$ is zero at every point except the line, and the total magnetic flux is fixed to Φ. The superfluid vortex circulation κ is analogous to the magnetic flux Φ, and lines of the super-current around the vortex resembles the distribution of the vector potential A around the AB–line. Furthermore, the wave equation for a phonon scattered by a vortex [4] can be presented in a form identical to the Schrödinger equation for a charge in the field of the AB–line. Of course, the analogy should not be extended too far. Unlike the purely gauge vector potential, $\operatorname{rot} A = 0$, which can be chosen almost arbitrary due to the gauge invariance, the superflow is an observable field and the region around the vortex is not force-free. Nevertheless, the analogy has proven to be rather useful in the vortex theory, in particular, for the understanding of the origin of the Iordanskii force [5] (see also references [4]). One may say that the analog to the Iordanskii force is the transverse force exerted by the AB–line, which is a quantum version of the classical Lorentz force. Surprisingly, the question of the existence of both, the Iordanskii and the Lorentz force, is still controversial [6,7] in spite of the fact that the exact solution to the Aharonov–Bohm scattering problem was found as early as 1959 [1], and the original Iordanskii's findings dates back to 1965 [5]. The purpose of the present chapter is to discuss the controversy and to attempt to resolve it. Since vortices and the Aharonov-Bohm lines are met in many contexts, from cosmic strings [8] to fractional statistics theories [9], the problem appears to be of general physical interest. This chapter deals with the Aharonov–Bohm magnetic flux line, and one can find a presentation with the emphasis on superfluidity in Chap. 8 of this volume or [10].

First, we explain what is meant here by "force". A scattering event changes the particle momentum from the initial value p_i to a certain final p_f. The net momentum transfer ΔP is found by averaging over all possible outcomes: $\Delta P = \langle (p_{\text{out}} - p_{\text{in}}) \rangle$. In the traditional terminology, the force $\mathcal{F}$ acting on the particle is $\mathcal{F} = \Delta P \times (\text{collision rate})$, being a close synonym to the momentum transfer. The term "Lorentz force" is understood here as the component of the total force, which has the symmetry of the classical

Lorentz force $\frac{e}{c}\boldsymbol{v}\times\boldsymbol{B}$: it is orthogonal to the velocity of the particle and changes its sign when either the velocity or the magnetic field (i.e. the flux Φ) are reversed.

To evaluate the momentum transfer, one uses essentially classical arguments saying that the probability of an event is proportional to the differential cross-section (see critics of the point below). Then the components of $\Delta\boldsymbol{P}$ or the force $\boldsymbol{\mathcal{F}}$, in the direction perpendicular to the initial velocity, $(\boldsymbol{\mathcal{F}})_\perp$, and parallel to it, $\boldsymbol{\mathcal{F}}_\|$, are proportional to the corresponding cross-sections $\sigma_\perp$ and $\sigma_\|$: $\boldsymbol{\mathcal{F}}_{\perp,\|} = p\sigma_{\perp,\|}J$ where p and J are the momentum and the current density in the incident wave, respectively. In the two-dimensional geometry, when scattering occurs in the plane orthogonal to the line, these effective cross-sections read

$$\sigma_\perp = \int_{-\pi}^{\pi} d\theta \sin\theta \left(\frac{d\sigma}{d\theta}\right), \quad \sigma_\| = \int_{-\pi}^{\pi} d\theta (1-\cos\theta)\left(\frac{d\sigma}{d\theta}\right), \qquad (9.1)$$

here θ is the scattering angle.

Aharonov and Bohm derived [1,2] the exact solution to the problem of scattering by a flux line and found the differential cross-section,

$$\frac{d\sigma}{d\theta} = \frac{1}{2\pi k}\frac{\sin^2\pi\alpha}{\sin^2\frac{\theta}{2}}, \quad \alpha = \frac{\Phi}{\Phi_0} \qquad (9.2)$$

where α is the flux in the units of $\Phi_0 = hc/e$. The cross-section $\sigma_\|$ can be easily calculated and the longitudinal force $\boldsymbol{\mathcal{F}}_\|$ does not pose any problems. The controversy centers on the transverse component.

The symmetry of the Aharonov–Bohm problem, where the time reversal $\mathcal{T}$ and the mirror symmetry $\mathcal{P}$ are broken by the flux line, does not forbid the existence of the transverse (Lorentz) force. However, the AB cross-section (9.2) is left-right symmetric, i.e. unchanged by $\theta \to -\theta$. Consequently, $\sigma_\perp$ in (9.1) is zero. By this simple and seemingly robust argument, many authors have come to the conclusion that the AB line does not exert the Lorentz force. In the context of superfluidity this would mean no Iordanskii's force.

The robustness of the argument is shattered when one calculates $\sigma_\perp$ using the partial waves expansion. The scattering amplitude is presented as $f(\theta) = \frac{1}{\sqrt{2\pi k}}\sum_{m=-\infty}^{\infty}(e^{2i\delta_m}-1)e^{im\theta}$, δ_m being the scattering phase shift, and $d\sigma/d\theta = |f|^2$. After apparently *identical* transformations of (9.1), $\sigma_\perp$ is written as

$$\sigma_\perp = \frac{1}{k}\sum_{m=-\infty}^{\infty}\sin 2(\delta_m - \delta_{m+1}). \qquad (9.3)$$

The phase shifts $\delta_m^{(AB)}$ are known to be $\delta_m^{(AB)} = \frac{\pi}{2}(|m|-|m-\alpha|)$. There is only one nonzero term in the sum, and one finds a *finite* value

$$\sigma_\perp = -\frac{1}{k}\sin(2\pi\alpha). \qquad (9.4)$$

It remains unclear how this result can be understood in view of the symmetry of the cross-section.

The force can be also calculated using the momentum balance arguments where it is found as the flux of the momentum-flow (stress) tensor (see e.g. [2] and below). The result [11,10,4] agrees with (9.4) as well as with earlier calculations of the Iordanskii force [5,12]. Again, the calculations seem to be incompatible with the fact that the scattering probability is insensitive to the direction of the magnetic field as it follows from the exact Aharonov–Bohm theory (9.2).

The fact that different approaches give contradicting results, of course, is due to the singular nature of the AB–line: The vector potential has infinitely long range and this leads to infinite scattering rate (9.2) in the forward direction $\theta = 0$. The integral for $\sigma_\perp$ is ill defined because of the divergence $d\theta/\theta$ at $\theta \to 0$. Obviously, by regrouping terms with different signs in a diverging sum or integral, one may assign any value to it. Grouping together terms with θ and $-\theta$, one assigns zero value to $\sigma_\perp$ in (9.1). In the partial wave representation (9.3), the difficulty is still present since the $\theta = 0$ divergence cannot be eliminated by truly identical transformations. Indeed, the sum allows different evaluations, as it is shown in [7] (see also [4]), because δ_m's do not tend to zero at $|m| \to \infty$. Choosing an arrangement of the terms, one may get either (9.4) or the result in [7] where $\sigma_\perp$ is proportional α^3 at $\alpha \ll 1$ instead of the linear dependence in (9.4). By these arguments, the validity of (9.4) is in question.

The ill-defined objects which are met in the theory call for regularization. In search for the regularization method, note that the forward scattering singularity is due to the long range of the vector potential. Thus, it cannot be eliminated by considering the line as a solenoid of a small but finite radius (see the corresponding solutions in [13,2]). In the gauge $\boldsymbol{A} = A(r)\boldsymbol{e}_\varphi$, the vector potential of the AB–line is $A = \frac{\Phi_0}{2\pi r}$. One might try the substitution $A \to A_\gamma = e^{-\gamma r}\frac{\Phi}{2\pi r}$, $\gamma \to 0$ to cut the potential at large distances (as is sometimes done in the Coulomb potential case). However, it appears that the vector potential cannot be changed from $1/r$ without changing physics: Indeed, A_γ is *not* a purely gauge field: the magnetic field, $B = \mathrm{rot}\boldsymbol{A}_\gamma = \gamma\frac{\Phi}{2\pi r}e^{-\gamma r}$, is finite in a large volume with the linear size $\propto \gamma^{-1}$. It is far from obvious that this limiting procedure would not influence the physics of the Aharonov–Bohm line.

In the approach taken in this chapter, the regularization is achieved by taking into consideration the fact that in any physical state the initial momentum has uncertainty, so that the wave function is a superposition of the AB–waves with different directions of the initial propagation. Then, the forward direction becomes blurred and the forward singularity is smeared. In other words, the incident wave is never an infinite coherent plane wave but is rather a beam of a finite width. In this picture, behaviour of the gauge potential at distances larger than the width of the beam (or the phase coherence length) becomes immaterial.

The plan of the chapter is the following: Taking an unbiased position, we first analyze an experiment, the goal of which is to answer the question about the existence of the Lorentz force. The analysis in Sects. 9.1.1 and 9.1.2 is based on a solution to the Schrödinger equation derived from scratch in the paraxial approximation. In Sect. 9.1.3, the exact solution to the scattering problem is built as a superposition of the AB–waves, and the expression for the scattering matrix is re-derived. In Sect. 9.1.4, it is argued that the scattering amplitude does not give a full description of scattering in the case of the AB–line. This means that the expressions for the effective cross-sections in (9.1) have to be re-derived. The derivation is presented in Sect. 9.2 and Appendices 9.4 and 9.5. The results are summarized in the last section.

9.1 The Magnetic Scattering

To get an answer to the question whether the transverse (Lorentz) force is finite, we analyze a "realistic" experiment where the force is measured. In the experiment, a particle of mass m, charge e and momentum $p = \hbar k$ moves on the $x-y$ plane in the x-direction from $-\infty$, and meets at $r = 0$ the AB–line, which goes in the z–direction. At some distance $x > 0$, there is a screen with detectors where the distribution with respect to the transverse coordinate y is measured (see Fig. 9.1). The averaged value $\bar{y}$ as a function of x defines the "trajectory" $\bar{y}(x)$ from which the deflection is extracted. The deflection is the measure of the transverse momentum transferred to the particle and, therefore, of the Lorentz force. To make the transverse coordinate y meaningful, the incident wave is beam-like with a finite transverse size W (e.g. a wave having passed through an aperture of the width W). The direction of propagation is well-defined if $kW \gg 1$.

First, we use the paraxial approximation which allows one to get the result in a most simple form.

9.1.1 Paraxial Solution

In the paraxial theory [14–16], the wave function of a particle with the energy $E = \hbar^2 k^2 / 2m$ moving at a small angle to the x−axis is presented as $\Psi = e^{ikx}\psi(x, y)$, where ψ is a slowly varying amplitude. Neglecting $\partial^2 \psi / \partial x^2$ in comparison with $k\partial \psi / \partial x$ in the stationary Schrödinger equation, one comes to the paraxial equation:

$$iv\partial_x \psi = -\frac{1}{2m}\partial_y^2 \psi \tag{9.5}$$

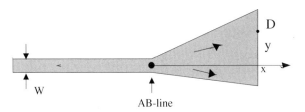

Fig. 9.1. Measuring the Lorentz force. The incident wave is a collimated beam, the profile of which is characterized by the width W. The beam is scattered by the AB–line shown as a black circle. For a fixed coordinate x, the detector D measures the intensity of the wave $|\psi(x,y)|^2$ as a function of the transverse coordinate y. The averaged value $\langle y \rangle$ defines the y–coordinate of the beam gravity centre as a function of x. The deflection angle, $\Delta\theta = \langle y \rangle / x$, $x \to \infty$ measures the Lorentz force.

where the velocity $v = \hbar k / m$ and $\boldsymbol{\partial} \equiv \hbar \boldsymbol{\nabla} - i\frac{e}{c}\boldsymbol{A}$. The conservation of the current, $J_x = v|\psi|^2$, allows one to impose the normalization condition

$$\int_{-\infty}^{\infty} dy |\psi(x,y)|^2 = 1 , \tag{9.6}$$

which fixes the total current in the x–direction to v.

Taking advantage of the gauge freedom, the vector potential of the AB–line with the flux $\alpha \Phi_0$ is conveniently chosen as

$$A_x = -\frac{\alpha}{2} \Phi_0 \delta(x) \operatorname{sgn} y , \quad A_y = 0 , \tag{9.7}$$

so that $\boldsymbol{A}$ is nonzero only on the line $x = 0$.

The back-scattering is absent in the paraxial approximation, and the wave function at $x < 0$ is a free incident wave. We will take the incoming wave $\psi(x < 0, y)$ to be a collimated beam (like that schematically shown in Fig. 1), and consider $\psi(x = -0, y) \equiv \psi_{\text{in}}(y)$ as the input specifying the profile of the beam. The paraxial approximation is applicable when $kW \gg 1$.

Solving (9.5), in the immediate vicinity of the line $x = 0$, where the vector potential (9.7) is concentrated, one finds

$$\psi(+0, y) = \psi_{\text{in}}(y) \exp\left(-i\pi\alpha \operatorname{sgn} y\right) . \tag{9.8}$$

The line $x = 0$ acts as a phase changing screen [17].

With the initial condition in (9.8), further propagation is free

$$\psi(x > 0, y) = \int_{-\infty}^{\infty} dy' \, G_0(y - y'; x) \psi(+0, y') \quad G_0(y; x) = \sqrt{\frac{k}{2\pi i x}} \, e^{\frac{ik}{2x} y^2} \tag{9.9}$$

where G_0 is the free propagator. Inserting $\psi(+0, y)$ from (9.8), one gets [15]

$$\psi(x > 0, y) = \cos \pi\alpha \; \psi_0(x, y)$$

$$+ \, \mathrm{i} \sin \pi \alpha \int_{-\infty}^{\infty} dy' \, G_0(y - y'; x) \frac{y'}{|y'|} \psi_{\mathrm{in}}(y') \,, \tag{9.10}$$

where $\psi_0(x, y) = \int_{-\infty}^{\infty} dy' \, G_0(y - y'; x)\psi_{\mathrm{in}}(y')$ is the free ($\alpha = 0$) incident wave having propagated to $x > 0$.

Berry [17] has derived the paraxial approximation from the exact Aharonov–Bohm solution. This gives confidence that the paraxial approximation is valid in the case of a highly singular potential of the AB–line.

Plane Incident Wave. In the simplest case of the plane incident wave when $\psi_{\mathrm{in}} = \psi_0(x, y) = 1$, one gets from (9.10)

$$\psi_{\mathrm{plane}}(x > 0, y) = \cos \pi \alpha + \mathrm{i} K(\tilde{y}) \sin \pi \alpha \,, \tag{9.11}$$

where $\tilde{y} = y/\sqrt{2kx}$ and $K(\tilde{y})$ is the Fresnel integral, $K(\tilde{y}) = \frac{2}{\sqrt{\mathrm{i}\pi}} \int_0^{\tilde{y}} dt' \, e^{\mathrm{i}t'^2}$. This expression is in agreement with the Aharonov–Bohm solution [1] at $kx \gg 1$ and $|\varphi| \ll 1$, the angle $\varphi = y/x$.

Note the anomalous behaviour in the forward direction: The AB–wave does not converge to the incident wave ($\psi = 1$) at large distances from the scatterer. Instead, $\psi(x, y = 0) = \cos \pi \alpha$ at *any* distance x from the line, whatever large [18]. In the near-forward directions $\tilde{y} \lesssim 1$, the anomaly persists at a progressively narrow region of angles $\varphi \sim 1/\sqrt{kx}$. Ultimately, this is the singularity that causes the problems.

The paraxial solution has a very simple representation on the $\mathrm{Re}\,\psi - \mathrm{Im}\,\psi$ plane: The complex function $\psi_{\mathrm{plane}}(x, y)$ (9.11) depends on x and y only in the combination $\tilde{y}$. Therefore, the points corresponding to $\psi(x, y)$ for x and y on the $x - y$ half-*plane* $x > 0$, draw a *line* on the complex $\mathrm{Re}\,\psi - \mathrm{Im}\,\psi$ plane. Up to a scaling and shift, the line is the Cornu spiral, well-known in the diffraction theory [19]. Unexpectedly, the Aharonov–Bohm wave can be mapped to the Cornu spiral not only at small angles as the paraxial theory suggests but for *any* angle (see [20] for detail).

9.1.2 Deflection of the Beam

Given the incoming wave $\psi_{\mathrm{in}}(y)$, equation (9.9) or (9.10) allows one to find the outgoing wave and to predict the result of the experiment shown in Fig. 9.1.

At a given x, the transverse position of the centre of gravity of the beam is $\bar{y}(x) = \int_{-\infty}^{\infty} dy \, y |\psi(x, y)|^2$. In the force free regions $x < 0$ (in) and $x > 0$ (out), the trajectories $\bar{y}(x)$ are straight lines at the angle $\varphi_{\mathrm{in},\,\mathrm{out}} = \left.\frac{d\bar{y}}{dx}\right|_{x<0,\,x>0}$ to the x–axis. Then, $\Delta\varphi = \varphi_{\mathrm{out}} - \varphi_{\mathrm{in}}$ denotes the deflection of the beam.

It follows from (9.5) and (9.7) that $d\bar{y}/dx = \langle \hat{p}_y \rangle / p$ where $\hat{p}_y = \frac{\hbar}{i}\frac{\partial}{\partial y}$. Thus, the angles $\varphi_{\text{in,out}}$ are proportional to the expectation values of the transverse (kinematical) momenta $\langle \hat{p}_y \rangle_{\text{in,out}}$. In the gauge (9.7),

$$\langle \hat{p}_y \rangle_{\text{out(in)}} = \int_{-\infty}^{\infty} dy\, \psi^*(x,y) \frac{\hbar}{i} \frac{\partial}{\partial y} \psi(x,y)\ , \quad x > 0\ (x < 0). \tag{9.12}$$

Denote the transverse momentum transfer $\Delta p_y = \langle \hat{p}_y \rangle_{\text{out}} - \langle \hat{p}_y \rangle_{\text{in}}$.

To evaluate $\langle \hat{p}_y \rangle_{\text{out}}$, note: (i) $\psi(x,y)$ on (9.10) is a smooth function of y and, therefore, one may safely substitute the derivative in (9.12) for a finite difference: $(\psi(y+a, x) - \psi(y-a, x))/2a$ where $a \to 0$; (ii) Since before and after the substitution, the integral does not depend on x (due to the conservation of p_y in the force free regions), the coordinate x may be chosen at $x = +0$ where ψ is given by (9.8); after that, one takes the limit $a \to 0$. The initial momentum, $\langle \hat{p}_y \rangle_{\text{in}}$, is evaluated in the incoming region at $x = -0$ where $\psi(x,y) = \psi_{\text{in}}(y)$. Simple calculations give the following result

$$\Delta p_y = -\hbar |\psi_{\text{in}}(0)|^2 \sin 2\pi\alpha\ , \tag{9.13}$$

where $\psi_{\text{in}}(0)$ is the amplitude of the normalized *incoming* wave at the position of the AB–line [15].

This result shows that the AB–line scatters particles asymmetrically: The beam as a whole gets deflected by the angle $\Delta\varphi = \Delta p_y/p$. As it should be, the deflection – to the right or to the left – is reversed if the magnetic flux $\Phi = \alpha\Phi_0$ changes its sign. The deflection of the charge by the AB–line may be attributed to the Lorentz force in its ultra-quantum version when the wave length considerably exceeds the size of the region where the field is present. A non-classical origin of the effect is revealed by its periodicity in the flux. Besides, the deflection (9.13) is finite only if $\psi_{\text{in}}(0) \neq 0$, i.e. when the incoming wave overlaps with the line. For instance, the deflection is zero if the line is in the shadow region in a double-slit geometry.

By order of magnitude, $|\psi_{\text{in}}(0)|^2 \sim 1/W$ as it follows from the normalization (9.6). From this estimate, the deflection, $\Delta\varphi \leq \hbar|\psi_{\text{in}}(0)|^2/p$, never exceeds the beam angular width $\varphi_0 \sim 1/kW$.

To understand this result in view of the left-right symmetry of the AB cross-section (9.2), it is instructive to consider an example. Let the incident wave $\psi_{\text{in}}(y) = \exp(-|y|/W)$. It follows from (9.8) and (9.9) that at large distances ($x \gg kW^2$) the solution behaves as a diverging spherical wave, i.e. $|\psi|^2 = P(\varphi)/x$, $\varphi = y/x$, with the angular distribution $P(\varphi)$,

$$P(\varphi) = \frac{2}{\pi k} \frac{(\varphi \sin \pi\alpha - \varphi_0 \cos \pi\alpha)^2}{(\varphi^2 + \varphi_0^2)^2}\ ,$$

$\varphi_0 = 1/kW$ being the beam angular width ($\varphi_0 \ll 1$ in the paraxial theory). One sees that indeed the distribution is not symmetric relative to $\varphi \leftrightarrow -\varphi$, and the beam deflection $\propto \int d\varphi\, \varphi P(\varphi)$ is nonzero. However, the asymmetry exists only at small angles $|\varphi| \lesssim \varphi_0$. Larger angles, where one recovers the symmetric AB cross–section, do not contribute to the deflection.

In (9.13), Δp_y is the momentum transfer per collision. Multiplying it by the collision rate $\dot{N}$, one gets the transverse force acting on the charge, $\mathcal{F}_y = \Delta p_y \dot{N}$. The collision rate is found as $\dot{N} = \int_{-\infty}^{\infty} dy\, J_x(x,y)$, J_x being the current density. For the wave normalized as in (9.6), the collision rate is presented by $\dot{N} = v$. Then, the force can be expressed via the current, $J_{\text{in}}(0)$, in the *incident* wave at the point of the line, namely, $\mathcal{F}_y = -\hbar \sin(2\pi\alpha) J_{\text{in}}(0)$. In the scattering experiments one usually assumes a homogeneous illumination where the incident wave is an incoherent mixture of beams with different impact parameters. Averaging with respect to the impact parameter, one gets $\mathcal{F}_y = -\hbar \sin(2\pi\alpha) J_{\text{in}}$. Here, J_{in} is the incident current density. From the identification $\mathcal{F}_y = p\sigma_\perp J_{\text{in}}$, one gets the transverse cross-section in (9.4). In this derivation, the physical origin of the transverse force, which is asymmetric scattering, is obvious.

The main conclusion from the analysis of the experiment is that the AB–line does exert the transverse force. The force and the net deflection of the beam come from small scattering angles $|\varphi| \sim \varphi_0$ of order of the angular width of the incoming wave. In this near-forward region where one cannot tell apart the scattered and unscattered waves, the conception of the differential cross-section is not applicable. As it is explained in more detail below, this is the reason why one is not able to find the Lorentz force with the help of the Aharonov–Bohm cross-section.

9.1.3 Exact Solution

In this section, we consider an exact solution to the scattering problem for a general incident wave. The goal is to see how the conclusions made in the previous section can be reconciled with the Aharonov–Bohm theory.

Denote by $\psi_{\text{AB}}(\mathbf{r})$ the AB wave function corresponding to the plane wave scattered by the flux line with the vector potential in the circular gauge

$$\mathbf{A} = A_\varphi \mathbf{e}_\varphi, \quad A_\varphi = \alpha \frac{\Phi_0}{2\pi r}. \tag{9.14}$$

For the incident wave propagating in the *positive* x-direction, the AB solution [1,2] reads

$$\psi_{\text{AB}}(\mathbf{r}) = \sum_{m=-\infty}^{\infty} \exp\left(-i\frac{\pi}{2}|m-\alpha| + im\pi + im\varphi\right) J_{|m-\alpha|}(kr), \tag{9.15}$$

r and φ are the cylindrical coordinates, and $J_\nu(z)$ is the Bessel function.

The rotation $e^{-i\phi \hat{L}_z} \psi_{\text{AB}}$, where $\hat{L}_z = -i\partial/\partial\varphi$, generates the wave which is also a solution to the scattering problem since in the gauge (9.14) $\hat{L}_z$ commutes with the Hamiltonian. In the new solution, the angle of incidence is ϕ.

Superimposing the rotated waves, one gets the general *scattering* solution:

$$\Psi = \int_{-\pi}^{\pi} d\phi \, \mu(\phi) \exp\left(-i\phi\left(\hat{L}_z - \alpha\right)\right) \psi_{AB} \, , \qquad (9.16)$$

where $\mu(\phi)$ is the amplitude of the AB–state with the initial wave vector $\mathbf{k}_\phi = \mathbf{e}_x k \cos\phi + \mathbf{e}_y k \sin\phi$. Asymptotically, $\Psi|_{r\to\infty} = e^{i\alpha\varphi}\Psi_{\rm in}(\mathbf{r})$ ($-\pi < \varphi < \pi$), where

$$\Psi_{\rm in}(\mathbf{r}) = \int_{-\pi}^{\pi} d\phi \, \mu(\phi) \exp\left(\mathbf{k}_\phi \cdot \mathbf{r}\right) \, . \qquad (9.17)$$

By its physical meaning, $\Psi_{\rm in}(\mathbf{r})$ is the incident wave, *i.e.* the wave in the absence of the line. As such, it may be considered at any point, including $\mathbf{r} = 0$.

In the scattering geometry, Fig. 9.1, $\Psi_{\rm in}$ is a collimated beam (in the x-direction for definiteness) which is characterized by its width W and the impact parameter b. The modulus $|\mu(\phi)|$ is concentrated at $|\varphi| \lesssim \phi_0$ where again $\varphi_0 \sim 1/kW$ is the angular width of the beam. The impact parameter b, that is the position of the beam centre in the y–direction, enters via the phase factor as $\mu(\phi) \propto e^{ikb\sin\phi}$.

Substituting (9.15) into (9.16), one gets the partial wave expansion [17]:

$$\Psi(\mathbf{r}) = \sum_{m=-\infty}^{\infty} \mu_m \exp\left(-i\frac{\pi}{2}|m-\alpha| + im\pi + im\varphi\right) J_{|m-\alpha|}(kr) \qquad (9.18)$$

where

$$\mu_m = \int_{-\pi}^{\pi} d\phi \, \tilde{\mu}(\phi) \exp(-i\phi m) \, , \quad \tilde{\mu}(\phi) = e^{i\alpha\phi}\mu(\phi) \, . \qquad (9.19)$$

Note that μ_m's tend to zero for $|m| > m_{\max}$, where $m_{\max} \sim \max[1/\varphi_0, 1/kb]$, so that the sums over m are effectively cut off at $|m| \sim m_{\max}$. Seeing that infinitely large $|m|$'s do not contribute to the sum, one is allowed to substitute μ_m for $\mu_m \cdot e^{-\epsilon|m|}$, $\epsilon \to +0$. The substitution facilitates further manipulations with the partial waves sums.

At large distances from the line, $r \gg b, W$, one can safely use the asymptotics $J_{|m-\alpha|}(kr) \approx \sqrt{\frac{2}{\pi kr}} \cos\left(kr - \frac{\pi}{2}|m-\alpha| - \frac{\pi}{4}\right)$ in (9.18) since kr is larger than $m_{\max}$. Asymptotically, the wave in (9.18) acquires the form,

$$\Psi(r, \varphi) \approx \sqrt{\frac{2\pi}{kr}} \left(e^{-i\left(kr-\frac{\pi}{4}\right)} \tilde{\mu}(\pi + \varphi) + e^{i\left(kr-\frac{\pi}{4}\right)} \hat{S}\tilde{\mu}(\varphi)\right) \, , \qquad (9.20)$$

which is well-known in the scattering theory [21]: Here, $\hat{S}$ is the scattering operator,

$$\hat{S}\tilde{\mu}(\varphi) = \int_{-\pi}^{\pi} d\varphi\, S(\varphi - \varphi')\tilde{\mu}(\varphi')\,,$$

where the kernel (i.e. the scattering matrix) reads ($0 < \alpha < 1$):

$$S(\theta) = \frac{1}{2\pi} \sum_{m=-\infty}^{\infty} S_m e^{im\theta} e^{-\epsilon|m|}\,,$$

$$S_m \equiv e^{2i\delta_m} = \begin{cases} e^{i\pi\alpha}\,, & m \geq 1\,; \\ e^{-i\pi\alpha}\,, & m \leq 0\,. \end{cases} \quad (9.21)$$

Summation of the geometrical progressions in (9.21) gives

$$S(\theta) = \frac{e^{i\theta}}{2\pi}\left(\frac{e^{i\pi\alpha}}{1 - e^{i(\theta+i\epsilon)}} - \frac{e^{-i\pi\alpha}}{1 - e^{i(\theta-i\epsilon)}}\right)\,, \quad \epsilon \to +0\,. \quad (9.22)$$

In a slightly different form, this expression can be found in [22] and [17]. At $\theta \ll 1$, one recovers the paraxial approximation [15].

Evaluating the angular distribution in the outgoing wave with the help of the exact theory, Berry [17] has calculated numerically the deflection of a beam with the Gaussian profile: $\Psi_{\text{in}}(x=0,y) \propto e^{-y^2/2W^2}$. For beams coming in the positive x-direction, the deflection D is defined in [17] as

$$D = \int_{-\frac{\pi}{2}}^{\frac{\pi}{2}} d\varphi\, \sin(\varphi)\left(|\hat{S}\tilde{\mu}(\varphi)|^2 - |\tilde{\mu}(\varphi)|^2\right) \bigg/ \int_{-\frac{\pi}{2}}^{\frac{\pi}{2}} d\varphi\, |\mu(\varphi)|^2\,.$$

In the approximation where $\sin(\varphi) \to \varphi$ and $\varphi = y/x$, the deflection D corresponds to the paraxial $\bar{y}/x = \langle p_y \rangle_{\text{out}}/p$. To check the paraxial result in (9.13) one compares the deflection D with $(\Delta p_y)/p$.

It turns out [17] that the paraxial approximation works very well for the paraxial beams with $kW \gg 1$. The quantitative agreement is found when the width W is as small as $2-3\,k^{-1}$. When the beam has the profile $\Psi_{\text{in}}(x=0,y) \propto y^2 e^{-y^2/2W^2}$, the deflection turns out to be almost zero, again confirming (9.13) with $|\psi_{\text{in}}(0)|^2 = 0$.

9.1.4 Scattering Amplitude

The usual way to proceed from (9.22) would be to single out the scattering amplitude from the S-matrix. The actual motivation for this would be the following: the S-matrix is expected to possess a trivial $\delta(\theta)$ singularity, the term which generates the unscattered part of the outgoing wave. Then, the *regular* part, that is $S(\theta) - \delta(\theta) = \sqrt{\frac{ik}{2\pi}} f(\theta)$, constitutes the scattering amplitude $f(\theta)$.

Physically, one is able to introduce the scattering amplitude if the scattered and unscattered waves can be separated. The formal prerequisite for this is that the forward singularity of the S-matrix is exhausted by the free δ-function. This requirement is not met in the AB–line case. Indeed, rewriting the S-matrix (9.22) in the form [22],

$$S(\theta) = \cos \pi \alpha \, \delta(\theta) - \frac{\sin \pi \alpha}{2\pi} \, \mathsf{P} \frac{e^{i\frac{1}{2}\theta}}{\sin \frac{\theta}{2}} \qquad (9.23)$$

(P stands for the principal value), one observes that the weight of the δ-function differs from 1. Consequently, $\hat{S} - \hat{1}$ is not regular: it still contains the δ-function, not mentioning a weaker singularity due to the principal value. Therefore, the usual procedure does *not* reach its goal in the AB–line case and there is no practical alternative other than to keep the S-matrix as a single object.

One comes to the same conclusion from another direction. Usually, $f(\theta)$ is read off from the solution corresponding to the plane incident wave

$$\Psi(r,\varphi)|_{r\to\infty} = e^{ikr\cos\varphi} + f(\varphi)\frac{e^{ikr}}{\sqrt{ikr}} \; . \qquad (9.24)$$

As shown by Berry et al. [18] and seen from the paraxial (9.11), the forward asymptotics does not obey (9.24), and again one is not able to define a reasonable scattering amplitude (in the near forward direction).

These arguments show that the standard theory should be applied with caution to the peculiar AB–scattering. Results, for which near-forward direction is of importance, $\sigma_\perp$ in (9.1) in particular, have to be reconsidered. In the next section, the expression for the momentum transfer is re-derived.

9.2 The Momentum Balance

It is convenient to perform the derivation in a time dependent theory where the momentum transfer $\Delta\boldsymbol{P}$ can be expressed with the help of the *identity*,

$$\Delta\boldsymbol{P} = \int dt \, \langle \psi(t)|\hat{\dot{\boldsymbol{P}}}|\psi(t)\rangle \; ,$$

via the expectation value of the quantum operator corresponding to $\dot{\boldsymbol{P}}$. This relation deals with the total momentum of the quantum state. As is well-known in the field theory, one may also introduce local variables which play the role of the momentum density. These quantities obey a local conservation law (see e.g. [2]): similar to hydrodynamics, the local momentum variation is related to the momentum-flow (stress) tensor and the force density. In relation to the AB–line, this method of evaluation of the momentum transfer has been used in [2,11,10]. The derivation of the conservation law in the present context is outlined here (see Appendix for details).

The k-th Cartesian component ($k = x, y, z$) of the momentum density at the point r is described by the operator $\hat{\pi}_k(r)$

$$\hat{\pi}_k(r) = \frac{1}{2}\left(\delta(r-\hat{r})\,\hat{p}_k + \hat{p}_k\,\delta(r-\hat{r})\right) \tag{9.25}$$

where $\hat{p}$ is the kinematical momentum operator,

$$\hat{p} = m\hat{v} = \frac{\hbar}{i}\nabla - \frac{e}{c}A(\hat{r})\ . \tag{9.26}$$

Here, r is the label attached to the operator $\hat{\pi}_k(r)$, whereas $\hat{r}$ and $\hat{p}$ are quantum operators.

Finding the time derivative of $\hat{\pi}_k$ with the help of (9.39) and taking the diagonal matrix element $\langle\Psi|\ldots|\Psi\rangle$ of (9.41), one comes to the following local relation,

$$\dot{\pi} + \nabla\cdot\overleftrightarrow{\mathcal{K}} = F \tag{9.27}$$

where $\pi = mJ(r)$, J being the current density, is the expectation value of $\hat{\pi}(r)$ and

$$\left(\overleftrightarrow{\mathcal{K}}(r)\right)_{kl} = \frac{1}{2m}\mathrm{Re}\left.\left((\hat{p}_k\Psi)^*(\hat{p}_l\Psi) + \Psi^*(\hat{p}_k\hat{p}_l\Psi)\right)\right|_r\ , \tag{9.28}$$

is a (symmetric) momentum-flow tensor, and F denotes the force density,

$$F(r,t) = e|\Psi(r,t)|^2 E(r,t) + \frac{e}{c}\left.(J\times B)\right|_{r,t}\ , \tag{9.29}$$

E and B being the electric and magnetic fields.

Integrating (9.27) over the volume V surrounded by the surface S, one gets

$$\frac{d}{dt}\int_V dV\ \pi + \int_S dS\cdot\overleftrightarrow{\mathcal{K}} = \int_V dV\ F(r,t), \tag{9.30}$$

that is the momentum balance equation.

In a time dependent scattering problem $\Psi(r,t)$ is a normalized wave packet and $\Psi(r\to\infty,t) = 0$ (see e.g. [23]). When $V\to\infty$, the surface term vanishes and (9.30) relates the increment of the total momentum $P = \langle\Psi|p|\Psi\rangle$,

$$\dot{P} = \mathcal{F}$$

to the total force $\mathcal{F} = \int dV\ F(r,t)$.

In the stationary state, when $\dot{\pi} = 0$, the total force $\mathcal{F}$ identically equals to the surface integral,

$$\mathcal{F} = \int_S dS\cdot\overleftrightarrow{\mathcal{K}} \tag{9.31}$$

9.2.1 The Force

To calculate the total force in the two-dimensional case under consideration, we choose the circle of radius R with the centre on the line as the integration "surface" in (9.31). The x- and y-components of the force are found by integrating the tensor components $\mathcal{K}_{r,x}$ and $\mathcal{K}_{r,y}$ respectively. It is convenient to build the combinations $\mathcal{F}_\nu = (\mathcal{F})_x + i\nu (\mathcal{F})_y$, $\nu = \pm 1$, so that $\mathcal{F}_x = \frac{1}{2}(\mathcal{F}_+ + \mathcal{F}_-)$, $\mathcal{F}_y = \frac{1}{2i}(\mathcal{F}_+ - \mathcal{F}_-)$. Also, $\mathcal{K}_{r,\nu} = \mathcal{K}_{r,x} + i\nu \mathcal{K}_{r,y}$. From (9.28), one sees that $\mathcal{K}_{r,\nu}$ contains operators $\hat{p}_r = \frac{\hbar}{i}\frac{\partial}{\partial r}$ as well as $\hat{p}_\nu = \hat{p}_x + i\nu \hat{p}_y$. The latter, has a simple form $\hat{p}_\nu \approx e^{i\nu\varphi}\hat{p}_r$ at $r \to \infty$ (in the circular gauge (9.14)). Taking advantage of these simplifications and using the representation in (9.20), one readily gets

$$\mathcal{F}_\nu = \frac{2\pi v p}{k}\int_{-\pi}^{\pi} d\phi\, e^{i\nu\phi}\left(|\hat{S}\tilde{\mu}(\phi)|^2 - |\tilde{\mu}(\phi)|^2\right). \qquad (9.32)$$

Introducing the collision rate $\dot{N}$ as the total probability flux $\int J_r\, r d\varphi$ in the converging spherical wave in (9.20),

$$\dot{N} = \frac{2\pi v}{k}\int_{-\pi}^{\pi} d\phi\, |\tilde{\mu}(\phi)|^2, \qquad (9.33)$$

and the momentum transfer ΔP_ν,

$$\Delta P_\nu = \int_{-\pi}^{\pi} d\phi\, p\, e^{i\nu\phi}\left(|\hat{S}\tilde{\mu}(\phi)|^2 - |\tilde{\mu}(\phi)|^2\right) \Big/ \int_{-\pi}^{\pi} d\phi\, |\tilde{\mu}(\phi)|^2,$$

one is able to write down (9.32) as

$$\mathcal{F}_\nu = \dot{N} \cdot \Delta P_\nu, \qquad (9.34)$$

so that indeed $\mathcal{F}$ possesses the expected structure.

Now, when the general expression for the force is available, we are in a position to discuss the physics behind the difficulties mentioned in the beginning of the chapter. The expressions for σ's in (9.1) are based on the assumption that the scattering can be described in terms of probabilities as in classical mechanics: the particle is either unscattered or scattered at an angle θ with the probability $\propto d\sigma/d\theta$. Then, the momentum transfer, which is due only to the scattered component, can be expressed via the scattering probabilities (the differential cross-section) as in (9.1). However, this simple picture does not necessarily follow from (9.32) where the outgoing momentum is calculated via the angle resolved *total* intensity of the outgoing wave without a separation into scattered and unscattered pieces.

To proceed, we let the incident beam propagate in the x-direction and, taking advantage of the unitarity of the S-matrix, write (9.32) in the form

$$\mathcal{F}_\nu = \frac{2\pi p v}{k} \int_{-\pi}^{\pi} d\phi \left(e^{i\nu\phi} - 1 \right) \left(|\hat{S}\tilde{\mu}(\phi)|^2 - |\tilde{\mu}(\phi)|^2 \right).$$

If the incident wave is wide enough so that $\mu(\phi)$ is concentrated at vanishingly small angles, the last term can be omitted, and

$$\mathcal{F}_\nu = \frac{2\pi p v}{k} \int_{-\pi}^{\pi} d\phi \left(e^{i\nu\phi} - 1 \right) |\hat{S}\tilde{\mu}(\phi)|^2.$$

Since the contribution of the small angles is suppressed by the factor $(e^{i\nu\phi}-1)$, one is normally allowed to substitute $S(\phi)$ for $\sqrt{\frac{ik}{2\pi}} f(\phi)$, $f(\phi)$ being the scattering amplitude, and come to the expressions (9.1). However, this procedure turns out to be wrong in the AB–line case. The reason is that in the near-forward direction, which is important for the calculation of the transverse force, the S-matrix is not reduced to the scattering amplitude. One sees again that it is impossible to give a full description of the AB–scattering in terms of only the scattering amplitude. Technically, the transverse component of the force in (9.37) comes as the interference of the two terms in the S-matrix (9.23). This means that if one insists on splitting the outgoing wave into scattered and unscattered parts, their interference must be taken into account.

9.2.2 The AB–Line

In the partial wave representation, (9.32) reads

$$\mathcal{F}_\nu = (2\pi)^2 \frac{vp}{k} \sum_m \mu^*_{m+\nu} \left(S^*_{m+\nu} S_m - 1 \right) \mu_m \qquad (9.35)$$

where μ_m is defined in (9.19). Equations (9.32), (9.35), and (9.34) are applicable for any scattering potential.

In the AB–line case, when the scattering matrix is given by (9.21), there is only one non-zero term in the sum (9.35) which corresponds to $m = 0$ for $\nu = +1$, and $m = -1$ for $\nu = -1$. One finds

$$\mathcal{F}_+ = (2\pi)^2 \frac{vp}{k} \left(e^{-2i\pi\alpha} - 1 \right) \mu^*_1 \mu_0, \quad \mathcal{F}_- = \mathcal{F}^*_+. \qquad (9.36)$$

This relation allows one to find the force for an arbitrary incoming wave. Leaving the general case to Section 9.5, we use here the approximation,

$(2\pi)^2 \mu_1^* \mu_0 = \frac{i}{k}(\nabla_+ \Psi_{\text{in}})^* \Psi_{\text{in}}|_{r=0}$, valid, when the incident wave is almost a plane wave (see Sect. 9.5 for detail). In this limit,

$$\mathcal{F} = -\hbar (1 - \cos(2\pi\alpha)) \boldsymbol{J}_{\text{in}} + \hbar \sin(2\pi\alpha)[\boldsymbol{J}_{\text{in}} \times \boldsymbol{e}_z] , \quad (9.37)$$

where $\boldsymbol{J}_{\text{in}}$ is the current in the incident wave on the line. The first term is the well-known longitudinal force $\mathcal{F}_\parallel$ proportional to $\sigma_\parallel$ in (9.1). The second term is the transverse force. Its value agrees with the paraxial (9.13) and the result first derived by Cleary [13]. (See Sect. 9.5 for the theory beyond the paraxial approximation.)

In the linear in α approximation, when the first term can be omitted and $\sin \pi\alpha \approx \pi\alpha$, (9.37) can be presented in a very familiar form,

$$\mathcal{F} = \frac{e}{c} \int dV \, [\boldsymbol{J}_{\text{in}} \times \boldsymbol{B}] , \quad (9.38)$$

where $\boldsymbol{B} = \Phi \delta(\boldsymbol{r}) \boldsymbol{e}_z$ is the magnetic induction of the AB–line, and $e\boldsymbol{J}_{\text{in}}(\boldsymbol{r})$ is the charge current density. One sees that AB–lines with a small flux create the same force density as the classical Lorentz force. (This is seen already from (9.29).)

9.3 Conclusions

The goal of this chapter has been to understand why different approaches to the scattering by the Aharonov–Bohm flux line give contradicting results. A short answer to the question is that the scattering theory in its text-book form is not applicable to the AB–line. The reason for this is that, unlike what is usually assumed, the potential of the AB–line perturbs the wave function at any distance. (This is most obvious in the sheet gauge (9.7) where the vector potential does not depend on the distance to the line.)

The singular behaviour of the potential at large distances translates to the forward scattering singularity. The singularity reveals itself in the Aharonov–Bohm solution where the wave does not converge at large distances to the incident wave and does not follow the usual asymptotics (9.24) in the near forward direction (most clearly it is seen in the Cornu spiral representation [20]). In the S-matrix equation (9.23), the δ-function contribution, which usually has the weight unity representing the unscattered part of the outgoing wave, turns out to be modified and, besides, there is another forward anomaly due to the principal value in (9.23). For this reason, the combination, $\hat{f} = \sqrt{\frac{2\pi}{ik}}(\hat{S} - \hat{1})$, which usually defines the scattering amplitude $f(\theta)$, is singular. What is even more important, the notion of the scattering amplitude appears to be obsolete in the AB–case because one is not able to give a full description of scattering in terms of only $|f|^2$, the transverse force gives an example. Therefore, it seems advisable to use only the S-matrix formalism and refrain from the usage of the scattering amplitude: Uncritical application of the textbook theory leads to the paradoxes when seemingly identical expressions for $\sigma_\perp$ in (9.1) and (9.3) return different values.

In the approach taken in this chapter, the difficulties related to the long range of the AB–potential, are avoided by taking the incident wave to be a coherent beam of a large but finite width W. The value of the vector potential in the region beyond the beam is essentially irrelevant, and the tail of the vector potential is effectively cut. Formulated differently, the regularization is achieved because the Heisenberg uncertainty in the momentum $\sim \hbar/W$ and the direction of propagation $\varphi_0 \sim \hbar/pW$, smears out the forward singularity.

The existence of the Lorentz force from the AB–line has been "verified" in the experiment Fig. 9.1 where a finite value of the beam deflection is found [15,17]. The deflection angle is of the order of the diffraction beam size φ_0, and this is why the consideration based on the notion of the differential cross-section $|f|^2$ misses this *qualitative* effect. (In passing, we note that problems with the non-analytical forward scattering amplitude and a divergent total cross-section exist for the Coulomb $1/r$ potential [21]. Unlike the AB–line case, no qualitative effects are related to the Coulomb forward singularity.)

Throughout the chapter we dealt with pure quantum states described by the wave function. The momentum transfer in (9.13) and the force in (9.37) depend on the wave intensity at the line, and, therefore, they are sensitive to the structure of the incident wave. In the transport-like theories, one may be interested in a situation where the incident particles are in a state which is an incoherent mixture of beams with evenly distributed impact parameters (a homogeneous flow). The final result in (9.37) can be used also for a partially coherent state described by the density matrix with the understanding that J_{in} is the total (incoming) current density created by all incoming particles at the point where they meet the flux line.

If one accepts the equivalence of the Aharonov–Bohm problem and the problem of phonon scattering by a vortex in superfluids [4,7,10], the Iordanskii force acting on the *vortex* from the phonons can be obtained from (9.38): One changes the sign and substitutes $\frac{e}{c}B$ for $\frac{\kappa}{c}\delta(\bm{r})$, $\kappa = \frac{h}{m}$ and c being the quantized vortex circulation and the sound velocity, respectively; besides, J_{in} must be understood as the momentum flow density. The fact that both the Iordanskii and Lorentz force can be derived from the same expression, allows one to argue that the controversial Iordanskii force is an analog to the more familiar Lorentz force.

Transport properties of charged particles in the field of many AB–lines can be considered with the help of the Boltzmann equation derived in [16]. As is no surprise, the collision integral which describes the scattering of the charges by the AB–lines, contains the Aharonov–Bohm cross-section (9.2). In addition, the lines enter the *field* part of the Boltzmann equation via the effective magnetic field $B_{\text{eff}} = \frac{1}{2\pi}\sin(2\pi\alpha)\Phi_0 \times$(the AB–line density). This shows that the transverse force is more a renormalization effect rather than scattering.

The final general remark is that the Aharonov–Bohm line may serve as a counterexample in quantum mechanics: It shows one more time that when conditions of the validity of a theory are not met (here, it is a finite radius of the potential in the scattering theory), an uncritical application of the theory may lead to wrong and self-contradicting results.

I would like to thank E. Sonin for numerous communications. I am grateful to M. Berry for discussions about the Aharonov–Bohm physics. I would like also to thank the organizers of the workshop in Dresden for their kind invitation. This chapter was completed during my stay at the Institut für Theorie der Kondensierten Materie, Universität Karlsruhe, and was partly financed by SFB 195 der Deutsche Forschungsgemeinschaft.

9.4 Appendix: The Momentum–Flow Tensor for the Schrödinger Equation

The momentum density operator $\hat{\pi}_k(\boldsymbol{r})$ is defined in (9.25) which we write as

$$\hat{\pi}_k(\boldsymbol{r}) = \frac{1}{2}\{\delta(\boldsymbol{r}-\hat{\boldsymbol{r}}), \hat{p}_k\}\ ,$$

here and below we denote $\{A, B\} = AB + BA$.

To avoid confusion, the coordinate $\boldsymbol{r}$ as well as the index k are the labels attached to the operator. On the contrary, $\hat{\boldsymbol{r}}$ is the quantum operator. For instance, $\langle \Psi|\delta(\boldsymbol{r}-\hat{\boldsymbol{r}}|\Psi\rangle \equiv \int d\boldsymbol{r}'\, \delta(\boldsymbol{r}-\boldsymbol{r}')|\Psi(\boldsymbol{r}')|^2 = |\Psi(\boldsymbol{r})|^2$. As another example, the expectation value

$$\langle \Psi|\hat{\boldsymbol{\pi}}(\boldsymbol{r})|\Psi\rangle = \mathrm{Re}\left(\Psi^*(r)\left(\frac{\hbar}{i}\frac{\partial}{\partial \boldsymbol{r}}-\frac{e}{c}\boldsymbol{A}(\boldsymbol{r})\right)\Psi(\boldsymbol{r})\right) = m\boldsymbol{J}(\boldsymbol{r})$$

gives the current density $\boldsymbol{J}(\boldsymbol{r})$ in the state Ψ at the point $\boldsymbol{r}$.

To find time dependent operators, one uses the usual formula

$$\dot{\hat{X}} = \frac{\partial \hat{X}}{\partial t} + \frac{i}{\hbar}[\hat{H},\hat{X}]\ , \tag{9.39}$$

with the Hamiltonian

$$\hat{H} = \frac{1}{2m}\hat{\boldsymbol{p}}^2 + V(\hat{\boldsymbol{r}})\ ,$$

where $\hat{\boldsymbol{p}}$ is the kinematical momentum (9.26), and V is the potential energy.

Using the following relations for the commutators: $[\hat{p}_l,\hat{p}_k] = ie\hbar\varepsilon_{lk\gamma}B_\gamma$ and $[\hat{p}_l,\delta(\hat{\boldsymbol{r}}-\boldsymbol{r})] = i\hbar\frac{\partial}{\partial \boldsymbol{r}}\delta(\hat{\boldsymbol{r}}-\boldsymbol{r})$, one can write $[\hat{H},\hat{\pi}_k]$ as

$$[\hat{H},\hat{\pi}_k] = \frac{\partial}{\partial x_l}\frac{i\hbar}{4m}\{\hat{p}_l,\{\delta(\hat{\boldsymbol{r}}-\boldsymbol{r}),\hat{p}_k\}\}$$
$$+\frac{ie\hbar}{2m}\varepsilon_{lk\gamma}\{\hat{p}_l,\delta(\hat{\boldsymbol{r}}-\boldsymbol{r})B_\gamma(\hat{\boldsymbol{r}})\} + e\frac{\hbar}{i}\hat{\boldsymbol{E}} \tag{9.40}$$

where $\hat{\boldsymbol{E}} = \boldsymbol{E}(\hat{\boldsymbol{r}})$ and $\hat{\boldsymbol{B}} = \boldsymbol{B}(\hat{\boldsymbol{r}})$ are the electric and magnetic fields (the operator nature of this objects does not pose any problems since they are diagonal in the coordinate representation).

Finally, one comes to the operator relation,

$$\hat{\pi}_k(\boldsymbol{r}) = -\frac{\partial}{\partial x_l}\hat{K}_{lk}(\boldsymbol{r}) + \hat{\boldsymbol{F}}(\boldsymbol{r}) ,\qquad(9.41)$$

where $\hat{K}$ corresponds to the momentum-flow tensor (stress tensor),

$$\hat{K}_{lk}(\boldsymbol{r}) = \frac{1}{2m}\{\hat{p}_l, \hat{\pi}_k(\boldsymbol{r})\} ,$$

and the force operator $\hat{\boldsymbol{F}}(\boldsymbol{r})$,

$$\hat{\boldsymbol{F}}(\boldsymbol{r}) = e\delta(\hat{\boldsymbol{r}} - \boldsymbol{r})\hat{\boldsymbol{E}} + e\frac{1}{2}\left(\left(\hat{\boldsymbol{v}}\times\hat{\boldsymbol{B}}\right)\delta(\hat{\boldsymbol{r}} - \boldsymbol{r}) - \delta(\hat{\boldsymbol{r}} - \boldsymbol{r})\left(\hat{\boldsymbol{B}}\times\hat{\boldsymbol{v}}\right)\right) .$$

All these operators are Hermitian, their expectation values are given by (9.28) and (9.29).

Taking the expectation value of (9.41) and integrating over the volume V, one gets the conservation law

$$\frac{d}{dt}\int_V dV\, \pi_k + \int_S dS_l K_k = \int_V dV\, F_k(\boldsymbol{r})$$

where the components of $\boldsymbol{\pi}$, K, and F denote the expectation value of the corresponding operators, e.g. $\boldsymbol{\pi}(\boldsymbol{r}) \equiv \langle\boldsymbol{\pi}(\boldsymbol{r})\rangle$. This expression is presented in vector form (9.30).

9.5 Appendix: The Force: Arbitrary Wave

The force $\mathcal{F}$ is found from (9.17) with $\mu_{0,1}$ defined in (9.19). In terms of a modified incoming wave $\tilde{\Psi}_{\text{in}}$,

$$\tilde{\Psi}_{\text{in}}(\boldsymbol{r}) = \int_{-\pi}^{\pi}\frac{d\phi}{2\pi}e^{i\alpha\phi}\mu(\phi)\exp\left(\boldsymbol{k}_\phi\cdot\boldsymbol{r}\right) ,$$

one can write $\mu_1 = \frac{1}{ik}\nabla_-\tilde{\Psi}_{\text{in}}|_{r=0}$, with $\nabla_\pm = \frac{\partial}{\partial x} \pm i\frac{\partial}{\partial y}$, and $\mu_0 = \tilde{\Psi}_{\text{in}}|_{r=0}$, and present (9.36) as

$$\mathcal{F}_+ = \frac{i\hbar^2}{2m}\left(e^{-2i\pi\alpha} - 1\right)\left(\nabla_+|\tilde{\Psi}_{\text{in}}|^2 + \frac{2m}{i\hbar}\tilde{J}_+^{(\text{in})}\right)$$

where $\tilde{J}_+^{(\text{in})} = \frac{i\hbar}{2m}\left(\left(\nabla_+\tilde{\Psi}_{\text{in}}^*\right)\tilde{\Psi}_{\text{in}} - \tilde{\Psi}_{\text{in}}^*\nabla_+\tilde{\Psi}_{\text{in}}\right)_{r=0}$. This expression is valid for a general incident wave. For a beam with a small angular width, $e^{i\alpha\phi}$ is effectively a constant, and, in the main approximation $\tilde{\Psi}_{\text{in}}$ differs from Ψ_{in}

only in a phase factor; the term $\nabla_+|\tilde{\Psi}_{in}|^2$ can be neglected. In this limit, one arrives at (9.37).

Alternatively, one formally expresses $\mu_{1,0}$ as

$$\mu_0 = \left(\frac{\partial_+}{ik}\right)^\alpha \Psi_{in}|_{r=0} \;,\; \mu_1 = \left(\frac{\partial_-}{ik}\right)^{1-\alpha} \Psi_{in}|_{r=0} \tag{9.42}$$

where $\partial_\pm = \partial_x \pm i\partial_y$. For a wave propagating mainly in the x-direction, $\Psi_{in}(r) = e^{ikx}\psi_{in}(r)$, equation (9.42) can be written a

$$\mu_0 = \left(1 + \frac{\partial_+}{ik}\right)^\alpha \psi_{in}|_{r=0} \;,\; \mu_1 = \left(1 + \frac{\partial_-}{ik}\right)^{1-\alpha} \psi_{in}|_{r=0} \;. \tag{9.43}$$

Since $|\partial_{x,y}\psi| \ll k$ for a beam with a narrow angular width, one can expand with respect to $\partial_\pm$ in (9.43). In the main approximation one arrives again at again (9.37). The corrections $\sim (kW)^{-2}$ can explain the small but noticeable deviations from the paraxial theory observed by Berry [17].

It seems to be of interest to consider the case when the incident wave does not overlap with the line and the paraxial force is zero. Take a beam for which $\psi_{in}(\mathbf{0}) = 0$ and also $\frac{\partial}{\partial y}\psi_{in}(\mathbf{0}) = 0$ (a symmetric beam $\Psi(x,y) = \Psi(x,-y)$). The free Schrödinger equation requires that $\left(-2ik\partial_x + \partial_x^2 + \partial_y^2\right)\psi_{in} = 0$. To simplify the calculations, we suppose also that $\frac{\partial}{\partial x}\psi_{in}(\mathbf{0}) = 0$ (as it is the case if $x = 0$ is the focus "plane" for the incident wave). Consequently, $\partial_x^2\psi_{in}(\mathbf{0}) = -\partial_y^2\psi_{in}(\mathbf{0})$. Besides, $\partial_x\partial_y\psi_{in}(x,y=0) = 0$ for a symmetric beam. After these simplifications,

$$\mu_0 \approx \mu_1 \approx -\frac{1}{k^2}\alpha(1-\alpha)\,\partial_y^2\psi_{in}(\mathbf{0})\;.$$

Expressed via $\mu(\phi)$, $\partial_y^2\psi_{in}(\mathbf{0}) = -k^2\int_{-\pi}^{\pi} d\phi\,\mu(\phi)\sin^2\phi$ Known $\mu_0^*\mu_1$, one finds the force (9.36) and, combining (9.33) and (9.34,) the deflection of the beam $\Delta\phi \equiv \frac{1}{p}\Delta P_y$,

$$\Delta\varphi = -2\pi\,\alpha^2(1-\alpha)^2\,\sin(2\pi\alpha)\frac{\left|\int_{-\pi}^{\pi} d\phi\,\mu(\phi)\sin^2\phi\right|^2}{\int_{-\pi}^{\pi} d\phi\,|\mu(\phi)|^2}\;.$$

In this case, the deflection scales as φ_0^5 with the angular width of the beam φ_0.

References

1. Y. Aharonov and D. Bohm, Phys. Rev. **115**, 485 (1959).
2. S. Olariu, I. I. Popescu, Rev. Mod. Phys. **57**, 339 (1985).

3. M. Peshkin, and A. Tanomura, "Aharonov–Bohm Effect", Lecture Notes in Physics **340**, Springer-Verlag, (1989).
4. E.B. Sonin, Chap. 8 of this book.
5. S.V. Iordanskii, Zh. Eksp. Teor. Fiz. **49**, 225 (1965) [Sov.Phys.-JETP **22**, 160 (1966)]; Ann. Phys. **29**, 335 (1964).
6. P. Ao, D.J. Thouless, Phys. Rev. Lett. **70**, 2158 (1993); D.J. Thouless et al., ibid, **76**, 3758 (1996); C. Wexler, ibid **79**, 1321 (1997).
7. D. J. Thouless et al., *The 9th International Conference on Recent Progress in Many-Body Theories, 21–25 July 1997, Sydney, Australia* (cond-mat/9709127).
8. G.E. Volovik, Pis'ma Zh. Eksp. Teor. Fiz. **67**, 841 (1998) [JETP Lett. **67**, 881 (1998)].
9. J.K. Jain, Adv. Phys. **41**, 105 (1992); J. March–Russel, F. Wilczek, Phys. Rev. Lett. **61** , 2066 (1988).
10. E. B. Sonin, Phys. Rev. **B 55** , 485 (1997).
11. M. Nielsen, P. Hedegård, Phys. Rev. **B51**, 7679 (1995).
12. E.B. Sonin, Zh. Eksp. Teor. Fiz. **69**, 921 (1975) [Sov.Phys.-JETP **42**, 469 (1976)].
13. R. M. Cleary, Phys. Rev. **175**, 587 (1968).
14. M.A. Leontovich, and V.A. Fock, Zh. Eksp. Teor. Fiz. **16** , 557 (1946).
15. A. Shelankov, Europhys. Lett, **43**, 623–628 (1998).
16. A. Shelankov, Phys. Rev. **B 62**, No. 5 (2000) (in print).
17. M. V. Berry, J. Phys. A: Math. Gen., **32**, 5627, (1999).
18. M. V. Berry, et al., 1980, Eur. J. Phys. **1**, 154 (1980)
19. M. Born, E. Wolf, Principles of Optics, (Pergamon Press), 1959, p.431.
20. M. V. Berry, and A.Shelankov, J. Phys. A: Math. Gen., **32**, L447–L455, (1999).
21. L.D. Landau and E.M. Lifshitz, *Quantum Mechanics* (Pergamon Press, New York, 1977).
22. S.N.M. Ruijsenaars, Ann. Phys. (NY), **146** 1, (1983); A. Moroz, Phys. Rev. **A 53**, 669 (1996).
23. D. Stelitano, Phys. Rev. **D 51** , 5876 (1995).

10 Relativistic Solution of Lordanskii Problem in Multi-Constituent Superfluid Mechanics

B. Carter, D. Langlois, and R. Prix

Summary. Flow past a line vortex in a simple perfect fluid or superfluid gives rise to a transverse Magnus force that is given by the well known Joukowski lift formula. The problem of generalising this to multiconstituent superfluid models has been controversial since it was originally posed by the work of Iordanski in the context of the Landau 2-constituent model for ^{4}He at finite temperature. The present work deals not just with this particular case but with the generic category of perfect multiconstituent models including the kind proposed for a mixture of ^{4}He and ^{3}He by Andreev and Bashkin. It is shown here (using a relativistic approach) that each constituent will provide a contribution proportional to the product of the corresponding momentum circulation integral with the associated asymptotic current density.

10.1 Introduction

For a simple perfect fluid with asymptotically uniform density $\bar{\rho}$ say, the Magnus effect of a uniform background flow with relative velocity v^i say in the rest frame of a vortex in the direction of a 3 dimensional unit vector ℓ^i results in a force per unit length given by the well known (non-relativistic) Joukowsky formula as

$$\mathcal{F}_i = \kappa \bar{\rho} \varepsilon_{ijk} \ell^j \bar{v}^k \,, \tag{10.1}$$

where κ is the relevant velocity circulation integral.

The question raised by Iordanskii [1] of how this formula should be generalised to the case of Landau's 2 constituent model for superfluid ^{4}He at finite temperature has been a subject of controversy: the most widely accepted [2,3] prescription is that of Sonin [4,5], but various alternatives have been proposed by other authors [6–8]. The present work clarifies the issue by demonstrating the existence of an elegant generalisation of the Joukowsky formula (10.1) for an extensive class of perfect multiconstituent fluid models, including, as well as the Landau model, the Andreev–Bashkin model [9] for a mixture of superfluid ^{4}He with ("normal") ^{3}He. (However our analysis does not cover the more complicated subject [2] of superfluid ^{3}He).

For technical convenience (not to mention the consideration that it is more accurate in contexts such as that of neutron star matter) the work is carried out using a (special) relativistic formulation.

10.2 Perfect Multiconstituent Fluid Dynamics

As well as including the original Landau model in the (not so simple) Galilean limit [10], two independent lines of early development of relativistic (and thus technically simpler) perfect multiconstituent fluid theory – using currents [11,12] and momenta [13] respectively as independent variables in (suitably constrained) variational formulations – were subsequently shown to be entirely equivalent[14]: the independent momentum covectors $\mu_\nu^{\rm X}$ of the latter approach are identifiable just as the dynamical conjugates of the independent conserved current vectors $n_{\rm X}^\nu$ on which the former approach was based. In the latter approach, which is the most convenient for our present purpose, the fundamental equation of state characterising a particular perfect multiconstituent fluid model is given by specifying the dependence of the relevant generalised pressure function Ψ on the independent momentum covectors $\mu_\nu^{\rm X}$, and the associated current vectors $n_{\rm X}^\nu$ are then obtained as the corresponding partial derivatives in the infinitesimal variation formula

$$\delta\Psi = -\sum_{\rm X} n_{\rm X}^\nu \delta\mu_\nu^{\rm X}\,. \tag{10.2}$$

For such a model, the complete set of equations of motion consists just of a set of particle and vorticity conservation laws of the form

$$\nabla_\nu n_{\rm X}^\nu = 0\,, \qquad n_{\rm X}^\nu w_{\nu\sigma} = 0\,. \tag{10.3}$$

where, for any particular constituent with label X the corresponding generalised vorticity 2-form is defined as the exterior derivative

$$w_{\nu\sigma}^{\rm X} = \nabla_\nu \pi_\sigma^{\rm X} - \nabla_\sigma \pi_\nu^{\rm X}\,, \qquad \pi_\nu^{\rm X} = \mu_\nu^{\rm X} + e^{\rm X} A_\nu \tag{10.4}$$

of the generalised momentum covector $\pi_\nu^{\rm X}$, whose specification[15] allows for the possibility of coupling to an electromagnetic field

$$F_{\nu\sigma} = \nabla_\nu A_\sigma - \nabla_\sigma A_\nu\,. \qquad \nabla_\sigma F^{\sigma\nu} = 4\pi J^\nu\,, \tag{10.5}$$

with current and electromagnetic stress tensor given by

$$J^\nu = \sum_{\rm X} e^{\rm X} n_{\rm X}^\nu\,, \qquad T_{{\rm F}\,\nu}^{\sigma} = \frac{1}{4\pi}\left(F^{\mu\nu}F_{\mu\sigma} - \frac{1}{4}F^{\mu\rho}F_{\mu\rho}g_\nu^\sigma\right), \tag{10.6}$$

where $e^{\rm X}$ is the electric charge, if any, per particle of the X th species.

In such a model a (flat space) conservation law of the usual form

$$\nabla_\nu T_\sigma^{\,\nu} = 0\,, \tag{10.7}$$

will be satisfied by the relevant total stress energy tensor, which takes the form

$$T_\nu^{\,\sigma} = \sum_{\rm X} n_{\rm X}^\sigma \mu_\nu^{\rm X} + \Psi g_\nu^\sigma + T_{{\rm F}\,\nu}^{\sigma} = \sum_{\rm X} n_{\rm X}^\sigma \pi_\nu^{\rm X} + \Psi g_\nu^\sigma - J^\sigma A_\nu + T_{{\rm F}\,\nu}^{\sigma}\,. \tag{10.8}$$

The transport law (10.3) for the vorticities is such that they will remain zero if they are zero initially. Thus we shall have

$$w^{\text{X}}_{\nu\sigma} = 0, \tag{10.9}$$

not just for cases of superfluidity or superconductivity (i.e. cases for which the momentum covector is the gradient of a condensate phase scalar) but even for "normal" constituents in configurations of the kind to be considered here, in which a perturbing vortex moves through an *asymptotically uniform medium* characterised by vanishing of the asymptotic background value (indicated here by an overhead bar) not just of the current (as is necessary for uniformity) but also of the electromagnetic field, and (in an appropriate gauge) of its vector potential,

$$\overline{J}^\nu = 0 \qquad \overline{F}_{\nu\sigma} = 0, \qquad \overline{A}_\sigma = 0. \tag{10.10}$$

This must necessarily be the case (the Meissner effect), with the implication that the uniform background value of the stress energy density tensor will be given simply by

$$\overline{T}^\sigma{}_\nu = \sum_{\text{X}} \overline{n}^\sigma_{\text{X}} \overline{\pi}^{\text{X}}_\nu + \overline{\Psi} g^\sigma_\nu, \tag{10.11}$$

whenever even just a single one of the uniform background constituents is superconducting (since $e^{\text{X}} F_{\nu\sigma} = \nabla_\nu \pi^{\text{X}}_\sigma - \nabla_\sigma \pi^{\text{X}}_\nu - w^{\text{X}}_{\nu\sigma}$).

10.3 Specification of Lift Force on Vortex

The subject of this investigation is an asymptotically uniform vortex configuration that is stationary with respect to a rest frame characterised by a uniform timelike unit symmetry generating vector field k^μ say, and that is aligned in the direction of a uniform orthogonal spacelike unit symmetry generating vector field ℓ^μ in a flat background spacetime using Minkowski coordinates. Provided the corresponding conditions of stationarity and longitudinal symmetry apply to $F_{\nu\sigma}$ not just outside but even within the vortex core region, they will be applicable to the potential A_σ in a suitable gauge, and hence also, not just to the gauge independent covectors μ^{X}_ν but to the corresponding generalised momenta π^{X}_ν as well. In view of the vanishing (10.9) of the vorticity vectors (10.4), the condition that the momentum covectors should be invariant with respect to the action of the the uniform symmetry generating vector fields k^ν and ℓ^ν can be seen to imply the uniformity of corresponding sets of generalised Bernouilli constants,

$$k^\nu \pi^{\text{X}}_\nu = k^\nu \overline{\pi}^{\text{X}}_\nu, \qquad \ell^\nu \pi^{\text{X}}_\nu = \ell^\nu \overline{\pi}^{\text{X}}_\nu. \tag{10.12}$$

The force per unit length, $\mathcal{F}_\nu$ acting on such a stationary longitudinally invariant vortex can be evaluated as the integral round a circuit s say surrounding the vortex in an orthogonal 2-plane in the form

$$\mathcal{F}_\nu = \oint f_\nu ds, \tag{10.13}$$

where ds is the proper distance element given by $ds^2 = g_{\nu\sigma}dx^\nu dx^\sigma$ and f_ν is the local force density that is given by

$$f_\nu = \nu_\sigma T^\sigma{}_\nu, \tag{10.14}$$

in terms of the unit normal covector ν_σ which will be given in terms of the antisymmetric background measure tensor $\varepsilon_{\lambda\mu\nu\sigma}$ by

$$\nu_\sigma ds = {}^*\varepsilon_{\sigma\nu}dx^\nu, \qquad {}^*\varepsilon_{\sigma\nu} = \ell^\rho\varepsilon_{\rho\sigma\nu}, \qquad \varepsilon_{\rho\sigma\nu} = k^\mu\varepsilon_{\mu\rho\sigma\nu}. \tag{10.15}$$

10.4 Generalised Joukowski Theorem

It is to be observed that, as a consequence of the conservation law (10.7), it makes no difference what circuit is employed for evaluating $\mathcal{F}_\nu$. We are thus allowed to choose a circuit sufficiently far out for reliability of our smoothed fluid description (whose physical validity might be questionable near the core) to be ensured, and also for the deviation from the uniform background value $\overline{T}^\sigma{}_\nu$ to be evaluated as a linear perturbation:

$$T^\sigma{}_\nu - \overline{T}^\sigma{}_\nu = \delta T^\sigma{}_\nu + O\{\delta^2\}. \tag{10.16}$$

Since the force integral for the unperturbed uniform background must evidently vanish, $\overline{\mathcal{F}}_\nu = 0$ by symmetry, the corresponding value in the presence of the vortex will be given by

$$\mathcal{F}_\nu = \delta\mathcal{F}_\nu + O\{\delta^2\}. \tag{10.17}$$

Using (10.2) and (10.11) it can immediately be seen that the required first order variation will be given by

$$\delta T^\sigma{}_\nu = \sum_{\text{X}} \left(\overline{\pi}^{\text{X}}_\nu \delta n^\sigma_{\text{X}} + \overline{n}^\sigma_{\text{X}} \delta \pi^{\text{X}}_\nu - g^\sigma_\nu \overline{n}^\rho_{\text{X}} \delta \pi^{\text{X}}_\rho \right). \tag{10.18}$$

Using the decomposition of the 4-dimensional spacetime metric in the form $g^\nu_\sigma = \eta^\nu_\sigma + \perp^\nu_\sigma$ as the sum of the (rank 2) operators of projection respectively parallel to and orthogonal to the vortex given by $\eta^\nu_\sigma = -k^\nu k_\sigma + \ell^\nu \ell_\sigma$, and $\perp^\nu_\sigma = {}^*\varepsilon^{\mu\nu}{}^*\varepsilon_{\mu\sigma}$ and using the possibility of taking the Bernouilli constants (produced by parallel projection) outside the integration, (10.18) provides a result expressible simply as

$$\delta\mathcal{F}_\nu = \sum_{\text{X}} \left(\overline{n}^\sigma_{\text{X}} {}^*\varepsilon_{\sigma\nu}\delta\mathcal{C}^{\text{X}} + \overline{\pi}^{\text{X}}_\nu \delta D_{\text{X}} \right), \tag{10.19}$$

where for each species X the corresponding momentum circulation integral $\mathcal{C}^{\text{X}}$ and current outflux integral D_{X} are defined by

$$\mathcal{C}^{\text{X}} = \oint \pi^{\text{X}}_\nu dx^\nu, \qquad D_{\text{X}} = \oint n^\sigma_{\text{X}} \nu_\sigma ds. \tag{10.20}$$

The irrotationality condition (10.9) ensures that $\mathcal{C}^{\rm X}$ is independent of the choice of circuit, and the current conservation law (10.3) ensures that the same will apply to $D_{\rm X}$, which furthermore will simply vanish, $D_{\rm X} = 0$, provided there is no current creation in the vortex core. Thus by the fact that the uniform background value of the circulation integrals must also vanish, $\overline{\mathcal{C}}^{\rm X} = 0$, and by taking the limit in which the circuit is taken to a very large distance outside, one obtains an exact net force formula of the simple form

$$\mathcal{F}_\nu = {}^\star\!\varepsilon_{\sigma\nu} \sum_{\rm X} \mathcal{C}^{\rm X} \overline{n}_{\rm X}^{\,\sigma} \,. \tag{10.21}$$

This result is the required (relativistic, multiconstituent) generalisation of Joukowsky's well known formula (10.1) for the single constituent case. What it means is that each constituent contributes an amount proportional to, but orthogonal to, its asymptotic current vector, with a coefficient given by the corresponding momentum circulation integral.

10.5 Application to the Landau Model

The particular example that motivated this work is that of superfluid ^{4}He at finite temperature, as described by the Landau model in terms of just two constituents with conserved 3-dimensional current densities $n_\alpha^i = n_\alpha v_\alpha^i$ and $n_\beta^i = n_\beta v_\beta^i$ of which the first represents Helium atoms, i.e. "dressed" alpha particles, characterised by a "rest mass" m_α, and the second represents units of entropy, characterised by a vanishing rest mass $m_\beta = 0$. As in the less specialised case of the 2 constituent (zero temperature) limit of the Andreev–Bashkin model for which the second constituent is ^{3}He with non vanishing rest mass, $m_\beta \simeq 3m_\alpha/4$, the Newtonian limit description can be formulated in terms of a total mass density and 3 dimensional mass current

$$\rho = \rho_\alpha + \rho_\beta \,, \qquad \rho^i = \rho_\alpha v_\alpha^i + \rho_\beta v_\beta^i \,, \tag{10.22}$$

where $\rho_\alpha = m_\alpha n_\alpha$ and $\rho_\beta = m_\beta n_\beta$, so that the latter vanishes in the particular case of the Landau model. The total mass current is identifiable with the total momentum density $\rho_i = n_\alpha \mu_i^\alpha + n_\beta \mu_i^\beta$, in which, due to the effect of "entrainment" (which is describable in terms of "effective masses" that are different from the bare masses) the vanishing of the second contribution to the mass current does not imply absence of the second momentum contribution proportional to μ_i^β.

The pseudo-velocity $v_{\rm S}^i$ that is commonly referred to as the "superfluid velocity" is defineable by $v_{{\rm S}i} = m_\alpha^{-1}\mu_i^\alpha$. In the Landau case (unlike the generic Andreev–Bashkin case) it is not possible to define an analogous pseudo velocity for the other consituent, because of the vanishing of m_β, and the quantity commonly denoted as $v_{\rm N}^i$ and known as the "normal" velocity is simply identifiable with the velocity of the entropy current, i.e. $v_{\rm N}^i = v_\beta^i$.

In a mass and momentum decomposition of the commonly used (effectively "mongrel") form

$$\rho = \rho_{\rm S} + \rho_{\rm N}\,, \qquad \rho^i = \rho_{\rm S} v_{\rm S}^i + \rho_{\rm N} v_{\rm N}^i\,, \tag{10.23}$$

the coefficients $\rho_{\rm S}$ and $\rho_{\rm N}$ must not be confused with ρ_α and ρ_β (of which the latter is zero in the Landau case characterised by by $\rho = \rho_\alpha$). In the Landau case, as well as in the generic Andreev–Bashkin case, there are two independently conserved momentum circulation integrals,

$$\mathcal{C}^\alpha = \oint \mu_i^\alpha \, dx^i\,, \qquad \mathcal{C}^\beta = \oint \mu_i^\beta \, dx^i\,, \tag{10.24}$$

of which, by the superfluidity property, the former is quantised, $\mathcal{C}^\alpha = h$. Since $m_\alpha \neq 0$ we can write

$$\mathcal{C}^\alpha = m_\alpha \kappa_{\rm S}\,, \quad \mathcal{C}^\beta = m_\beta \kappa_{\rm S} + \oint \frac{\rho_{\rm N}}{n_\beta}(v_{{\rm N}i} - v_{{\rm S}i})dx^i\,, \quad \kappa_{\rm S} = \oint v_{{\rm S}i}\,dx^i\,, \tag{10.25}$$

in terms of the pseudo-velocity circulation integral $\kappa_{\rm S}$, which has no "normal" analogue in the Landau case, because of the vanishing of m^β. Thus (in the rest frame of the vortex) using the notation ${}^\star\varepsilon_{ij} = \varepsilon_{ijk}\ell^k$, one finally obtains a non-relativistic force formula of the form

$$\mathcal{F}_i = {}^\star\varepsilon_{ij}\left(\kappa_{\rm S}\overline{\rho}_\alpha \overline{v}_\alpha^j + \mathcal{C}^\beta \overline{n}_\beta \overline{v}_\beta^j\right) = \kappa_{\rm S}\overline{\rho}_{\rm S}\,{}^\star\varepsilon_{ij}\overline{v}_{\rm S}^j + \mathcal{F}_{{\rm I}i}\,, \tag{10.26}$$

where in this last version the first term is what is commonly referred to as the "superfluid Magnus force" contribution, while the remaining "Iordanskii" correction term is found to be given by

$$\mathcal{F}_{{\rm I}i} = \left(\mathcal{C}^\beta \overline{n}_\beta + \kappa_{\rm S}(\overline{\rho}_{\rm N} - \overline{\rho}_\beta)\right){}^\star\varepsilon_{ij}\overline{v}_{\rm N}^j\,. \tag{10.27}$$

The third term in this expression is needed for the generic Andreev–Bashkin case, but drops out for the special Landau case characterised by $\rho_\beta = 0$.

If, as well as setting ρ_β to zero, one adopts the plausible supposition that the "normal" circulation will vanish, $\mathcal{C}^\beta = 0$, then the first term also drops out so that our formula will reduce to a form that is in exact agreement with the result that was derived by Sonin[4,5] and confirmed, on the basis of a more rigorous microscopic analysis of phonon dynamics, by Stone[3].

This widely accepted conclusion has however been vigorously contested by Thouless and coworkers [7,8] who have used a more sophisticated – though not obviously more reliable – kind of microscopic analysis to argue that the Iordanskii force term $\mathcal{F}_{{\rm I}i}$ vanishes, leaving just the purely "superfluid" term (namely ${}^\star\varepsilon_{ij}\kappa_{\rm S}\overline{\rho}_{\rm S}\,\overline{v}_{\rm S}^j$) in (10.26). As *prima facie* evidence in favour of this dissident conclusion, it is to be observed that in the limit when there is no relative flow at all (i.e. $\overline{v}_{\rm S}^i = \overline{v}_{\rm N}^i = 0$) then – as a requirement for compatibility with strict stationarity – the long term effect of the small "normal" viscosity contribution that was neglected in the preceeding analysis will impose a rigidity condition to the effect that $v_{\rm N}^i = 0$ throughout. This imperative entails small deviations from strict irrotationality of the normal constituent except in the incompressible case for which the ratio $\rho_{\rm N}/n_\beta$ is exactly uniform, and it ensures in any case by (10.25) that the normal momentum circulation round a circuit at large distance will be given by the formula $\mathcal{C}^\beta = \left(m_\beta - \overline{\rho}_{\rm N}/\overline{n}_\beta\right)\kappa_{\rm S}$ whose substitution in (10.27) does indeed give, $\mathcal{F}_{1i} = 0$. However this simple counter argument is inconclusive because – as shown every time an ordinary light aircraft takes off – the stationary circulation value due to the long term effect of slight deviations from strictly inviscid behaviour will change as a function of the relative flow velocity.

To sum up, the present work shows how the Iordanskii force is given simply as a function of the (in the inviscid limit conserved) "normal" momentum circulation integral $\mathcal{C}^\beta$, but the issue of the appropriate value for this parameter in a realistic steady flow confguration is beyond the scope of a perfectly conducting fluid treatment such as is provided here. Experience with the analogous aerofoil problem in the context of aircraft engineering suggests that the final resolution of this issue may involve subtleties that have have eluded even the most sophisticated analysis available so far.

We wish to thank Uwe Fischer, Michael Stone, and Grigori Volovik for instructive conversations.

References

1. S.V. Iordanskii, *Sov. Phys. J.E.T.P.* **22**, 160 (1966).
2. N.B. Kopnin, G.E. Volovik, U. Parts, *Europhys. Lett.* **32**, 651–656 (1995).
3. M. Stone, "Iordanskii force and gravitational Aharonov–Bohm effect on a moving vortex". [cond-mat/9909313].
4. E.B. Sonin, *Sov. Phys. J.E.T.P.* **42**, 469 (1976).
5. E.B. Sonin, *Phys. Rev.* **B55**, 485 (1997) [cond-mat/9606099].
6. C.F. Barenghi, R.J. Donelly, W.F. Vinen, *J. Low. Temp. Phys.* **52**, 189 (1983).
7. D.J. Thouless, Ping Ao, Quian Niu *Phys. Rev. Letters* **76**, 3758–3760 (1996).
8. C. Wexler, D.J. Thouless, *Phys. Rev. B58*, 8897 (1998) [cond-mat/9804118].
9. A.F. Andreev, E.P. Bashkin, *Sov. Phys. J.E.T.P.,* **42**, 164–646 (1975).
10. B. Carter, I.M. Khalatnikov, *Rev. Math. Phys.* **6**, 277–305 (1994).
11. B. Carter, in Journées Relativistes 1979, ed. I. Moret–Bailly & C. Latremolière, 166–182 (Faculté des Sciences, Angers, 1979).
12. B. Carter, in *A random walk in Relativity and Cosmology* ed. N. Dadhich, J. Krishna Rao, J.V. Narlikar, C.V. Visveshwara, 48–62 (Wiley Eastern, Bombay, 1985).
13. V.V. Lebedev, I.M. Khalatnikov, *Sov. Phys. J.E.T.P.* **56**, 923–930 (1982).
14. B. Carter, I.M. Khalatnikov, *Phys. Rev.* **D45**, 4536–4544 (1992).
15. B. Carter, D. Langlois, "Relativistic models for superconducting superfluid mixtures", *Nuclear Phys.* **B531**, 478–504 (1998) [gr-qc/9806024].

11 Vortex Core Structure and Dynamics in Layered Superconductors

M. Eschrig, D. Rainer, and J. A. Sauls

Summary. We investigate the equilibrium and nonequilibrium properties of the core region of vortices in layered superconductors. We discuss the electronic structure of singly and doubly quantized vortices for both *s*-wave and *d*-wave pairing symmetry. We consider the intermediate clean regime, where the vortex-core bound states are broadened into resonances with a width comparable to or larger than the quantized energy level spacing, and calculate the response of a vortex core to an a.c. electromagnetic field for vortices that are pinned to a metallic defect. We concentrate on the case where the vortex motion is nonstationary and can be treated by linear response theory. The response of the order parameter, impurity self energy, induced fields and currents are obtained by a self-consistent calculation of the distribution functions and the excitation spectrum. We then obtain the dynamical conductivity, spatially resolved in the region of the core, for external frequencies in the range, $0.1\Delta < \hbar\omega \lesssim 3\Delta$. We also calculate the dynamically induced charge distribution in the vicinity of the core. This charge density is related to the nonequilibrium response of the bound states and collective mode, and dominates the electromagnetic response of the vortex core.

11.1 Introduction

Vortex motion is the principal mechanism for resistive losses in type II superconductors. Vortices also provide valuable information about the nature of low lying excitations in the superconducting state. In clean *s*-wave BCS superconductors the low-lying excitations in the core are the bound states of Caroli, de Gennes and Matricon [1]. These excitations have superconducting as well as normal metallic properties. For example, these states are the source of circulating supercurrents in the equilibrium vortex core, and they are strongly coupled to the condensate by Andreev scattering [2,3]. Furthermore, the response of the vortex core states to an electromagnetic field is generally very different from that of normal electrons. However, in the dirty limit, $\hbar/\tau \gg \Delta$, the the Bardeen–Stephen model [4] of a normal-metal spectrum with the local Drude conductivity in the core provides a reasonable description of the dissipative dynamics of the vortex core. The opposite extreme is the "superclean limit", $\hbar/\tau \ll \Delta^2/E_f$, in which the quantization of the vortex-core bound states must be taken into account. In this limit a single impurity and its interaction with the vortex core states must be considered.

The a.c. electromagnetic response is then controlled by selection rules governing transition matrix elements for the quantized core levels and the level structure of the core states in the presence of an impurity [5–7]. In the case of d-wave superconductors in the superclean limit, *nodes* in the spectrum of bound states lead to a finite dissipation from Landau damping for $T \to 0$ [8].

The superclean limit is difficult to achieve even for short coherence length superconductors; weak disorder broadens the vortex core levels into a quasi-continuum. We investigate the intermediate-clean regime, $\Delta^2/E_f \ll \hbar/\tau \ll \Delta$, where the discrete level structure of the vortex-core states is broadened and the selection rules are broken due to strong overlap between the bound state wave functions. However, the vortex core states remain well defined on the scale of the superconducting gap, Δ. In this regime we can take advantage of the power of the quasiclassical theory of nonequilibrium superconductivity [9–13].

The energy required to maintain a net charge density of the order of an elementary charge per particle within a coherence volume (or coherence area in two dimensions) is much larger than the condensation energy. Thus, charge accumulation in the vortex core is strongly suppressed. In order to reduce the Coulomb energy associated with the charge accumulation, an internal electrochemical potential, $\Phi(\boldsymbol{R};t)$, develops in response to an external electric field. This potential produces an internal electric field, $\boldsymbol{E}^{\text{int}}(\boldsymbol{R};t)$, which is the same order of magnitude as the external field. Even though the external field may vary on a scale that is large compared to the coherence length, ξ_0, the internal field develops on the coherence length scale. The source of the internal field is a charge density that accumulates inhomogeneously over length scales of the order of the coherence length. It is necessary to calculate the induced potential self consistently from the spatially varying order parameter, spectral function and distribution function for the electronic states in the vicinity of the vortex core. An order of magnitude estimate shows that to produce an induced field of the order of the external field, the dynamically induced charge is of the order of $e\,(\Delta/E_f)(\delta v_\omega/\Delta)$, where $\delta v_\omega \sim eE^{\text{ext}}/\xi_0\omega$ is the typical energy scale set by the strength of the external field. This charge density accumulates predominantly in the vortex core region and creates a dipolar field around the vortex core. For a pinned vortex the charge accumulates near the interface separating the metallic inclusion from the superconductor.

Disorder plays a central role in the dissipative dynamics of the mixed state of type II superconductors. Impurities and defects are a source of scattering that limits the mean free path of carriers, thus increasing the resistivity. Defects also provide 'pinning sites' that inhibit vortex motion and suppresses the flux-flow resistivity. However, for a.c. fields even pinned vortices are sources for dissipation. The magnitude and frequency dependence of this dissipation depends on the electronic structure and dynamics of the core states of the pinned vortex. In the analysis presented below we consider vortices in the presence of pinning centers. We model a pinning center as a normal metallic inclusion which is coupled to the electronic states of the superconductor through a highly transmitting interface. In this model the charge dynamics of the electronic states near the interface, between the pinning region and superconductor, plays an important role in the electromagnetic response of the core.

In the next section we provide a short summary of the nonequilibrium quasiclassical equations, including the transport equations for the quasiparticle distribution and spectral functions, constitutive equations for the order parameter, impurity self-energy and electromagnetic potentials. In Sect. 11.3 we present calculations for the the electronic structure of vortices for superconductors with both s-wave and d-wave pairing symmetry. The results are based on self-consistent calculations of the order parameter and impurity self energy. For s-wave superconductors, impurity scattering leads to inhomogeneous broadening of the vortex core bound states, as well as bands of impurity states within the gap. In the case of d-wave pairing the core states are further broadened by coupling between bound states and the continuum states through impurity scattering. We also discuss the structure of doubly quantized vortices and vortices bound to mesoscopic size metallic inclusions. In the case of the doubly quantized vortex there are two branches of zero-energy bound states centered at a finite impact parameter from the vortex center. This leads to a unique signature of a doubly quantized vortex: currents in the core circulate opposite to the supercurrents outside the core. Section 11.4 summarizes calculations of the vortex core dynamics for s-wave vortices in the presence of impurity scattering. We describe the charge dynamics of the vortex core for both pinned and unpinned vortices, and calculate the local a.c. conductivity that results from the coupled dynamics of the order parameter collective mode and the quasiparticle bound states in the vortex core. We discuss energy transport by the core states and the absorption features in the conductivity spectrum, which we interpret in terms of absorption within the bound-state band centered at the Fermi level and resonant transitions involving the bound and continuum states.

11.2 Nonequilibrium Transport Equations

The quasi-classical theory describes equilibrium and nonequilibrium properties of superconductors on length scales that are large compared to microscopic scales (i.e. the lattice constant, Fermi wavelength, k_f^{-1}, Thomas–Fermi screening length, etc.) and energies that are small compared to the atomic scales (e.g. Fermi energy, E_f, plasma frequency, conduction band width, etc.). Thus, there are small dimensionless parameters that define the limits of validity of the quasi-classical theory. In particular, we require $k_f \xi_0 \gg 1$, $k_B T_c / E_f \ll 1$ and $\hbar \omega \ll E_f$, where the a.c. frequencies of interest are typically of the order of $\Delta \sim T_c$, or smaller, and the length scales of interest are of the order of the coherence length, $\xi_0 = \hbar v_f / 2\pi k_B T_c$, or longer. Hereafter we use units in which $\hbar = k_B = 1$, and adopt the sign convention $e = -|e|$ for the electron charge.

In quasi-classical theory, quasiparticle wavepackets move along nearly straight, classical trajectories at the Fermi velocity. The classical dynamics of the quasiparticle excitations is governed by semi-classical transport equations for their phase-space distribution function. The quantum mechanical degrees of freedom are the "spin" and "particle-hole degree of freedom", described by 4×4 density matrices (Nambu matrices). The quantum dynamics is coupled to the classical dynamics of the quasiparticles in phase space through the matrix structure of the quasi-classical transport equations.

The nonequilibrium quasi-classical transport equations [9–13] are formulated in terms of a quasi-classical Nambu–Keldysh propagator $\check{g}(\mathbf{p}_f, \mathbf{R}; \varepsilon, t)$, which is a matrix in the combined Nambu–Keldysh space, and is a function of position $\mathbf{R}$, time t, energy ε, and momenta $\mathbf{p}_f$ on the Fermi surface.[1] We denote Nambu–Keldysh matrices by a "check", and their 4×4 Nambu submatrices of advanced (A), retarded (R) and Keldysh-type (K) propagators by a "hat". The Nambu–Keldysh matrices for the quasi-classical propagator and self-energy have the form,

$$\check{g} = \begin{pmatrix} \hat{g}^R & \hat{g}^K \\ 0 & \hat{g}^A \end{pmatrix}, \quad \check{\sigma} = \begin{pmatrix} \hat{\sigma}^R & \hat{\sigma}^K \\ 0 & \hat{\sigma}^A \end{pmatrix}, \quad (11.1)$$

where $\hat{g}^{R,A,K}$ are the retarded (R), advanced (A) and Keldysh (K) quasi-classical propagators, and similarly for the self-energy functions. Each of these components of $\check{g}$ and $\check{\sigma}$ are 4×4 Nambu matrices in combined particle-hole-spin space. For a review of the methods and an introduction to the notation we refer to [15,14,16]. In the compact Nambu–Keldysh notation the transport equations and the normalization conditions read

$$\left[(\varepsilon + \frac{e}{c}\mathbf{v}_f \cdot \mathbf{A})\check{\tau}_3 - eZ_0\Phi\check{1} - \check{\Delta}_{mf} - \check{\nu}_{mf} - \check{\sigma}_i, \check{g}\right]_\otimes + i\mathbf{v}_f \cdot \nabla \check{g} = 0, (11.2)$$

$$\check{g} \otimes \check{g} = -\pi^2 \check{1}, \quad (11.3)$$

where the commutator is $[\check{A}, \check{B}]_\otimes = \check{A} \otimes \check{B} - \check{B} \otimes \check{A}$,

$$\check{A} \otimes \check{B}(\varepsilon, t) = e^{\frac{i}{2}(\partial_\varepsilon^A \partial_t^B - \partial_t^A \partial_\varepsilon^B)} \check{A}(\varepsilon, t)\check{B}(\varepsilon, t). \quad (11.4)$$

The vector potential, $\mathbf{A}(\mathbf{R}; t)$, includes $\mathbf{A}_0(\mathbf{R})$ which generates the static magnetic field, $\mathbf{B}_0(\mathbf{R}) = \nabla \times \mathbf{A}_0(\mathbf{R})$, as well as the non-stationary vector potential describing the time-varying electromagnetic field; $\check{\Delta}_{mf}(\mathbf{p}_f, \mathbf{R}; t)$ is the mean-field order parameter matrix, $\check{\nu}_{mf}(\mathbf{p}_f, \mathbf{R}; t)$ describes diagonal mean fields due to quasiparticle interactions (Landau interactions), and $\check{\sigma}_i(\mathbf{p}_f, \mathbf{R}; \varepsilon, t)$ is the impurity self-energy. The electrochemical potential $\Phi(\mathbf{R}; t)$

---

[1] In quasi-classical theory the description in terms of the variables $(\varepsilon, \mathbf{p}_f, \mathbf{R})$ is related to the phase-space description in $\mathbf{p} - \mathbf{R}$ space by a transformation, $g(\mathbf{p}_f, \varepsilon; \mathbf{R}, t) = f(\mathbf{p}, \mathbf{R}; t)$, with $\varepsilon = \varepsilon(\mathbf{p}, \mathbf{R}; t) - \mu$ and $\hat{\mathbf{p}} = \hat{\mathbf{p}}_f$ [14].

includes the field generated by the induced charge density, $\rho(\boldsymbol{R};t)$. The coupling of quasiparticles to the external potential involves virtual high-energy processes, which result from polarization of the non-quasiparticle background. The interaction of quasiparticles with both the external potential Φ and the polarized background can be described by coupling to an effective potential $Z_0\Phi$ [14]. The high-energy renormalization factor Z_0 is defined below in (11.12). The coupling of the quasiparticle current to the vector potential in (11.2) is given in terms of the quasiparticle Fermi velocity. No additional renormalization is needed to account for the effective coupling of the charge current to the vector potential because the renormalization by the non-quasiparticle background is accounted for by the effective potentials that determine the band structure, and therefore the quasiparticle Fermi velocity.

11.2.1 Constitutive Equations

Equations (11.2–11.3) must be supplemented by Maxwell's equations for the electromagnetic potentials, and by self-consistency equations for the order parameter and the impurity self-energy. We use the weak-coupling gap equation to describe the superconducting state, including unconventional pairing. The mean field self energies are then given by,

$$\hat{\Delta}^{R,A}_{mf}(\boldsymbol{p}_f,\boldsymbol{R};t) = N_f \int_{-\varepsilon_c}^{+\varepsilon_c} \frac{d\varepsilon}{4\pi i} \langle V(\boldsymbol{p}_f,\boldsymbol{p}'_f) \hat{f}^K(\boldsymbol{p}'_f,\boldsymbol{R};\varepsilon,t) \rangle \,, \tag{11.5}$$

$$\hat{\nu}^{R,A}_{mf}(\boldsymbol{p}_f,\boldsymbol{R};t) = N_f \int_{-\varepsilon_c}^{+\varepsilon_c} \frac{d\varepsilon}{4\pi i} \langle A(\boldsymbol{p}_f,\boldsymbol{p}'_f) \hat{g}^K(\boldsymbol{p}'_f,\boldsymbol{R};\varepsilon,t) \rangle \,, \tag{11.6}$$

$$\hat{\Delta}^{K}_{mf}(\boldsymbol{p}_f,\boldsymbol{R};t) = 0 \,, \qquad \hat{\nu}^{K}_{mf}(\boldsymbol{p}_f,\boldsymbol{R};t) = 0 \,. \tag{11.7}$$

The impurity self-energy,

$$\check{\sigma}_i(\boldsymbol{p}_f,\boldsymbol{R};\varepsilon,t) = n_i \, \check{t}(\boldsymbol{p}_f,\boldsymbol{p}_f,\boldsymbol{R};\varepsilon,t) \,, \tag{11.8}$$

is specified by the impurity concentration, n_i, and impurity scattering t-matrix, which is obtained from the the self-consistent solution of the t-matrix equations,

$$\check{t}(\boldsymbol{p}_f,\boldsymbol{p}''_f,\boldsymbol{R};\varepsilon,t) = \check{u}(\boldsymbol{p}_f,\boldsymbol{p}''_f) + N_f \langle \check{u}(\boldsymbol{p}_f,\boldsymbol{p}'_f)$$
$$\otimes \check{g}(\boldsymbol{p}'_f,\boldsymbol{R};\varepsilon,t) \otimes \check{t}(\boldsymbol{p}'_f,\boldsymbol{p}''_f,\boldsymbol{R};\varepsilon,t) \rangle \,. \tag{11.9}$$

The Nambu matrix $\hat{f}^K$ is the off-diagonal part of $\hat{g}^K$, while $\hat{g}^K$ is the diagonal part in particle-hole space. The Fermi surface average is defined by

$$\langle \ldots \rangle = \frac{1}{N_f} \int \frac{d^2 \boldsymbol{p}'_f}{(2\pi)^3 \, |\boldsymbol{v}'_f|} (\ldots) \,, \quad N_f = \int \frac{d^2 \boldsymbol{p}'_f}{(2\pi)^3 \, |\boldsymbol{v}'_f|} \,, \tag{11.10}$$

where N_f is the average density of states on the Fermi surface. The other material parameters that enter the self-consistency equations are the dimensionless pairing interaction, $N_f V(\boldsymbol{p}_f, \boldsymbol{p}'_f)$, the dimensionless Landau interaction, $N_f A(\boldsymbol{p}_f, \boldsymbol{p}'_f)$, the impurity concentration, n_i, the impurity potential, $\breve{u}(\boldsymbol{p}_f, \boldsymbol{p}'_f)$, and the Fermi surface data: $\boldsymbol{p}_f$ (Fermi surface), $\boldsymbol{v}_f(\boldsymbol{p}_f)$ (Fermi velocity). We eliminate both the magnitude of the pairing interaction and the cut-off, ε_c, in favor of the transition temperature, T_c, using the linearized equilibrium form of the mean-field gap equation (11.5).

The quasi-classical equations are supplemented by constitutive equations for the charge density, the current density and the induced electromagnetic potentials. The formal result for the non-equilibrium charge density, to linear order in Δ/E_f, is given in terms of the Keldysh propagator by

$$\rho^{(1)}(\boldsymbol{R};t) = \mathrm{e}N_f \int_{-\varepsilon_c}^{+\varepsilon_c} \frac{d\varepsilon}{4\pi \mathrm{i}} \left\langle Z(\boldsymbol{p}'_f) \,\mathrm{Tr}\left[\hat{g}^{\mathrm{K}}(\boldsymbol{p}'_f, \boldsymbol{R}; \varepsilon, t)\right] \right\rangle$$
$$- 2\mathrm{e}^2 N_f Z_0 \Phi(\boldsymbol{R}; t) \,, \tag{11.11}$$

with the renormalization factors given by

$$Z(\boldsymbol{p}_f) = 1 - \langle A(\boldsymbol{p}'_f, \boldsymbol{p}_f) \rangle \,, \quad Z_0 = \langle Z(\boldsymbol{p}_f) \rangle \,. \tag{11.12}$$

The high-energy renormalization factor is related to an average of the scattering amplitude on the Fermi surface by a Ward identity that follows from the conservation law for charge [14]. The charge current induced by $\boldsymbol{A}(\boldsymbol{R};t)$, calculated to leading order in Δ/E_f, is also obtained from the Keldysh propagator,

$$\boldsymbol{j}^{(1)}(\boldsymbol{R};t) = \mathrm{e}N_f \int \frac{d\varepsilon}{4\pi \mathrm{i}} \mathrm{Tr} \langle \boldsymbol{v}_f(\boldsymbol{p}'_f) \hat{\tau}_3 \hat{g}^{\mathrm{K}}(\boldsymbol{p}'_f, \boldsymbol{R}; \varepsilon, t) \rangle \,. \tag{11.13}$$

There is no additional high-energy renormalization of the coupling to the vector potential because the quasiparticle Fermi velocity already includes the high-energy renormalization of the charge-current coupling in (11.13). Furthermore, the self-consistent solution of the quasi-classical equations for $\hat{g}^{\mathrm{K}}$ ensures the continuity equation for charge conservation,

$$\partial_t \, \rho^{(1)}(\boldsymbol{R};t) + \boldsymbol{\nabla} \cdot \boldsymbol{j}^{(1)}(\boldsymbol{R};t) = 0 \,, \tag{11.14}$$

is satisfied.

An estimate of the contribution to the charge density from the integral in (11.11) leads to the condition of "local charge neutrality" [17,18]. A charge density given by the elementary charge times the number of states within an energy interval Δ around the Fermi surface implies $\rho^{(1)} \sim 2\mathrm{e}N_f \Delta$. Such a charge density cannot be maintained within a coherence volume because of the cost in Coulomb energy. The Coulomb energy is suppressed by requiring the leading order contribution to the charge density vanish: i.e. $\rho^{(1)}(\boldsymbol{R};t) = 0$.

Thus, the spatially varying renormalized electro-chemical potential, $Z_0 \Phi$, is determined by

$$2eZ_0\Phi(\boldsymbol{R};t) = \int_{-\varepsilon_c}^{+\varepsilon_c} \frac{d\varepsilon}{4\pi \mathrm{i}} \mathrm{Tr} \langle Z(\boldsymbol{p}'_f) \hat{g}^{\mathrm{K}}(\boldsymbol{p}'_f, \boldsymbol{R}; \varepsilon, t) \rangle . \tag{11.15}$$

The continuity equation implies $\boldsymbol{\nabla} \cdot \boldsymbol{j}^{(1)}(\boldsymbol{R};t) = 0$. We discuss violations of the charge neutrality condition (11.15), which are of a higher order in Δ/E_f, in Sect. 11.4.1. Finally, Ampere's equation, with the current given by (11.13), determines the vector potential in the quasi-classical approximation,

$$\boldsymbol{\nabla} \times \boldsymbol{\nabla} \times \boldsymbol{A}(\boldsymbol{R};t) = \frac{8\pi e N_f}{c} \int \frac{d\varepsilon}{4\pi \mathrm{i}} \mathrm{Tr} \langle \boldsymbol{v}_f(\boldsymbol{p}'_f) \hat{\tau}_3 \hat{g}^{\mathrm{K}}(\boldsymbol{p}'_f, \boldsymbol{R}; \varepsilon, t) \rangle . \tag{11.16}$$

Equations (11.2–11.9) and (11.15–11.16) constitute a complete set of equations for calculating the electromagnetic response of vortices in the quasi-classical limit. For high-κ superconductors we can simplify the self-consistency calculations to some degree. Since quasiparticles couple to the vector potential via $\frac{e}{c} \boldsymbol{v}_f \cdot \boldsymbol{A}$, (11.16) shows that this quantity is of the order of $8\pi e^2 N_f v_f^2/c^2 = 1/\lambda^2$, where λ is the magnetic penetration depth. Thus, for $\kappa = \lambda/\xi_0 \gg 1$, as in the layered cuprates, the feedback effect of the current density on the vector potential is smaller by a factor of $1/\kappa^2$.

11.2.2 Linear Response

For sufficiently weak fields we can calculate the electromagnetic response to linear order in the external field. The propagator and the self-energies are separated into unperturbed equilibrium parts and terms that are first-order in the perturbation,

$$\check{g} = \check{g}_0 + \delta \check{g}, \quad \check{\Delta}_{mf} = \check{\Delta}_0 + \delta \check{\Delta}_{mf}, \quad \check{\sigma}_i = \check{\sigma}_0 + \delta \check{\sigma}_i , \tag{11.17}$$

and similarly for the electromagnetic potentials, $\boldsymbol{A} = \boldsymbol{A}_0 + \delta \boldsymbol{A}$, $\Phi = \delta\Phi$. The equilibrium propagators obey the matrix transport equation,

$$\left[(\varepsilon + \frac{e}{c} \boldsymbol{v}_f \cdot \boldsymbol{A}_0) \check{\tau}_3 - \check{\Delta}_0 - \check{\sigma}_0 , \check{g}_0 \right] + \mathrm{i} \boldsymbol{v}_f \cdot \boldsymbol{\nabla} \check{g}_0 = 0 . \tag{11.18}$$

These equations are supplemented by the self-consistency equations for the mean fields, (11.5–11.6), the impurity self energy, (11.8–11.9), the local charge-neutrality condition for the scalar potential, (11.15), Ampère's equation for the vector potential, (11.16), the equilibrium normalization conditions,

$$\check{g}_0^2 = -\pi^2 \check{1} , \tag{11.19}$$

and the equilibrium relation between the Keldysh function and equilibrium spectral density,

$$\hat{g}_0^{\mathrm{K}} = \tanh\left(\frac{\varepsilon}{2T}\right) \left[\hat{g}_0^R - \hat{g}_0^A \right] . \tag{11.20}$$

The first-order correction to the matrix propagator obeys the linearized transport equation,

$$\left[\left(\varepsilon + \frac{e}{c}\boldsymbol{v}_f \cdot \boldsymbol{A}_0\right)\check{\tau}_3 - \check{\Delta}_0 - \check{\sigma}_0, \delta\check{g}\right]_\otimes + i\boldsymbol{v}_f \cdot \boldsymbol{\nabla}\delta\check{g}$$
$$= [\delta\check{\Delta}_{mf} + \delta\check{\sigma}_i + \delta\check{v}, \check{g}_0]_\otimes, \quad (11.21)$$

with source terms on the right-hand side from both the external field ($\delta\check{v}$) and the internal fields ($\delta\check{\Delta}_{mf}$, $\delta\check{\sigma}_i$). In addition, the first-order propagator satisfies the "orthogonality condition",

$$\check{g}_0 \otimes \delta\check{g} + \delta\check{g} \otimes \check{g}_0 = 0. \quad (11.22)$$

obtained from linearizing the full normalization condition.[2] The system of linear equations are supplemented by the equilibrium and first-order self-consistency conditions for the order parameter,

$$\hat{\Delta}_0^{R,A}(\boldsymbol{p}_f, \boldsymbol{R}) = N_f \int_{-\varepsilon_c}^{+\varepsilon_c} \frac{d\varepsilon}{4\pi i} \langle V(\boldsymbol{p}_f, \boldsymbol{p}'_f) \hat{f}_0^K(\boldsymbol{p}'_f, \boldsymbol{R}; \varepsilon) \rangle, \quad (11.23)$$

$$\delta\hat{\Delta}_{mf}^{R,A}(\boldsymbol{p}_f, \boldsymbol{R}; t) = N_f \int_{-\varepsilon_c}^{+\varepsilon_c} \frac{d\varepsilon}{4\pi i} \langle V(\boldsymbol{p}_f, \boldsymbol{p}'_f) \delta\hat{f}^K(\boldsymbol{p}'_f, \boldsymbol{R}; \varepsilon, t) \rangle, \quad (11.24)$$

and the impurity self-energy,

$$\check{\sigma}_0(\boldsymbol{p}_f, \boldsymbol{R}; \varepsilon) = n_i \check{t}_0(\boldsymbol{p}_f, \boldsymbol{p}_f, \boldsymbol{R}; \varepsilon), \quad (11.25)$$

$$\check{t}_0(\boldsymbol{p}_f, \boldsymbol{p}''_f, \boldsymbol{R}; \varepsilon) = \check{u}(\boldsymbol{p}_f, \boldsymbol{p}''_f)$$
$$+ N_f \langle \check{u}(\boldsymbol{p}_f, \boldsymbol{p}'_f) \check{g}_0(\boldsymbol{p}'_f, \boldsymbol{R}; \varepsilon) \check{t}_0(\boldsymbol{p}'_f, \boldsymbol{p}''_f, \boldsymbol{R}; \varepsilon) \rangle, \quad (11.26)$$

$$\delta\check{\sigma}_i(\boldsymbol{p}_f, \boldsymbol{R}; \varepsilon, t) = n_i N_f \langle \check{t}_0(\boldsymbol{p}_f, \boldsymbol{p}'_f, \boldsymbol{R}; \varepsilon)$$
$$\otimes \delta\check{g}(\boldsymbol{p}'_f, \boldsymbol{R}; \varepsilon, t) \otimes \check{t}_0(\boldsymbol{p}'_f, \boldsymbol{p}_f, \boldsymbol{R}; \varepsilon) \rangle. \quad (11.27)$$

In general the diagonal mean fields also contribute to the response. However, we do not expect Landau interactions to lead to qualitatively new phenomena for the vortex dynamics, so we have neglected these interactions in the following analysis and set $A(\boldsymbol{p}_f, \boldsymbol{p}'_f) = 0$ (i.e. $\check{v}_{mf} = 0$). As a result the local charge neutrality condition for the electro-chemical potential becomes,

$$2e\delta\Phi(\boldsymbol{R}; t) = \int_{-\varepsilon_c}^{+\varepsilon_c} \frac{d\varepsilon}{4\pi i} \text{Tr} \langle \delta\hat{g}^K(\boldsymbol{p}_f, \boldsymbol{R}; \varepsilon, t) \rangle. \quad (11.28)$$

In what follows, we work in a gauge in which the induced electric field, $\boldsymbol{E}^{\text{ind}}(\boldsymbol{R}; t)$, is obtained from $\delta\Phi(\boldsymbol{R}; t)$ and the uniform external electric field, $\boldsymbol{E}_\omega^{\text{ext}}(t)$, is determined by the vector potential $\delta\boldsymbol{A}_\omega(t)$. For $\lambda/\xi_0 \gg 1$ we can

---
[2] Note that the convolution product between an equilibrium and a nonequilibrium quantity simplifies after Fourier transforming $t \to \omega$: $\check{A}_0 \otimes \delta\check{B}(\varepsilon, \omega) = \check{A}_0(\varepsilon + \omega/2)\check{B}(\varepsilon, \omega)$, $\check{B}(\varepsilon, \omega) \otimes \check{A}_0 = \check{B}(\varepsilon, \omega)\check{A}_0(\varepsilon - \omega/2)$.

safely neglect corrections to the vector potential due to the induced current. Thus, in the Nambu–Keldysh matrix notation the electromagnetic coupling to the quasiparticles is given by

$$\delta \check{v} = -\frac{e}{c} \bm{v}_f \cdot \delta \bm{A}_\omega(t) \check{\tau}_3 + e\delta\Phi(\bm{R};t) \check{1} \ . \tag{11.29}$$

The validity of linear response theory requires the external perturbation $\delta\check{v}$ be sufficiently small and that the induced vortex motion responds to the external field at the frequency set by the external field. At very low frequencies, frictional damping of the vortex motion, arising from the finite mean free path of quasiparticles scattering from impurities, gives rise to a nonlinear regime in the dynamical response of a vortex. This regime is discussed extensively in the literature [13], and is not subject of our study. However, for sufficiently small field strengths the vortex motion is nonstationary over any time interval, although it may be regarded as quasi-stationary at low enough frequencies. The nonstationary motion of the vortex can be described by linear response theory if $\delta\check{v} \ll 1/\tau$ for $\omega \lesssim 1/\tau$, and $\delta\check{v} \ll \omega$ for $\omega \gtrsim 1/\tau$. Note that the frequency of the perturbation, ω, is not required to be small compared to the gap frequency; it is only restricted to be small compared to atomic scale frequencies, e.g $\omega \ll E_f/\hbar$.

Self-consistent solutions of (11.24), (11.27) and (11.28) for the self-energies and scalar potential are fundamental to obtaining a physically sensible solution for the electromagnetic response. The dynamical self-energy corrections are equivalent to "vertex corrections" in the Kubo formulation of linear response theory. They are particularly important in the context of nonequilibrium phenomena in inhomogeneous superconductors.[3] In our case these corrections are of vital importance; the self-consistency conditions enforce charge conservation. In particular, (11.25–11.27) imply charge conservation in scattering processes, whereas (11.23) and (11.24) imply charge conservation in particle–hole conversion processes; any charge which is lost (gained) in a particle–hole conversion process is compensated by a corresponding gain (loss) of condensate charge. It is the coupled quasiparticle and condensate dynamics which conserves charge in superconductors. Neglecting the dynamics of either component, or using a non-conserving approximation for the coupling leads to unphysical results.

Self-consistent calculations for the equilibrium order parameter, impurity self-energy and local excitation spectrum (spectral density) are necessary inputs to the linearized transport equations for the dynamical response of a vortex. The equilibrium spectral function also provides key information for the interpretation of the dynamical response. Because of particle–hole coherence, the spectral density is sensitive to the phase winding and symmetry

---

[3] Vertex corrections usually vanish in homogeneous superconductors because of translational and rotational symmetries. Inhomogeneous states break these symmetries and typically generate non-vanishing vertex corrections.

of the order parameter, as well as material properties such as the transport mean-free path and impurity cross-section. In the following section we present results for the low-energy excitation spectra of singly- and doubly-quantized vortices in layered superconductors with s-wave and d-wave pairing symmetry.

11.3 Electronic Structure of Vortices

The local density of states for excitations with Fermi momentum $\boldsymbol{p}_f$ is obtained from the retarded and advanced quasi-classical propagators,

$$N(\boldsymbol{p}_f, \boldsymbol{R}; \varepsilon) = N_f \frac{1}{4\pi\mathrm{i}} \mathrm{Tr} \left[\hat{\tau}_3 \hat{g}_0^A(\boldsymbol{p}_f, \boldsymbol{R}; \varepsilon) - \hat{\tau}_3 \hat{g}_0^R(\boldsymbol{p}_f, \boldsymbol{R}; \varepsilon) \right] . \tag{11.30}$$

This function measures the local density of the quasiparticle states with energy ε at the point $\boldsymbol{p}_f$ on the Fermi surface. The local density of states (LDOS) is obtained by averaging this quantity over all momentum directions of the quasiparticles,

$$N(\boldsymbol{R}; \varepsilon) = \left\langle N(\boldsymbol{p}'_f, \boldsymbol{R}; \varepsilon) \right\rangle . \tag{11.31}$$

The product of the angle-resolved density of states and the Fermi velocity, $\boldsymbol{v}_f$, determines the current density carried by these states; $\boldsymbol{v}_f$ also defines the direction of a quasi-classical trajectory passing through the space point $\boldsymbol{R}$. We introduce the angle-resolved *spectral current density* [3],

$$\boldsymbol{j}(\boldsymbol{p}_f, \boldsymbol{R}; \varepsilon) = 2e\boldsymbol{v}_f(\boldsymbol{p}_f) N(\boldsymbol{p}_f, \boldsymbol{R}; \varepsilon), \tag{11.32}$$

which measures the current density carried by quasiparticle states with energy ε moving along the trajectory defined by $\boldsymbol{v}_f$. The local spectral current density is then defined as

$$\boldsymbol{j}(\boldsymbol{R}; \varepsilon) = \left\langle \boldsymbol{j}(\boldsymbol{p}'_f, \boldsymbol{R}; \varepsilon) \right\rangle , \tag{11.33}$$

and the total current density is obtained by summing over the occupied states for each trajectory,

$$\boldsymbol{j}(\boldsymbol{R}) = \int d\varepsilon \, f(\varepsilon) \, \boldsymbol{j}(\boldsymbol{R}; \varepsilon), \tag{11.34}$$

where $f(\varepsilon) = 1/(1+e^{\beta\varepsilon})$. Self-consistent calculations of the equilibrium structure and spectral properties of vortices are relatively straight-forward computations. Below we present results for s-wave and d-wave pairing symmetry with impurity scattering included.

The calculations reported are carried for a circular Fermi surface, with an isotropic Fermi momentum $\boldsymbol{p}_f$ and Fermi velocity $\boldsymbol{v}_f$. The elastic scattering rate is chosen to represent the intermediate-clean regime, $\Delta^2/E_f < \hbar/\tau \ll \Delta$. The pairing potential can be represented as a sum over invariant products of basis functions $\{\eta_{\Gamma,i}(\mathbf{p}_f)|i = 1\ldots d_\Gamma\}$ for the irreducible representations of the crystal point group labeled by Γ,

$$N_f V(\mathbf{p}_f, \mathbf{p}'_f) = \sum_{\Gamma,i} v_\Gamma \eta^*_{\Gamma i}(\mathbf{p}_f) \eta_{\Gamma i}(\mathbf{p}'_f). \qquad (11.35)$$

The pairing interaction, v_Γ, and the cutoff, ε_c, are eliminated in favor of the instability temperature for pairing in symmetry channel Γ. We limit the discussion here to even-parity, one-dimensional representations, which for tetragonal symmetry includes the 's-wave' (identity) representation, A_{1g}, two 'd-wave' representations, B_{1g} and B_{2g}, and a 'g-wave' representation, A_{2g}. The corresponding basis functions we use are listed in Table 11.1.

Table 11.1. Symmetry classes and model basis functions for the 1D even-parity representations of D_{4h}. The angle ϕ is the angular position of $\mathbf{p}_f$ on the Fermi surface with respect to the crystallographic a-axis ($= x$-axis).

Pairing Symmetry	Representation [Γ]	Basis Function [η_Γ]
s-wave	A_{1g}	1
d-wave	B_{1g}	$\sqrt{2}\cos(2\phi)$
d'-wave	B_{2g}	$\sqrt{2}\sin(2\phi)$
g-wave	A_{2g}	$\sqrt{2}\sin(4\phi)$

The results for the order parameter, impurity self energy and spectral properties of vortices that follow are calculated self consistently in the t-matrix approximation for point impurities (pure s-wave scattering), i.e. $\breve{u}(\mathbf{p}_f, \mathbf{p}'_f) = u_0 \breve{1}$. The quasiparticle scattering rate, $1/2\tau$, and normalized impurity cross section, $\bar{\sigma}$, are then given by,

$$\frac{1}{2\tau} = \frac{n_i}{\pi N_f} \bar{\sigma}, \quad \bar{\sigma} = \frac{(\pi N_f u_0)^2}{1 + (\pi N_f u_0)^2}. \qquad (11.36)$$

The Born limit corresponds to $\bar{\sigma} \ll 1$, while unitary scattering corresponds to $|u_0| \to \infty$ or $\bar{\sigma} \to 1$. In the calculations that follow the temperature is set at $T = 0.3 T_c$, and the mean free path is is chosen to represent the intermediate-clean regime; $\ell = 10\,\xi_0$.

11.3.1 Singly Quantized Vortices for S-Wave Pairing

For isotropic s-wave pairing the equilibrium order parameter for an isolated vortex with winding number p has the form,

$$\Delta(\mathbf{p}_f, \mathbf{R}) = |\Delta(\mathbf{R})| e^{ip\varphi}, \qquad (11.37)$$

where the amplitude $|\Delta(\boldsymbol{R})|$ is isotropic and φ is the azimuthal angle of $\boldsymbol{R}$ in the plane.

The angle-resolved local density of states spectra for a singly quantized vortex is shown in Fig. 11.1 for space points $\boldsymbol{R} = (0, y)$ along the y-axis with a spacing of $\delta y = \sqrt{3}\pi/4\xi_0 \simeq 1.36\xi_0$, and for trajectories parallel to the x-axis, $\boldsymbol{v}_f = v_f \boldsymbol{e}_x$. The vortex center is at $y = 0$, and the phase winding is such that the direction of the superflow is in x-direction for spectra with negative y coordinate.

Far outside the core the angle-resolved density of states resembles the BCS density of states with a gap in the spectrum roughly between $\varepsilon = \pm\Delta$, and peaks in the spectrum near the continuum edges. Careful inspection of Fig. 11.1 shows that the coherence peak for positive energy at $y = -8\delta y \simeq -11\xi_0$ is not at $\varepsilon = \Delta$, but is shifted to higher energy by the Doppler effect, $\Delta\varepsilon = \boldsymbol{v}_f \cdot \boldsymbol{p}_s$, where $\boldsymbol{p}_s = \frac{1}{2}\hbar\boldsymbol{\nabla}\vartheta - \frac{e}{c}\boldsymbol{A}$ is the condensate momentum [3]. In a homogeneous superflow field the spectrum is the Doppler-shifted BCS spectrum; the Doppler shift increases with condensate momentum until pair-breaking sets in at the bulk critical momentum, $p_c = \Delta/v_f$. However, in the vortex core *nonlocal effects* associated with the inhomogeneous flow field lead to a redistribution of the spectral weight near the gap edge. The positive energy continuum edge is broadened considerably compared to the square-root singularity in the absence of the Doppler effect. The continuum starts at $+\Delta$ even for the Doppler-shifted spectra near the maximal current regions. In contrast, the negative energy continuum edge shows sharp structures due to the accumulation of spectral weight in the region between $-\Delta$ and $-\Delta + \boldsymbol{v}_f \cdot \boldsymbol{p}_s$. The sharp structure corresponds to a bound state that is separated from the continuum edge. The density of states at the continuum edge drops precipitously at $\varepsilon = -\Delta$. We emphasize that nonlocal effects lead to qualitative differences in the spectrum near the gap edges compared to the widely used approximation of Doppler-shifted quasiparticles in a locally homogeneous superflow [19]. In clean superconductors nonlocal effects dominate the spectrum.

For a homogeneous superflow the current is carried mainly by the states that are Doppler shifted the region between $-\Delta$ and Δ. The spectral current density shows that contributions to the current density from states outside of this region nearly cancel. In the case of a vortex, the bound state that splits off from the continuum not only robs the Doppler-shifted continuum edge of its spectral weight, but the bound state also carries most of the supercurrent in the vortex core region. At distances approaching the vortex center the bound state is clearly resolved and disperses through $\varepsilon = 0$ at zero impact parameter. As shown in the left panel of Fig. 11.1, the bound state also broadens considerably for Born scattering as it disperses towards the Fermi level, but remains a sharp resonance in the limit of unitary scattering, as shown in Fig. 11.1.

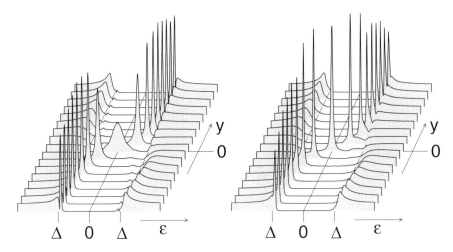

Fig. 11.1. Angle-resolved density of states for an s-wave vortex for quasiparticles propagating along trajectories parallel to the x-direction, at points along the y-axis spaced by $1.38\,\xi_0$. The left panel is for Born scattering, and the right panel is for unitary scattering with the same mean free path of $\ell = 10\xi_0$. The temperature is $T = 0.3 T_{\rm c}$.

The coherence peaks are completely suppressed at the vortex center, both in the Born and unitary limits. Note the difference in the evolution of the spectral weight for the bound and continuum states: for positive energies the spectral weight of the coherence peak is shifted to higher energies as one approaches the vortex core, with a continuously diminishing intensity at the coherence peak. In contrast, the spectral weight of the negative energy coherence peak is transferred to the bound state which splits off from the continuum. In principle there could be additional, secondary bound states, which would split off from the continuum if the vortex core were wider. However, a self-consistent calculation of the order parameter suppresses the secondary bound state in the vortex core. We return to the discussion of secondary bound states when we discuss the spectrum of pinned vortices. For trajectories with $y > 0$ the structure of the spectrum for positive and negative energies is reversed because the superflow is now counter to the quasiparticle velocity, leading to negative Doppler shifts and bound states at positive energies below the continuum edge. The small spectral features that appear at the energies corresponding to the *negative* of the bound state energies are due to mixing of trajectories of opposite velocity by backscattering from impurities.

Finally, consider the differences in the spectral features for unitary versus Born scattering. In Fig. 11.2 we show the local density of states at the center of the vortex for several scattering cross sections ranging from the Born limit ($\bar{\sigma} \ll 1$) to the unitary limit ($\bar{\sigma} = 1$). In the Born limit the broadening is a maximum. For a small, but finite cross section, two bands of impurity bound

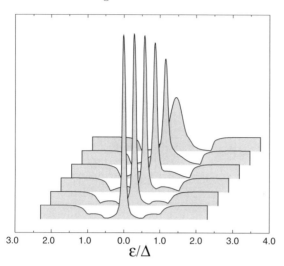

Fig. 11.2. Local density of states (LDOS) in the center of an s-wave vortex for different effective impurity scattering cross sections $\bar{\sigma}$ from Born to unitary limit, with a fixed mean free path $\ell = 10\xi_0$, and temperature $T = 0.3T_c$. From top to bottom: $\bar{\sigma} = 0.0, 0.2, 0.4, 0.6, 0.8, 1.0$.

states split off from the zero-energy resonance and remove spectral weight from the central peak. The impurity bands evolve towards the continuum edges as the cross-section increases and merge with them in the unitary limit. When the impurity bound-state bands no longer overlap the central peak, the zero-energy resonance sharpens dramatically with the width remaining constant as the unitary limit is approached. The overlap between the impurity bands and the continuum edges in the unitary limit is determined by the scattering rate, and is increasing with increasing scattering rate, $1/\tau$. As shown in the right panel of Fig. 11.1, the impurity bands are localized in the vortex core region; their existence depends on impurity scattering in a region where the phase changes rapidly over length scales of the order of the coherence length.

11.3.2 Singly Quantized Vortices for D-Wave Pairing

For d-wave pairing symmetry the order parameter has the form,

$$\Delta(\mathbf{p}_f, \mathbf{R}) = \eta_{B_{1g}}(\mathbf{p}_f)\,\Delta(\mathbf{R}), \tag{11.38}$$

where $\eta_{B_{1g}}(\mathbf{p}_f)$ changes sign on the Fermi surface at the points, $\hat{\mathbf{p}}_{fx} = \pm\hat{\mathbf{p}}_{fy}$. These nodal points lead to strong anisotropy and gapless excitations in the quasiparticle spectrum, which feeds back to produce anisotropy in the spatial structure of the order parameter in the core region of of vortex. The spatial part of the order parameter,

$$\Delta(\mathbf{R}) = |\Delta(\mathbf{R})|\,e^{i\vartheta(\mathbf{R})}, \tag{11.39}$$

for a vortex, at distances far from the core, approaches that of an isotropic vortex: $|\Delta(\mathbf{R})| \to \Delta(T)$ and $\vartheta \to p\varphi$. However, in the core region the current density is comparable to the critical current density and develops a four-fold anisotropy as a result of the backflow current concentrated near the nodes [20]. This current-induced pairbreaking effect is dominant for flow parallel to the nodal directions and leads to weak anisotropy of the order parameter in the core region.

The electronic structure of the d-wave vortex, presented in Fig. 11.3, shows distinct differences from that of an s-wave vortex. The resonances in the vortex core are broader than the core states of vortex in an s-wave superconductor. Mixing with extended states in the nodal direction broadens the peaks near the continuum edges. Also note the effect of impurity scattering on the spectra far from the vortex core region, which show a broadened continuum rather than a sharp continuum edge. In the right panel of Fig. 11.3 we show the angle-resolved density of states for a d-wave vortex for unitary impurity scattering. As in the s-wave case there is sharpening of the zero resonance, but note that the width of the resonance is much broader than in the case of s-wave pairing because impurity scattering provides a coupling and mixing of the bound state resonance with the low-energy extended states for momenta near the nodal directions, even for a trajectory in the antinodal direction.

In Fig. 11.4 we show the local density of states (LDOS) for both s-wave (top two panels) and d-wave pairing (bottom two panels) as a function of position $\mathbf{R}$ along an anti-nodal direction (bottom left) and along a nodal direction (bottom right). The LDOS is obtained by averaging the angle-resolved density of states over the Fermi surface at a particular space point $\mathbf{R}$. The averaging, together with the dispersion of the bound states and resonances as a function of angle, leads to one-dimensional bands with characteristic

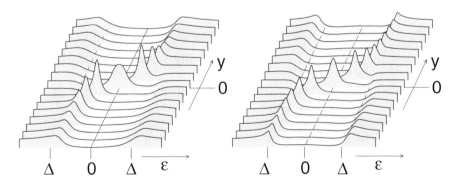

Fig. 11.3. Angle resolved density of states for a d-wave vortex for a quasiparticle trajectory along the x-direction (antinodal) as a function of impact parameter (y-direction). The impact parameter spacing is $\delta y = 1.36 \xi_0$. The left panel corresponds to impurity scattering in the Born limit, and the right panel is for the unitary limit. The temperature is $T = 0.3 T_c$.

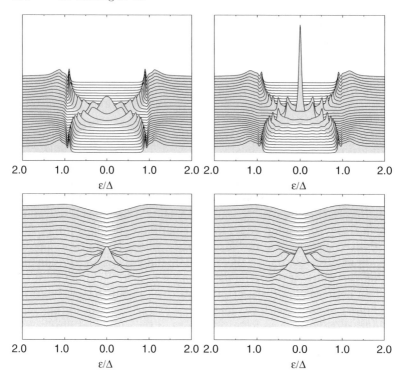

Fig. 11.4. Local density of states for an s-wave vortex (top panels) and a d-wave vortex (bottom panels), as a function of distance from the vortex center with a spacing $\delta y = 0.79\xi_0$. For s-wave pairing the Born limit is shown on the left and unitary scattering is on the right. For d-wave pairing the LDOS is shown only for Born scattering, but along axes parallel to a anti-nodal direction (left) and parallel to the nodal (right). The temperature is $T = 0.3T_c$.

Van Hove singularities, which are clearly visible for s-wave symmetry, but considerably smeared and broadened for d-wave symmetry. The LDOS for a vortex with s-wave pairing symmetry is shown for impurity scattering in both the Born (top left) and unitary limits (top right). The spectra show the characteristic bound state bands, Van Hove singularities and the dramatic reduction in the width of the zero bias resonance for unitary scattering.

Calculations of the LDOS in the superclean limit for a vortex with d-wave pairing are discussed by Schopohl and Maki [21] and by Ichioka et al. [22]. The self-consistent calculations shown in Fig. 11.4 include impurity scattering in the Born limit for space points, $\boldsymbol{R}$, along two different directions; the left panel corresponds the LDOS measured as a function of distance along the anti-nodal direction and the right panel is the LDOS measured along the nodal direction. The nodes of the order parameter for d-wave pairing lead to continuum states with energies down to the Fermi level. These states are visible in the LDOS as the smooth background extending to zero energy from both positive and negative energies, even for distances far from the core. In the vortex core region several broad peaks disperse as a function of distance from the vortex center. These peaks correspond to broadened Van Hove singularities resulting from averaging the vortex core resonances over the Fermi surface for at a fixed position $\boldsymbol{R}$. The differences in the d-wave spectra for the two directions reflects the weak fourfold anisotropy of the LDOS around the vortex at fixed energy.

11.3.3 Vortices Pinned to Mesoscopic Metallic Inclusions

The calculations discussed above describe the average effects of atomic scale impurity disorder on the spectral properties of a vortex. A specific defect can also act as a pinning site for a vortex. We model such a defect as a mesoscopic size, normal metallic inclusion in the superconductor. The defect is assumed to be a circular inclusion, with a radius, ξ_{pin}, of the order of the coherence length of the superconductor. For simplicity we assume that the metallic properties of the inclusion, e.g. Fermi surface parameters, are the same as those of the normal-state of the host superconductor. The calculations presented below neglect normal reflection processes at the interface between the inclusion and the host metal, but include Andreev reflection. The analysis and calculations can be generalized for more detailed models of a pinning center.

Figures 11.5 and 11.6 show the angle-resolved local density of states and the LDOS for a vortex pinned on a metallic inclusion of radius $\xi_{\text{pin}} = \pi\xi_0$. The order parameter, impurity self energy and spectral densities were calculated self-consistently for impurity scattering in the Born limit. Qualitative changes resulting from the inclusion occur inside the pinning center. The shape of the bound state resonance lines are asymmetric in energy. The asymmetry arises from multiple Andreev reflection processes at the interface between the pinning center and the superconductor, which leads to additional bands of resonances that overlap the vortex core resonances.

In addition to the asymmetry in the linewidth of the resonances, the zero-energy bound state at the vortex center has a peculiar spectral shape, shown in more detail in Fig. 11.7 for s-wave pairing symmetry, but also visible in the right panel of Fig. 11.6 for d-wave pairing symmetry as well. In contrast to the spectra for a vortex without a pinning center, the coherence peaks at the continuum edges are present at the vortex center.

Figure 11.7 shows the peaks near the continuum edges in the vortex center for a pinned vortex. The spectral weight near the continuum edge is taken from the resonance at the Fermi level. The enhanced weight at the continuum edges is a precursor to the formation of a secondary bound state that splits off from the continuum. This can be seen in the evolution of the spectral density at the vortex center as a function of the radius of the pinning center. There is a zero-energy resonance for all pinning radii, however, increasing the

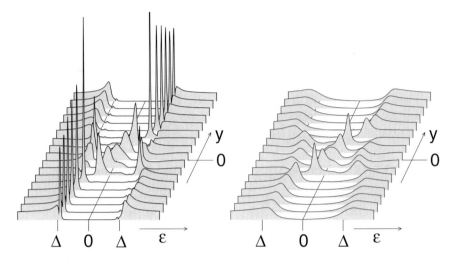

Fig. 11.5. Angle resolved density of states for a quasiparticle trajectory along the x-direction, as a function of impact parameter (y-axis) with a spacing of $\delta y = 1.36\xi_0$ for a vortex centered on a pinning center of radius of $\xi_{\rm pin} = \pi\xi_0$. The left panel is for s-wave symmetry and right panel is for d-wave symmetry. Impurity scattering is included for Born scattering with $\ell = 10\xi_0$. The temperature is $T = 0.3T_{\rm c}$.

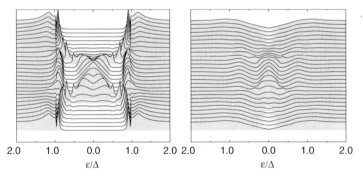

Fig. 11.6. Local density of states as a function of distance from the vortex center, with a spacing of $\delta y = 0.79\xi_0$, for a vortex pinned to an inclusion of radius of $\xi_p = \pi\xi_0$. The left panel corresponds to s-wave symmetry and the right panel is for d-wave symmetry along the anti-nodal direction. The temperature is $T = 0.3T_{\rm c}$.

radius of the pinning center transfers spectral weight from the zero-energy resonance to the continuum edge. A coherence peak develops, splits off from the continuum edge, strengthens and evolves to energies within the gap as the pinning radius changes from $\xi_{\rm pin} = 0.79\xi_0$ to $\xi_{\rm pin} = 4.71\xi_0$. Thus, the appearance of the coherence peak for pinning centers the size of a coherence length or so is a precursor to the formation of a secondary bound state within the gap. The spectral weight comes at the expense of states just above the

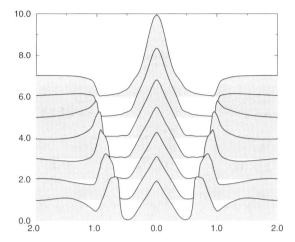

Fig. 11.7. Local density of states at the center of a vortex pinned in the center of a normal inclusion. The spectra (from top to bottom) correspond to different pinning radii: $\xi_{pin} = 0$, $0.79\xi_0$, $1.57\xi_0$, $2.36\xi_0$, $3.14\xi_0$, $3.93\xi_0$, and $4.71\xi_0$. The temperature is $T = 0.3T_c$.

continuum edge and the zero-energy bound state, which is diminished in intensity with increasing pinning radius.

11.3.4 Doubly Quantized Vortices

Vortices with winding numbers larger than $p = 1$ generally have line energies per unit winding number that are larger than that of a singly quantized vortex.[4] Nevertheless, doubly quantized vortices once formed are metastable and in principle it should be possible to observe the rare doubly quantized vortex using an atomic probe such as a scanning tunneling microscopic [25,26]. The spectrum of a doubly quantized vortex differs in a fundamental way from that of a singly quantized vortex. The singly quantized vortex, has a single branch of states that disperse through zero energy at the vortex center. The zero mode is guaranteed in the quasi-classical limit by π change along trajectories that pass through the vortex center. In contrast, there is no phase change along a trajectory through the vortex center for a doubly quantized vortex, and thus no topological requirement enforcing a zero energy bound state at the center of a doubly quantized vortex. Nevertheless, there is a spectrum of bound states in the cores of doubly quantized vortices which lead to characteristic structures in the LDOS and current spectral density of a doubly quantized vortex.

---

[4] There are counter examples for unconventional pairing with a multi-component order parameter in which the lowest energy vortex states are doubly quantized vortices [23,24].

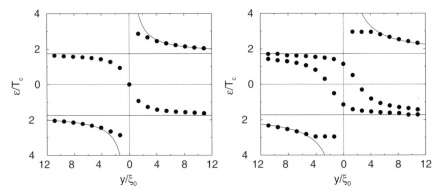

Fig. 11.8. Dispersion of the bound states (below the continuum edge) and of the coherence peaks (above the continuum edge) for quasiparticle trajectories in x-direction as a function of y, for a singly quantized vortex (left) and for a doubly quantized vortex (right). The center of the vortex is at $y = 0$. Born impurity scattering with a mean free path of $\ell = 10\xi_0$ is assumed. The temperature is $T = 0.3T_c$. The thin lines show the energies of the continuum edges and the Doppler shifted energies of the coherence peaks, assuming the London form for the condensate momentum $\boldsymbol{p}_s = p\boldsymbol{e}_\phi/2r$, where p is the winding number. The dispersion is shown for s-wave pairing; the corresponding data for d-wave pairing is similar.

This structure was discussed for a doubly quantized vortex in the superclean limit in [3]. Figure 11.9 shows the angle-resolved density of states for a doubly quantized vortex in an s-wave superconductor for trajectories parallel to the x-axis at different impact parameters along the y-direction. *Two* branches of vortex bound states cross the Fermi level at distances of the order of a coherence length from the vortex center. Thus, zero-energy bound states exist in the core but they are localized (for s-wave pairing symmetry) on trajectories at finite impact parameter from the vortex center. The locus of these trajectories forms a circle of radius $r_{bs} \simeq 2.5\xi_0$ around the vortex center. Also note that the Doppler shift of the continuum spectrum is twice that for singly quantized vortices.

This can be seen by comparing the spectra far from the core in Figs. 11.9 and 11.1. The distinctive features in the spectrum of a doubly quantized vortex are present for either s- and d-wave pairing symmetry. The doubling of bound state branches that cross the Fermi level is observed for both pairing symmetries. The main difference in the spectra, as in the case of singly quantized vortices, is in the width of the resonances.

The most distinguishing feature of doubly quantized vortices is that the supercurrents near the center of the vortex flow *counter* to the asymptotic superflow associated with the phase winding around the vortex [3]. The counter circulating currents in the core, shown in the right panel of Fig. 11.11, are due to the bound states interior to the radius defined by the zero-energy bound state (Note that the zero-energy bound state itself does not carry current).

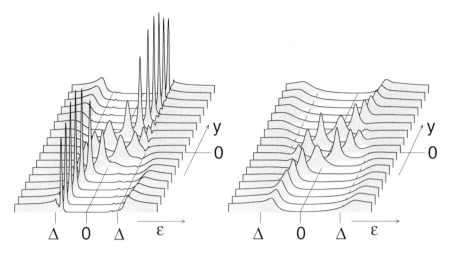

Fig. 11.9. Angle resolved density of states for a doubly quantized vortex along quasiparticle trajectories parallel to the x-direction, as a function of impact parameter, y, with a spacing $\delta y = 1.36\xi_0$. The left panel corresponds to s-wave symmetry and the right panel is for d-wave symmetry. Impurity scattering is included self-consistently in the Born limit with a mean free path $\ell = 10\xi_0$. The temperature is $T = 0.3T_c$.

This structure is revealed in Figs. 11.10 and 11.11, which show the LDOS and spectral current density for a doubly quantized vortex with s-wave and d-wave pairing symmetry. The left panel of Fig. 11.11 also shows the cumulative spectral current density and the reversal of the current for as one branch of bound states disperses below the Fermi level for $r < r_{\text{bs}}$.

These spectra show that the current density near the vortex core is carried mainly by the bound states, and that the reversal of the current direction near the vortex center is due to the branch of counter-flowing bound states dispersing below the Fermi level for impact parameter $r < r_{\text{bs}}$. In this region of the core the co-moving bound state is above the Fermi level so the asymmetry in the level occupation produces a counter flowing current.

11.4 Nonequilibrium Response

The dynamics of the electronic excitations of the vortex core play a key role in the dissipative processes in type II superconductors. Except in the dirty limit, $\ell \ll \xi_0$, the response of the core states to an electromagnetic field is generally very different from that of normal electrons. It is energetically unfavorable to maintain a charge density of the order of an elementary charge over a region with diameter of the order of the coherence length. Instead, an electrochemical potential is induced which ensures that almost no net charge accumulates in the core region. However, a dipolar-like charge distribution develops which

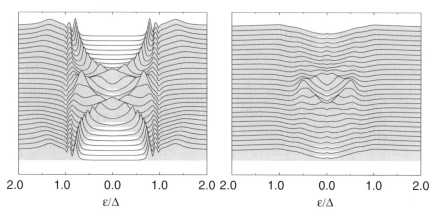

Fig. 11.10. Local density of states for doubly quantized vortices along trajectories parallel to the x-axis as a function of impact parameter along the y-axis with a spacing of $\delta y = 1.36\xi_0$. The left panel corresponds to s-wave pairing symmetry, and the right panel is for d-wave pairing along the anti-nodal direction. Impurity scattering is included in the Born limit with a mean free path of $\ell = 10\xi_0$ and the temperature is $T = 0.3T_c$.

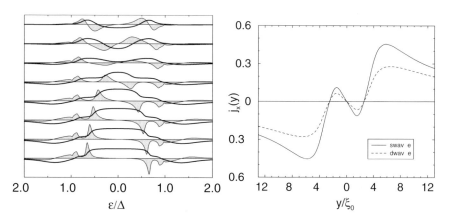

Fig. 11.11. The spectral current density for a doubly quantized vortex is shown along a trajectory parallel to the x-axis as a function of impact parameter (y-axis) with a spacing $\delta y = 1.36\xi_0$, starting at $y = 1.36\xi_0$. The thick curve is the cumulative spectral current density as a function of ε, obtained by integrating $j_x(\varepsilon; y)$ from $-\infty$ to ε. The right panel is the x-component of the current density as a function of impact parameter for s-wave (solid line) and d-wave (dashed line) pairing.

generates an internal electric field in the core. The internal field varies on the scale of the coherence length. This leads to a nonlocal response of the quasiparticles to the total electric field, even when the applied field varies on a much longer length scale and can be considered to be homogeneous. The dynamical response of the vortex core includes the collective mode of the inhomogeneous order parameter. This mode couples to the electro-chemical potential, $\delta\Phi$, in the vortex core region. This potential is generated by the charge dynamics of vortex core states and gives rise to internal electric fields which in turn drive the current density and the order parameter near the vortex core region. The induced electric fields in the core are the same order of magnitude as the external field. The dynamics of the core states are strongly coupled to the charge current and collective mode of the order parameter. Thus, the determination of the induced order parameter, as well as the spectrum and distribution function for the core states and non-equilibrium impurity scattering processes, requires dynamical self-consistency. Numerous calculations of the a.c. response neglect the self-consistent coupling of the collective mode and the spectral dynamics, or concentrate on the $\omega \to 0$ limit [27–30]. Quasi-classical theory is the only formulation of the theory of nonequilibrium superconductivity presently capable of describing the nonlocal response of the order parameter and quasiparticle dynamics in the presence of mesoscopic inhomogeneities and disorder. The numerical solution to the self-consistency problem was presented for unpinned vortices in [31]. Here we report results for the the electromagnetic response of isolated vortices bound to a pinning center in a superconductor with s-wave pairing symmetry.

11.4.1 Dynamical Charge Response

The charge dynamics of layered superconductors has two distinct origins. The c-axis dynamics is determined by the Josephson coupling between the conducting planes. Here we are concerned with the in-plane electrodynamics associated with the response of the order parameter and quasiparticle states bound to the vortex core. We assume strong Josephson coupling between different layers, and neglect variations of the response between different layers. This requires that the polarization of the electric field be in-plane, so that there is no coupling of the in-plane dynamics to the Josephson plasma modes. The external electromagnetic field is assumed to be long wavelength compared to the size of the vortex core, $\lambda_{\mathrm{EM}} \gg \xi_0$. In this limit we can assume the a.c. electric field to be uniform and described by a vector potential, $\mathbf{E}_\omega(t) = -\frac{1}{c}\partial_t \mathbf{A}_\omega$. We can also neglect the response to the a.c. magnetic field in the limit $\lambda \gg \xi_0$. In this case the spatial variation of the induced electric field occurs mainly within each conducting layer on the scale of the coherence length, ξ_0. Poisson's equation implies that induced charge densities are of the order of $\delta\Phi/\xi_0^2$, where $\delta\Phi$ is the induced electrochemical potential in the core. This leads to a dynamical charge of the order of $e(\Delta/E_f)$ in the vortex core. Once the electrochemical potential is calculated from (11.28) we can calculate the charge density fluctuations of the order of $(\Delta/E_f)^3$ from Poisson's equation,

$$\rho^{(3)}(\boldsymbol{R};t) = -\frac{1}{4\pi}\boldsymbol{\nabla}^2\,\Phi(\boldsymbol{R};t)\,. \tag{11.40}$$

In Fig. 11.12 we show the total electric field (external plus induced) acting on the quasiparticles in units of the external field. For $\omega \gtrsim 2\Delta$ the total field is approximately equal to the external field. However, at frequencies $\omega < 2\Delta$ an out-of-phase response develops. For $\omega \ll \Delta$, the total field in the pinning region approaches twice the external field, and the out-of-phase component vanishes. In the intermediate frequency region, $0.1\Delta \lesssim \omega \lesssim \Delta$, both in-phase and out-of-phase components are comparable. The right panel of Fig. 11.12 shows the charge distribution for $\omega = 0.1\Delta$ which oscillates out of phase with the external field. A dipolar charge distribution accumulates at the interface between the superconductor and the normal inclusion, oscillating at the frequency of the external field. At the center of the pinning site the out-of-phase component of the field is nearly zero at low frequencies (see also the left panel). The induced charge which accumulates is of the order of $e\Delta/E_f$ within a region of the order of ξ_0^2 in each conducting layer. This

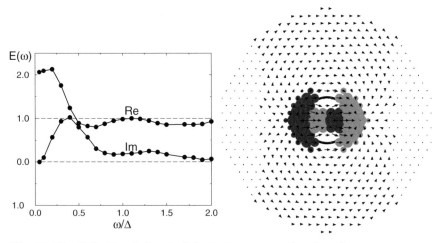

Fig. 11.12. Left: Total electric field in the center of a pinned s-wave vortex as a function of frequency ω of an external a.c. electric field with polarization vector in the x-direction and wavelength large compared to ξ_0. Right: The corresponding in-phase charge response around the pinning site for frequency $\omega = 0.1\Delta$. Gray corresponds to negative charge and black to positive charge. The arrows denote the total electric field vectors. The pinning center is a circular normal metallic inclusion with a radius $\xi_p = 1.57\xi_0$, shown as the black circle. Impurity scattering is included in the Born limit with a mean free path of $\ell = 10\xi_0$. The temperature is $T = 0.3T_c$ and the calculations are carried out in the high-κ limit.

charge is a factor of E_f/Δ larger than the static charge of a vortex that arises from particle–hole asymmetry [32–34].

11.4.2 Local Dynamical Conductivity

Because the total electric field varies on the scale of a coherence length, the current response expressed in terms of the the total field is nonlocal in the intermediate clean regime. However, we can define a local conductivity tensor in terms of the response to the external field, provided the external field varies on a length scale large compared to the coherence length,

$$J_\mu(\mathbf{R},\omega) = \sigma_{\mu\nu}(\mathbf{R},\omega) E_\nu^{\text{ext}}(\omega). \tag{11.41}$$

Figures 11.13 and 11.14 show results for the conductivity, $\sigma_{||}$, in the vortex core region as a function of frequency for both unpinned and pinned vortices. For the pinned vortex the radius of the pinning center is $\xi_{\text{pin}} = 1.57\xi_0$.

First consider the unpinned vortex. The absorptive part of the conductivity (right panel of Fig. 11.13) is strongly enhanced at the vortex center compared to the normal-state Drude conductivity. The reactive response exhibits a maximum at a frequency determined by the impurity scattering rate. A few coherence lengths away from the vortex center the real part of the conductivity changes sign at low frequencies. This is a signature that energy is transported by vortex-core excitations away from the vortex center producing "hot spots" outside the core. The net dissipation is determined by inelastic scattering processes in the region around the vortex core. At distances of the order of a coherence length or so from the vortex center, there is also structure in the conductivity spectrum at higher frequencies, reflecting absorptive transitions between quasiparticle excitations with energies corresponding to the Van Hove singularities in the local density of states. The maxima in the absorptive part of the conductivity at $y = 1.36\xi_0$ and $2.72\xi_0$ from the center correspond to the energy level separation between the Van Hove peaks above and below the Fermi level shown in Fig. 11.4.

The conductivity spectra for a vortex pinned by a metallic inclusion at its center is shown in Fig. 11.14. The most significant difference compared to the unpinned vortex occurs at frequencies $\omega < 1/\tau$. The absorptive part of the conductivity (right panel) is reduced compared to that of the unpinned case at low frequencies. The three broad peaks in the absorption spectrum correspond to scattering and dissipation within the zero-energy resonance (the dominant low-frequency peak), transitions between the zero-energy resonance and the continuum (the peak near $\omega \sim 1.5\Delta$), and pair-breaking transitions from the negative energy to positive energy continuum states (broad peak above 2Δ). The other notable feature is the reactive response at low frequencies which becomes *negative* in the low frequency limit, corresponding to superflow in the core that is counter to the induced supercurrent outside the vortex core and pinning center. This counterflow is required in order to

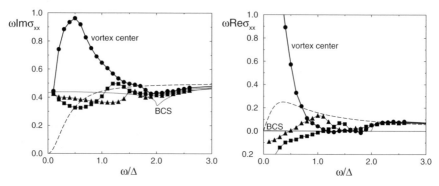

Fig. 11.13. Imaginary (left) and real part (right) of the local conductivity (multiplied with ω for convenience) in the center of an unpinned s-wave vortex (circles), and at different distances from the center along the y-axis: $1.1\xi_0$ (squares), and $2.2\xi_0$ (triangles), as a function of frequency ω of an external a.c. electric field with polarization vector in x-direction and wavelength large compared to ξ_0. Impurity scattering is taken into account in Born limit with a mean free path of $\ell = 10\xi_0$. The temperature is $T = 0.3T_c$. Calculations are done in the high-κ limit. Also shown are the response for a homogeneous s-wave superconductor (dotted, denoted 'BCS'), and the Drude conductivity of the normal metal (dashed).

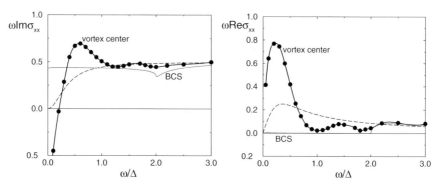

Fig. 11.14. Like Fig. 11.13 for a pinned vortex with a pinning radius $\xi_p = 1.57\xi_0$. As pinning site, a circular normal conducting inclusion located at the center of the vortex was assumed (see Fig. 11.12).

satisfy the conductivity sum rule. The counterflow in the vortex center is also present for unpinned vortices, but at smaller frequencies for this particular impurity scattering rate. The low-frequency counterflow is also related to characteristic current patterns associated with low-frequency vortex dynamics which we discuss below.

11.4.3 Induced Current Density

Results for the a.c. component of the current density near a pinned vortex are shown in Fig. 11.15 for $\omega = 0.1\Delta$. In addition to the a.c. current there is the time-independent circulating supercurrent around the vortex center which adds to the current shown in Fig. 11.15. The current response shows a dipolar pattern, which is also observed for unpinned vortices. The in-phase current response (right panel) indicates a region of strong absorption within the pinning region ($\boldsymbol{j} \| \mathbf{E}^{\text{ext}}$), and emission ($\boldsymbol{j} \cdot \mathbf{E}^{\text{ext}} < 0$) in the region roughly perpendicular to the direction of the applied field several coherence lengths away from the pinning center. Calculations of the energy transport current show that energy absorbed in the core is transported away from the vortex center by the vortex core excitations in directions predominantly perpendicular to the applied field. The net absorption is ultimately determined by inelastic scattering and requires integrating the local absorption and emission rate over the vortex array. Note that the long-range dipolar component does not contribute to the total dissipation. Far from the vortex core the current response is out of phase with the electric field and predominantly a non-dissipative supercurrent. Also note that at low frequency we clearly observe the counterflowing supercurrent within the pinning center.

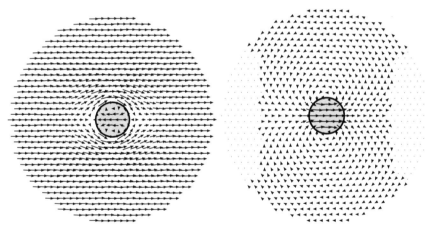

Fig. 11.15. Alternating current density pattern for a pinned s-wave vortex with a pinning radius $\xi_p = 1.57\xi_0$. As pinning site, a circular normal conducting inclusion located at the center of the vortex was assumed. The border of the pinning center (gray) is shown as black circle. The frequency of the external electric field with polarization vector in x-direction is chosen $\omega = 0.1\Delta$. Left picture for out-of-phase (reactive) response, right picture for in-phase (absorptive) response. Impurity scattering is taken into account in Born limit with a mean free path of $\ell = 10\xi_0$. The temperature is $T = 0.3T_c$. Calculations are done in the high-κ limit.

11.4.4 Summary

The electrodynamics of the vortex state in the intermediate-clean regime is nonlocal and largely determined by the response of the vortex-core states. Transitions involving the vortex-core states, and their coupling to the collective motion of the condensate, requires dynamically self-consistent calculations of the order parameter, self-energies, induced fields, excitation spectra and distribution functions. The results of these calculations provide new insight into the dynamics of vortex cores in conventional and unconventional superconductors.

Acknowledgements. This work was supported in part by the NSF through grant DMR-9972087 (M.E., J.A.S.), and in part by the U.S. Department of Energy, Basic Energy Sciences, Contract No. W-31-109-ENG-38 (M.E.).

References

1. C. Caroli, P. G. deGennes, and J. Matricon, Phys. Lett. **9**, 307 (1964).
2. J. Bardeen, R. Kümmel, A. E. Jacobs, and L. Tewordt, Phys. Rev. **187**, 556 (1969).
3. D. Rainer, J. A. Sauls, and D. Waxman, Phys. Rev. B **54**, 10094 (1996).
4. J. Bardeen and M. J. Stephen, Phys. Rev. **140**, A1197 (1965).
5. A. I. Larkin and Y. N. Ovchinnikov, Phys. Rev. B **57**, 5457 (1998).
6. A. A. Koulakov and A. I. Larkin, Phys. Rev. B **59**, 12021 (1999).
7. W. A. Atkinson and A. H. MacDonald, Phys. Rev. B **60**, 9295 (1999).
8. N. B. Kopnin and G. E. Volovik, Phys. Rev. Lett. **79**, 1377 (1997).
9. G. Eilenberger, Zeit.f. Physik **214**, 195 (1968).
10. A. I. Larkin and Y. N. Ovchinnikov, Sov. Phys. JETP **28**, 1200 (1969).
11. G. M. Eliashberg, Sov. Phys. JETP **34**, 668 (1972), [zetf, 61, 1254 (1971)].
12. A. Larkin and Y. Ovchinnikov, Sov. Phys. JETP **41**, 960 (1976).
13. A. Larkin and Y. Ovchinnikov, Sov. Phys. JETP **46**, 155 (1977).
14. J. W. Serene and D. Rainer, Phys. Rep. **101**, 221 (1983).
15. A. I. Larkin and Y. N. Ovchinnikov, in *Modern Problems in Condensed Matter Physics*, edited by D. Langenberg and A. Larkin (Elsevier Science Publishers, Amsterdam, 1986).
16. D. Rainer and J. A. Sauls, in *Superconductivity: From Basic Physics to New Developments*, edited by P. N. Butcher and Y. Lu (World Scientific, Singapore, 1995), pp. 45–78.
17. L. Gorkov and N. Kopnin, Sov. Phys. -Usp. **18**, 496 (1976).
18. S. Artemenko and A. Volkov, Sov. Phys. -Usp. **22**, 295 (1979).
19. M. Franz and Z. Tesanovic, Phys. Rev. B **60**, 3581 (1999).
20. S. K. Yip and J. A. Sauls, Phys. Rev. Lett. **69**, 2264 (1992).
21. N. Schopohl and K. Maki, Phys. Rev. B **52**, 490 (1995).
22. M. Ichioka, N. Hayashi, N. Enomoto, and K. Machida, Phys. Rev. B **53**, 15316 (1996).
23. T. Tokuyasu, D. Hess, and J. A. Sauls, Phys. Rev. B **41**, 8891 (1990).
24. A. S. Mel'nikov, Sov. Phys. JETP **74**, 1059 (1992).

25. H. Hess et al., Phys. Rev. Lett. **62**, 214 (1989).
26. C. Renner, A. D. Kent, P. Niedermann, and Ø. Fischer, Phys. Rev. Lett. **70**, 3135 (1991).
27. B. Jankó and J. Shore, Phys. Rev. B **46**, 9270 (1992).
28. T. Hsu, Phys. Rev. B **52**, 9178 (1995).
29. N. B. Kopnin, Sov. Phys. JETP Lett. **27**, 391 (1978).
30. N. B. Kopnin and A. V. Lopatin, Phys. Rev. B **51**, 15291 (1995).
31. M. Eschrig, J. A. Sauls, and D. Rainer, Phys. Rev. B **60**, 10447 (1999).
32. D. I. Khomskii and A. Freimuth, Phys. Rev. Lett. **75**, 1384 (1995).
33. M. V. Feigel'man, V. B. Geshkenbein, A. I. Larkin, and V. M. Vinokur, Sov. Phys. JETP Lett. **62**, 834 (1995).
34. G. Blatter et al., Phys. Rev. Lett. **77**, 566 (1996).

Part III

Fermion Zero Modes on Vortices

12 Band Theory of Quasiparticle Excitations in the Mixed State of d-Wave Superconductors

Alexander S. Mel'nikov

Summary. This review contains an account of recent advances in understanding of the electronic structure of the mixed state in superconductors with gap nodes. We discuss a number of fundamentally new features of low energy quasiparticle (QP) excitations in these systems as compared to the case of s-wave compounds, where low-lying QP states are bound to vortex cores. The vanishing pair potential in the nodal directions results in an extremely large extension of QP wavefunctions which are sensitive to both the long-range magnetic field effects and the superfluid velocity (V_s) fields of all vortices. The periodic V_s field is responsible for splitting of cyclotron orbits and induces a band structure in the spectrum. The distinctive features of the band spectrum, eigenfunctions and density of states are discussed for the particular case of d-wave compounds. Recent theoretical predictions are compared with experimental data for high-T_c superconductors.

12.1 Introduction

Understanding the nature of low energy quasiparticle (QP) excitations in isolated vortices and vortex lattices of type-II superconductors is of considerable importance, since at low temperatures these excitations impact on various static and dynamic properties. The peculiarities of the electronic structure of vortices are responsible for the magnetic field dependence of the density of states (DOS), which can be probed, e.g., by scanning tunneling spectroscopy and specific heat measurements. Low-lying QP states are also known to strongly influence both the electrical and thermal transport properties of the mixed state. The QP contributions to the vortex mobility and pinning characteristics play an essential role in the vortex dynamics and, thus, are important for the analysis of experimental data on flux-flow resistivity and Hall effect. For conventional s-wave superconductors these issues have been studied for several decades and now one may conclude that the physical picture of the electronic structure of the mixed state in s-wave systems is rather clear. The low-lying QP states are bound to the vortex cores as was first predicted by Caroli, de Gennes, and Matricon [1]. For an isolated vortex line the eigenvalues for these localized states can be written as follows: $E_\mu \sim \mu\Delta/(k_\mathrm{F}\xi)$, where Δ is the gap value far from the vortex axis, ξ is the coherence length, k_F is the Fermi momentum, and the angular momentum quantum number μ is half an odd integer. These states are only weakly perturbed by the presence of the neighboring vortices in a vortex lattice, at least, for magnetic fields H much less than the upper critical field H_{c2}. The small overlapping of wavefunctions results in splitting of discrete levels into narrow energy bands [2,3,30,5–8]. As for the spatially averaged DOS at the Fermi level, it appears to be proportional to the vortex concentration, i.e. to the average magnetic field. This physical picture is based on the assumption of the isotropic gap function and, thus, is no more valid for superconductors with gap nodes at the Fermi surface (FS). The vanishing pair potential in the nodal directions results in essential changes in quantum mechanical motion of QPs. The interest to these fundamental issues is stimulated by recent experimental observations of unconventional behavior of QP excitations in the mixed state of high-T_c compounds where the dominating order parameter is believed to be of d-wave symmetry. Perhaps the most notable examples of such surprising behavior are the following: (i) scaling in the variable $T/\sqrt{H}$ in low temperature specific heat [9–14]; (ii) the behavior of the longitudinal thermal conductivity, $\kappa(H)$, at ultralow temperatures (an increase in $\kappa(H)$ with an applied magnetic field) is in sharp contrast to the one observed at temperatures above a few Kelvin (an initial decrease and a high-field plateau in the dependence $\kappa(H)$) [15,16]; (iii) peculiarities of the local DOS detected by scanning tunneling spectroscopy [17–19,9]. The main goal of this review is to focus on the qualitatively new features of the electronic structure of the mixed state in d-wave systems as compared to the case of s-wave compounds. I am going to analyse recent theoretical results and compare them with the existing experimental data for high-T_c cuprates.

One of the most successful predictions in this field was made by Volovik [11] who showed that, in contrast to conventional superconductors, the DOS at low energies ε in d-wave systems is dominated by contributions which come from the regions far from the vortex cores, and depends essentially on the intervortex distance R_v. This unusual fact is a direct consequence of the vanishing pair potential in the directions of the gap nodes. The spatially averaged residual DOS was found to have the form $\bar{N}(\varepsilon=0) \propto N_F\sqrt{H/H_{c2}}$, where N_F is the DOS at the Fermi level in a normal state. The Volovik-type description as well as more complicated approaches allowed to obtain general scaling relations in the variable $T/\sqrt{H}$ for low temperature thermodynamic and transport characteristics [11,22–27], which are consistent, in particular, with the recent specific heat measurements. The Volovik-type description mentioned above is based on the semiclassical approach which takes account of the Doppler shift of the QP energy by the local superfluid velocity $\boldsymbol{V}_\mathrm{s}$:

$$\varepsilon^{\pm}(\boldsymbol{k},\boldsymbol{r}) = \pm\sqrt{\hbar^2 V_F^2(k-k_F)^2 + \Delta^2(\boldsymbol{k})} + \hbar\boldsymbol{k}\boldsymbol{V}_\mathrm{s}(\boldsymbol{r}) , \qquad (12.1)$$

where V_F is the Fermi velocity. Hereafter we assume FS to be two-dimensional (2D), which is appropriate to high-T_c superconductors, and take the gap function in the form $\Delta_d = 2\Delta_0 k_x k_y/k_F^2$ (the x axis is taken along the [110] crystal direction and thus makes an angle $\pi/4$ with the a axis of the CuO_2 planes).

We restrict ourselves to the study of a vortex lattice under a magnetic field applied parallel to the c axis and, as a consequence, the vector $\boldsymbol{V}_\mathrm{s}$ is a periodic function of x, y with the periodicity of the vortex lattice. The semiclassical model (12.1) is essentially based on two approximations: (i) first, we assume that the order parameter and the superfluid velocity are modulated on a scale which is much larger than the minimum spatial extension of wave packets made with excitations (12.1); (ii) second, we neglect the noncommutability of momentum components, which is associated with a nonzero magnetic field and, thus, we do not take account of the Landau-type quantization. For s-wave systems in a magnetic field $H \ll H_{c2}$ the first approximation breaks down only in the vortex core region and, as a consequence, the semiclassical model surely fails to describe the localized Caroli–de Gennes–Matricon states. For delocalized QP states above the gap the model (12.1) is valid though in the extremely clean limit the Landau-type quantization for QPs precessing along the whole FS can be essential (see [28] and below). To examine the validity of the semiclassical model for the d-wave case let us start from the homogeneous state and consider a Dirac spectrum for the low-lying excitations which are close to one of the gap nodes at $\boldsymbol{k}_i$ ($i = 1, 2, 3, 4$):

$$\varepsilon_i^\pm = \pm\hbar\sqrt{V_\mathrm{F}^2 q_\perp^2 + V_\Delta^2 q_\|^2} \,, \tag{12.2}$$

where $(q_\|, q_\perp)$ defines a coordinate system whose origin is at the node, with $q_\perp$ ($q_\|$) normal (tangential) to the FS. The V_Δ value determines the gap slope near the node and for our choice of the gap function is given by the expression $V_\Delta = 2\Delta_0/(\hbar k_\mathrm{F})$. One can distinguish the following length scales for QP wavefunctions: an atomic length scale k_F^{-1} and two characteristic wavelengths of a slowly varying envelope $l_\perp \sim q_\perp^{-1} \sim \hbar V_\mathrm{F}/\varepsilon$, $l_\| \sim q_\|^{-1} \sim \hbar V_\Delta/\varepsilon$. The length scales $l_\perp$ and $l_\|$ determine the size of the semiclassical wave packet, which appears to diverge in the low energy limit. For $\varepsilon \lesssim \hbar V_\mathrm{F}/R_v$ the $l_\perp$ value becomes even larger than the intervortex distance R_v (the characteristic length of the superfluid velocity variation). Such a divergence is responsible for the extremely important role of quantum mechanical effects in QP motion, which can not be treated within the standard semiclassical approach. To develop a description which would provide a possibility to describe true quantum mechanical motion we need more powerful methods using either the Bogolubov–de Gennes (BdG) equations or Green's-function techniques. Direct numerical solutions based on the BdG theory were previously presented in [29,41] for specific lattice models and in [31,15] for the continuum limit. The microscopic analysis can be essentially simplified in the quasiclassical limit ($k_\mathrm{F}\xi \gg 1$) which is known to be represented by the Andreev's or Eilenberger equations. Within the usual quasiclassical theory one should solve the one-dimensional (1D) quantum mechanical problem for the particle motion along the quasiclassical trajectory which is characterized by the impact parameter $b = \mu/k_\mathrm{F}$ and the angle θ in the $x - y$ plane [11,22,33–38]. Using the Bohr–Sommerfeld quantization rule for the angular momentum $\mu(\theta, \varepsilon)$

one can in principle determine the quantized energy spectrum (see [22,33–35]). For the d-wave case the main difficulty of the quantization method mentioned above is connected with the correct description of quasiclassical trajectories with momenta close to the nodal directions, which is of considerable importance since it is this angular interval that determines the true quantum levels according to the Bohr–Sommerfeld quantization rule. From this point of view there is an important difference between the quantization of the energy spectrum in the mixed state of s- and d-wave superconductors: for s-wave systems the energy quantization and corresponding localized states exist even in a single isolated vortex line (and they are weakly influenced by the presence of neighboring vortices at least for $R_v \gg \xi$), while for the d-wave case the low-lying energy spectrum may be quantized only due to the finite intervortex distance (see [22,24,28,33–35,39–47]). The latter conclusion is proved by the numerical solution of the BdG equations [15] which shows that there are no truly localized states for a single isolated vortex in a pure d-wave superconductor. Due to the extremely large extension of wavefunctions for the trajectories close to the nodal directions one should take account of two important facts: (i) first, one can not neglect the long-range magnetic field effects (i.e., the finite trajectory curvature caused by the momentum precession); (ii) second, for the trajectories which make an angle $\theta < \xi/R_v$ with the gap node directions the extension of the wavefunction exceeds the R_v value and QP states are sensitive to the superfluid velocity fields of all vortices. The modification of the quasiclassical theory, which takes account of the finite trajectory curvature and formation of cyclotron orbits has been recently discussed in [28,42,45,47]. The periodic potential associated with a nonzero superfluid velocity is responsible for splitting of these cyclotron orbits and removes the degeneracy of the spectrum with respect to the cyclotron orbit (CO) center [28,39–41,47]. As a consequence, we obtain the band spectrum with the energy scales determined by the interplay between the shape of the CO and the vortex lattice geometry. This interplay results in a strong dependence of the behavior of energy branches on the anisotropy of the Dirac cone $\alpha = V_F/V_\Delta$ [39–41,43,46].

12.2 Basic Equations

In this section we give a brief outline of the general BdG equations for superconductors with anisotropic pairing and discuss a model Hamiltonian for QPs confined near gap nodes.

12.2.1 BdG Equations for a Spin Singlet Unconventional Superconductor in a Magnetic Field

The BdG equations for spin singlet unconventional superconductors can be written as:

$$\hat{h}_0\left(-i\nabla + \frac{\pi}{\phi_0}\boldsymbol{A}\right)u(\boldsymbol{r}) + \int D(\boldsymbol{r},\boldsymbol{r}')v(\boldsymbol{r}')d\boldsymbol{r}' = \varepsilon u(\boldsymbol{r}) \tag{12.3}$$

$$-\hat{h}_0\left(i\nabla + \frac{\pi}{\phi_0}\boldsymbol{A}\right)v(\boldsymbol{r}) + \int D^*(\boldsymbol{r},\boldsymbol{r}')u(\boldsymbol{r}')d\boldsymbol{r}' = \varepsilon v(\boldsymbol{r}) , \tag{12.4}$$

where u, v are the particle-like and hole-like parts of the QP wavefunction and $\phi_0 = \pi\hbar c/|e|$ is the flux quantum. The one-particle Hamiltonian $\hat{h}_0$ in the most simple isotropic case takes the form $\hat{h}_0(\boldsymbol{k}) = \hbar^2\boldsymbol{k}^2/(2M) - E_F$, where E_F is the Fermi energy and M is the electron effective mass. The selfconsistency condition for D reads:

$$D(\boldsymbol{r},\boldsymbol{r}') = -K(\boldsymbol{r},\boldsymbol{r}')$$
$$\times \sum_p (u_p(\boldsymbol{r})v_p^*(\boldsymbol{r}')(-n_f(\varepsilon_p)) + u_p(\boldsymbol{r}')v_p^*(\boldsymbol{r})(1 - n_f(\varepsilon_p))) ,$$
$$\tag{12.5}$$

where $K(\boldsymbol{r},\boldsymbol{r}')$ is the pairing interaction, n_f is the Fermi function, and the sum over p runs over positive energy eigenstates. To simplify the nonlocal off-diagonal terms we use the following procedure: (i) first, we eliminate the order parameter phase χ in D by an appropriate gauge transformation $u = \tilde{u}\exp(i\chi/2)$, $v = \tilde{v}\exp(-i\chi/2)$, $D(\boldsymbol{r},\boldsymbol{r}') = \tilde{D}(\boldsymbol{r},\boldsymbol{r}')\exp(i(\chi(\boldsymbol{r}) + \chi(\boldsymbol{r}'))/2)$; (ii) second, we rewrite $\tilde{D}(\boldsymbol{r},\boldsymbol{r}')$ in terms of the center of mass $\boldsymbol{R} = (\boldsymbol{r} + \boldsymbol{r}')/2$ and relative coordinates $\boldsymbol{\rho} = \boldsymbol{r} - \boldsymbol{r}'$ and introduce the gap function as a Fourier transform with respect to $\boldsymbol{\rho}$: $\tilde{D}(\boldsymbol{k},\boldsymbol{R}) = \Delta(\boldsymbol{k})f(\boldsymbol{R})$ (here the function $\Psi(\boldsymbol{R}) = f\exp(i\chi)$ is the order parameter used in the Ginzburg–Landau theory); (iii) third, we restrict ourselves to the quasiclassical limit (i.e. assume the characteristic length scale of the order parameter f variation to be much larger than an atomic length scale k_F^{-1}) and neglect the gap dependence on distance to the FS. Introducing single-valued wavefunctions $U = \tilde{u}\exp(-i\chi/2)$ and $V = \tilde{v}\exp(-i\chi/2)$ we obtain:

$$\hat{h}_0\left(\hat{\boldsymbol{k}} + \frac{2M}{\hbar}\boldsymbol{V}_s\right)U + \left.\left(\Delta\left(\hat{\boldsymbol{k}}' + \frac{M}{\hbar}\boldsymbol{V}_s(\boldsymbol{r}')\right)f(\boldsymbol{R})V(\boldsymbol{r}')\right)\right|_{\boldsymbol{r}'=\boldsymbol{r}} = \varepsilon U \tag{12.6}$$

$$-\hat{h}_0(\hat{\boldsymbol{k}})V + \left.\left(\Delta^*\left(\hat{\boldsymbol{k}}' + \frac{M}{\hbar}\boldsymbol{V}_s(\boldsymbol{r}')\right)f(\boldsymbol{R})U(\boldsymbol{r}')\right)\right|_{\boldsymbol{r}'=\boldsymbol{r}} = \varepsilon V , \tag{12.7}$$

where $\boldsymbol{V}_s = \frac{\hbar}{2M}\left(\nabla\chi + \frac{2\pi}{\phi_0}\boldsymbol{A}\right)$ is the superfluid velocity, $\hat{\boldsymbol{k}}' = -i\nabla' - \frac{\pi}{\phi_0}\boldsymbol{A}(\boldsymbol{r}')$, $\hat{\boldsymbol{k}} = -i\nabla - \frac{\pi}{\phi_0}\boldsymbol{A}(\boldsymbol{r})$. The final transformation of wavefunctions ($U = u\exp(-i\chi)$, $V = v$) is single-valued and coincides with the one introduced previously in [39–42]. Note that other possible versions of single-valued gauge transformations were discussed in [43,46]. If we consider a particular case of pure d-wave superconductivity and neglect the effects connected with a possible formation of states of mixed symmetry (with coexisting s- and d-wave or $d_{x^2-y^2}$ and d_{xy} order parameter components), the gap operators in the off-diagonal terms of equations (12.6,12.7) take the form:

$$\frac{\Delta_0}{2k_F^2} \left\{ \hat{k}_x + \frac{M}{\hbar} V_{sx}, \left\{ \hat{k}_y + \frac{M}{\hbar} V_{sy}, f \right\} \right\} , \qquad (12.8)$$

where V_{sx} and V_{sy} are the components of the superfluid velocity $\boldsymbol{V}_s = V_{sx}\boldsymbol{x}_0 + V_{sy}\boldsymbol{y}_0$, and $\boldsymbol{x}_0$, $\boldsymbol{y}_0$, $\boldsymbol{z}_0$ are the unit vectors of the coordinate system. Here we use the notation $\{\hat{A}, \hat{B}\}$ for the anti-commutator of two operators $\hat{A}$ and $\hat{B}$.

12.2.2 Quasiparticles Confined Near Gap Nodes

These general BdG equations may be simplified, if we consider only the energy interval $|\varepsilon| \ll \Delta_0$ when QPs are locked to the nodes. Let us search for the solution in the form $U = \tilde{U}\exp(\mathrm{i}\boldsymbol{k}_F\boldsymbol{r})$, $V = \tilde{V}\exp(\mathrm{i}\boldsymbol{k}_F\boldsymbol{r})$, i.e. divide out the fast oscillations on a scale k_F^{-1}. Choosing one of four gap nodes on a 2D FS (e.g., $\boldsymbol{k}_{F1} = (k_F, 0)$) we consider the BdG equations linearized in gradient terms: $\hat{H}\hat{g} = \varepsilon\hat{g}$, where

$$\hat{H} = \hat{H}_0 + \hat{H}' , \qquad (12.9)$$

$$\hat{H}_0 = \hbar V_F \hat{\sigma}_z \hat{k}_x + \hbar V_\Delta \hat{\sigma}_x \{\hat{k}_y, f\}/2 , \qquad (12.10)$$

$$\hat{H}' = M V_F V_{sx}(1 + \hat{\sigma}_z) + M V_\Delta V_{sy}\hat{\sigma}_x , \qquad (12.11)$$

$\hat{\sigma}_x, \hat{\sigma}_y, \hat{\sigma}_z$ are the Pauli matrices and $\hat{g} = (\tilde{U}, \tilde{V})$. We follow here the treatment in [24] and neglect the curvature of the FS. For an isotropic FS, such an approximation is valid only for $\varepsilon \ll \Delta_0/(k_F \xi)$ (see [48,49]). However, as mentioned in [24,49], the range of validity of these equations may be even larger if the FS is somewhat flattened at the nodes. The superfluid velocity $\boldsymbol{V}_s(\boldsymbol{r})$ can be written as a superposition of contributions from individual vortices situated at points $\boldsymbol{r}_i$:

$$\boldsymbol{V}_s(\boldsymbol{r}) = \frac{\hbar}{2M} \sum_i \frac{[\boldsymbol{z}_0, \boldsymbol{r} - \boldsymbol{r}_i]}{\lambda_L |\boldsymbol{r} - \boldsymbol{r}_i|} K_1\left(\frac{|\boldsymbol{r} - \boldsymbol{r}_i|}{\lambda_L}\right) ,$$

where $[\boldsymbol{a}, \boldsymbol{b}]$ is a vector product of vectors $\boldsymbol{a}$ and $\boldsymbol{b}$, λ_L is the London penetration depth, K_1 is the Mcdonald function (modified Bessel function) of the first order. Note that in this expression we neglect the effects of nonlocal electrodynamics [50] which can not perturb the wavefunction $\hat{g}$ essentially in the limit $\varepsilon < \hbar V_\Delta/\xi$ when the minimum characteristic wavelength $l_{\min} \sim \hbar V_\Delta/\varepsilon$ exceeds the coherence length ξ. Outside the vortex cores the Hamiltonian $\hat{H}$ can be rewritten in the following form:

$$\hat{H} = -\hbar V_F \hat{\sigma}_z \left(\mathrm{i}\frac{\partial}{\partial x} - \frac{1}{2}\frac{\partial \chi}{\partial x}\right) - \hbar V_\Delta \hat{\sigma}_x \left(\mathrm{i}\frac{\partial}{\partial y} - \frac{1}{2}\frac{\partial \chi}{\partial y}\right) + M V_F V_{sx} . \qquad (12.12)$$

One can see that our equations are analogous to the ones describing the quantum mechanical motion of a massless Dirac particle in the 'vector potential' $-\phi_0 \nabla\chi/(2\pi)$ of Aharonov–Bohm solenoids and the scalar potential $MV_\mathrm{F} V_{sx}/|e|$ of 2D 'electrical dipoles' screened at a length scale λ_L [43,46,51]. Both the solenoids and the dipoles are positioned at $\boldsymbol{r}_i$. Each solenoid carries the flux quantum ϕ_0.

12.3 A Single Isolated Vortex Line: Aharonov–Bohm Effect for Quasiparticles

Let us start with the case of a single isolated vortex line and focus on the low energy limit when QPs are confined to the nodes. The analysis of the Hamiltonian (12.12) shows that the low energy QP states near each node are strongly influenced by the 'vector potential' $-\phi_0 \nabla\chi/(2\pi)$ due to the Aharonov–Bohm scenario [52] (for vortices in s-wave systems this mechanism was discussed, e.g., by Cleary [53] and Sonin [40]). Indeed, if we neglect the potential of a 2D 'electric dipole' in (12.12), the scattering cross section of a Dirac fermion in the Aharonov–Bohm potential appears to diverge for $\varepsilon \to 0$. Such a divergence directly follows from the results obtained in [55,56,24] for the particular case of the isotropic Dirac cone with $V_\mathrm{F} = V_\Delta$ (generalization for an anisotropic spectrum can be carried out using scale transformation). Introducing a polar coordinate system (r, θ) with the origin at the vortex center, one can write the solutions in the form:

$$\hat{g}_m \propto (1 + \mathrm{i}\hat{\sigma}_x)(\mathrm{e}^{\mathrm{i}m\theta} J_{m+1/2}(kr), \mathrm{sgn}\varepsilon \mathrm{e}^{\mathrm{i}(m+1)\theta} J_{m+3/2}(kr)) \,, \tag{12.13}$$

$$\hat{f}_m \propto \mathrm{i}\hat{\sigma}_y \mathrm{e}^{-\mathrm{i}\theta} \hat{g}_m^* \,, \tag{12.14}$$

where J_ν is the ν-th Bessel function, $V_\mathrm{F} = V_\Delta = V_i$, and $\hbar V_i k = \varepsilon$. The wavefunctions $\hat{g}_m$ and $\hat{f}_m$ with $m \geq 0$ are regular at the origin. The local DOS corresponding to a set of these regular solutions vanishes in the region $r < \hbar V_i/|\varepsilon|$ and saturates at the value $N_\infty \propto |\varepsilon|$ for $r \gg \hbar V_i/|\varepsilon|$ (N_∞ is the DOS in the absence of vortices). For low energies $\varepsilon \ll \hbar V_i/\lambda_\mathrm{L}$ these solutions are only weakly influenced by the potential of the screened 'electric dipole'. The wavefunctions (12.13,12.14) with negative m diverge at $r \to 0$ and are responsible for the formation of a nonzero DOS near the vortex inside the domain $r < \hbar V_i/|\varepsilon|$. The divergence should be cut off due to the matching with the solution inside the core, which results in a strong mixing of QP wavefunctions corresponding to all four nodes. In the limit $\varepsilon \to 0$ the angular harmonics g_{-1} and f_{-1} appear to decay most slowly from the vortex center (as $r^{-1/2}$). For distances $r < \lambda_\mathrm{L}$ ($\varepsilon \ll \hbar V_i/\lambda_\mathrm{L}$) these slowly decaying solutions take the form:

$$\hat{g}_{-1} \propto (1 + \mathrm{i}\hat{\sigma}_x) r^{-1/2} (\mathrm{e}^{-\mathrm{i}\theta} \cos(\cos\theta/2), -\mathrm{i}\sin(\cos\theta/2)) \,, \tag{12.15}$$

$$\hat{f}_{-1} \propto \mathrm{i}\hat{\sigma}_y \mathrm{e}^{-\mathrm{i}\theta} \hat{g}_{-1}^* \,. \tag{12.16}$$

These states are not localized in the core (in agreement with the numerical analysis [15]) and provide the following contribution to the residual local DOS: $N(\varepsilon = 0) \sim (\hbar V_i r)^{-1}$. For the general case $V_F \neq V_\Delta$ the expression for this DOS contribution (taking account of all four nodes) reads [58]:

$$N(\varepsilon = 0) \sim \hbar^{-1}((V_\Delta^2 x^2 + V_F^2 y^2)^{-1/2} + (V_F^2 x^2 + V_\Delta^2 y^2)^{-1/2}) \;.$$

Note that the influence of the Aharonov–Bohm field on the QP states can not be described within the standard quasiclassical approach (see Introduction) based on consideration of the wave packets propagating along the straight trajectories.

12.4 Quasiparticle States in Vortex Lattices

We now proceed with the analysis of the QP spectrum in vortex lattices and consider the quantization effects associated with a finite intervortex distance. At intermediate magnetic fields $H_{c1} \ll H \ll H_{c2}$ ($R_v \ll \lambda_L$) we can assume $\boldsymbol{H} = -H\boldsymbol{z}_0$ to be homogeneous and take the gauge $\boldsymbol{A} = Hy\boldsymbol{x}_0$.

12.4.1 Cyclotron Orbits in the Mixed State

Neglecting the terms proportional to $\boldsymbol{V}_s$ and the suppression of the order parameter in vortex cores (i.e., assuming $f = 1$) in (12.6,12.7), one can obtain a rather general scheme for description of the Landau-type quantization in superconductors with an arbitrary anisotropic gap function [28,42,47]. In the classical limit the QP motion occurs along the COs defined by the following expression:

$$\varepsilon^2 = \hbar^2 V_F^2 (k - k_F)^2 + \Delta^2(\boldsymbol{k}) = \text{const}. \tag{12.17}$$

The quasiclassical quantization rule reads:

$$S(\varepsilon) = \frac{2\pi^2 H}{\phi_0}(n + \gamma) \;, \tag{12.18}$$

where $S(\varepsilon)$ is the CO area in the $\boldsymbol{k}$- space, n is an integer and γ is of the order unity. Let us now use this general approach to analyse the peculiarities of the cyclotron motion in the d-wave case. Due to the gap anisotropy there appears an energy threshold Δ_0 separating the QP states with a qualitatively different structure of COs. For energies above this threshold ($\varepsilon > \Delta_0$) QP momentum precesses along the whole FS and the Larmor orbits in $\boldsymbol{k}$- space are defined by expression (12.17) with $\Delta(\boldsymbol{k}) = \Delta_d = 2\Delta_0 k_x k_y/k_F^2$. Integrating the classical equation of motion one can observe that $\boldsymbol{k} = \pi[\boldsymbol{r}, \boldsymbol{H}]/\phi_0$ and, thus, the orbits in the $\boldsymbol{r}$- space have the same shape as the ones in the $\boldsymbol{k}$- space but are rotated through $\pi/2$. These Larmor orbits are qualitatively similar to the ones for a normal metal (though in a superconductor these orbits are surely distorted due to the gap anisotropy and exhibit a fourfold symmetry even for an isotropic FS). The general quantization rule takes the following simple form:

$$\frac{4k_F\varepsilon}{\Delta_0\xi} E\left(\frac{\pi}{2}, \frac{\Delta_0}{2\varepsilon}\right) = \frac{2\pi^2 H}{\phi_0}(n+\gamma) ,\qquad(12.19)$$

where $E(\varphi, p)$ is the elliptic integral of the second kind. In the limit $\varepsilon \gg \Delta_0$ we obtain the usual equidistant spectrum: $\varepsilon = \hbar\omega_c(n + \gamma)$, where $\omega_c = eH/(Mc)$ is the cyclotron frequency. Below the energy threshold ($\varepsilon < \Delta_0$) excitations are confined to the nodes due to the Andreev reflection process (see the discussion in [42]). For each of the four gap nodes we obtain a set of anisotropic COs elongated along the FS (for $V_F > V_\Delta$) and a corresponding quantized energy spectrum:

$$\frac{k_F}{\xi}\left(E\left(\frac{\pi}{2}, \frac{2\varepsilon}{\Delta_0}\right) - \left(1 - \frac{4\varepsilon^2}{\Delta_0^2}\right)F\left(\frac{\pi}{2}, \frac{2\varepsilon}{\Delta_0}\right)\right) = \frac{2\pi^2 H}{\phi_0}(n+\gamma) ,\quad(12.20)$$

where $F(\varphi, p)$ is the elliptic integral of the first kind. In the low energy limit $\varepsilon \ll \Delta_0$ this expression can be simplified:

$$\varepsilon_n = \pm\sqrt{2\Delta_0\hbar\omega_c(n+\gamma)} .\qquad(12.21)$$

Such a case may be analysed even beyond the simple scheme (12.17,12.18) discussed in this section, if we neglect the FS curvature and start from the BdG equations with the Dirac Hamiltonian $\hat{H}_0$ ($f = 1$). One can see that these equations can be solved exactly in terms of harmonic oscillator eigenfunctions [28,42,45]. The energy spectrum is given by (12.21) with $\gamma = 0$.

12.4.2 Quasiparticle Band Spectrum in Vortex Lattices

Hereafter we focus on the analysis of the low energy part of the QP spectrum taking account of the periodic potentials produced by the superfluid velocity field. Our consideration is based on the BdG Hamiltonian (12.9) for QPs confined near the node $\mathbf{k}_{F1} = (k_F, 0)$ (the generalization for other gap nodes is straightforward). The periodic potential $\hat{H}'$ removes the degeneracy of the Landau levels with respect to the CO center and induces a band structure in the spectrum [39–41,43,44,46,47]. This potential cannot be considered as a small perturbation and, therefore, expression (12.21) does not provide an adequate description of the low energy spectrum.

For a vortex lattice with primitive translations $\mathbf{a}_1, \mathbf{a}_2$ the superfluid velocity may be written in the form:

$$\mathbf{V}_s = \frac{i\pi\hbar H}{M\phi_0} \sum_{\mathbf{b}\neq 0} \frac{[\mathbf{b}, \mathbf{z}_0]}{(\mathbf{b}, \mathbf{b})} e^{i\mathbf{b}\mathbf{r}} ,\qquad(12.22)$$

where $\mathbf{b} = n\mathbf{b}_1 + m\mathbf{b}_2$ and $\mathbf{b}_1 = 2\pi H[\mathbf{a}_2, \mathbf{z}_0]/\phi_0$, $\mathbf{b}_2 = 2\pi H[\mathbf{z}_0, \mathbf{a}_1]/\phi_0$ are the primitive translations in the reciprocal lattice, n and m run over all possible integers. Contrary to the conventional isotropic superconductors (where a hexagonal flux lattice appears to be energetically favorable), for d-wave compounds the previous theoretical works predicted a rich phase diagram, containing triangular, centered rectangular and square lattices with various orientations relative to the ionic lattice, as a function of magnetic field and temperature [50,59–61]. For the sake of simplicity we follow the treatment in [39–41] and restrict ourselves to the study of the energy spectra for two particular types of lattices with primitive translations: (I) a rectangular lattice with $\boldsymbol{a}_1 = a\boldsymbol{x}_0$, $\boldsymbol{a}_2 = \sigma a \boldsymbol{y}_0$, $H\sigma a^2 = \phi_0$; centered rectangular lattice with (II) $\boldsymbol{a}_1 = a\boldsymbol{x}_0$, $\boldsymbol{a}_2 = a(\boldsymbol{x}_0/2 - \sigma \boldsymbol{y}_0)$ (vortices in the unit cell form the shape of an isosceles triangle with the base along the x axis). In particular, we include in our consideration the square lattice tilted by $\pi/4$ from the a axis (type I, $\sigma = 1$). Such a lattice (though elongated in the a direction) is close to the one observed experimentally in twinned YBaCuO monocrystals by small angle neutron scattering [62] and scanning tunneling microscopy [17]. Note that our choice of lattice geometry influences strongly the symmetry properties of QP wavefunctions and spectrum. In particular, the existence of a center of inversion in both types of lattices results in the exact particle–hole symmetry of the linearized BdG Hamiltonian (12.9) near each node [46]. For more complicated structures such a symmetry should hold only on the whole spectrum of (12.6,12.7).

The Hamiltonian (12.9) commutes with operators of magnetic translations by periods of the vortex lattice. As a result, the general solution can be written in the form of a magnetic Bloch wave:

$$\hat{g} = \sum_n e^{ix(q_x + 2\pi n/a) + 2in\sigma q_y a} \hat{G}(y - 2n\sigma a, \boldsymbol{q}) , \qquad (12.23)$$

where n is an integer, and $\boldsymbol{q}$ is the quasimomentum lying within the first magnetic Brillouin zone (MBZ): $-\pi/(2a) < q_x < \pi/(2a)$, $-\pi/(2\sigma a) < q_y < \pi/(2\sigma a)$. The wavefunction $\hat{G}(y, \boldsymbol{q})$ is localized in the domain with the size L determined by $\boldsymbol{q}$ and energy values. The potential $\hat{H}'$ causes splitting of the CO near MBZ boundaries (see Fig. 12.1) and the spectrum consists of branches which correspond to the splitted portions of the CO.

Large Dirac Cone Anisotropy Limit: 1D and 2D Regimes in the Band Spectrum. For large Dirac cone anisotropy $\alpha = V_F/V_\Delta \gg 1$ one can distinguish two regimes with a qualitatively different behavior of energy branches. For $\varepsilon < 0.5\varepsilon^*$ ($\varepsilon^* = \pi\hbar V_F/a \sim \Delta_0 \sqrt{H/H_{c2}}$) the overlapping of the harmonics in (12.23) is small ($L < 2\sigma a$) and one can replace $\hat{H}'$ by the potential $\langle \hat{H}' \rangle_x$ averaged in the x direction (see [39–41]). Such a simplification is a natural consequence of a small size of the CO (CO1 in Fig. 12.1) in the nodal direction as compared to the size of the MBZ. The energy branches in such 1D regime are characterized by a negligible dispersion in the q_y ($q_\parallel$) direction. The latter conclusion agrees well with the numerical simulations [43,46] in the large-α limit. The crossover between 1D and 2D regimes in the band spectrum occurs at $\varepsilon_c \sim 0.5\varepsilon^*$, when the CO size in the q_x ($q_\perp$) direction becomes larger than the size of the first MBZ (CO2 in Fig. 12.1).

Let us now discuss the peculiarities of the eigenfunctions and band spectrum in the 1D regime in more detail. It is convenient to introduce $\hat{F} = (\hat{\sigma}_x + \hat{\sigma}_z)\hat{G}/2$ and dimensionless values $z = y/R_y$, $Q = q_x a/\pi$, $E = \varepsilon/\varepsilon^*$, where R_y (R_x) is the distance between the lines parallel to the x (y) axis which pass through the vortex centers. For type I (II) lattices we have $R_x = a$, $R_y = \sigma a$ ($R_x = a/2$, $R_y = \sigma a$). At the m-th interval ($m < z < m+1$) the equation for $\hat{F}$ reads:

$$-i\lambda\hat{\sigma}_z \frac{\partial \hat{F}}{\partial z} + \left(\frac{1}{2}\Phi(z) - E\right)\hat{F} + q_m \hat{\sigma}_x \hat{F} = 0 , \qquad (12.24)$$

where $q_m = Q - m - 1/2$ and $\lambda = (\pi\sigma a)^{-1}$ is a dimensionless wavelength, $\Phi(z) = 2z - (2m+1)$ for $m < z < m+1$, m is an integer. We omitted here the small corrections of the order ξ/a, which are connected with suppression of the order parameter f in the vortex cores. Inside the m-th interval equations (12.24) are equivalent to the ones describing the interband tunneling [63] or the 1D motion of a Dirac particle in a uniform electric field and can be solved exactly in terms of parabolic cylinder functions. The solution can be simplified using the standard 1D quasiclassical approach which is valid for $\lambda \ll 1$. This condition is easily met for a rather large Dirac cone anisotropy, if the σ parameter is not extremely small. Substituting $\hat{F} \propto \exp(iS(z))$ in equation (12.24) we obtain the classically allowed (CA) regions ($|\Phi(z)/2 - E| > |q_m|$) which shift in the z direction with a change in the Q value (dashed regions in Fig. 12.2). Thus, for the low energy (1D) regime the wavefunction G is locked between the vortex planes. Let us now consider the domain $|q_m| < 1$ and continue with the calculation of the quasiclassical energy levels corresponding to the QP motion at the m-th interval which contains two CA regions. The quantization rules for regions A and B can be written as follows:

$$\left(\frac{1}{2} \pm E_{A,B}\right)\sqrt{\left(\frac{1}{2} \pm E_{A,B}\right)^2 - q_m^2} - q_m^2 \cosh^{-1}\left(\frac{\frac{1}{2} \pm E_{A,B}}{|q_m|}\right)$$
$$= 2\pi\lambda(N_{A,B} + \gamma_{A,B}) , \qquad (12.25)$$

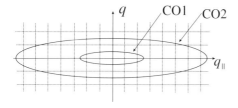

Fig. 12.1. Cyclotron orbits (CO1, CO2) and MBZ boundaries for a square lattice. ($q_\parallel, q_\perp$) is a coordinate system whose origin is at the node, with $q_\perp$ ($q_\parallel$) normal (tangential) to the FS.

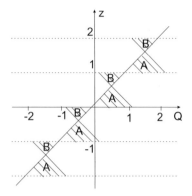

Fig. 12.2. Classically allowed regions A and B

where $N_{A,B}$ is an integer, the $\gamma_{A,B}$ values are of the order unity and should be determined from the boundary conditions at the vortex planes. These boundary conditions are influenced by the overlapping of harmonics in (12.23) (which was neglected above) and, thus, cannot be found by the direct matching of the solutions of (12.24) in classically allowed and forbidden regions. The energy spectrum consists of branches $\varepsilon_n(q_x) = \varepsilon^* E_n(Q, \sigma\alpha)$, which are displayed in Fig. 12.3 in the first MBZ for $\pi\sigma\alpha = 50$ and $\pi\sigma\alpha = 100$. The number of energy branches which cross the Fermi level can be determined as follows: $N \sim 2q_\parallel^*/\delta q_\parallel \sim 2\sqrt{\pi\sigma\alpha}$, where $q_\parallel^*$ is the minimum possible size of the CO in the $q_\parallel$ direction and $\delta q_\parallel$ is the distance between the MBZ boundaries. The spectrum appears to be equidistant only at the Brillouin zone boundaries ($Q = \pm 1/2$), and the distance between the energy levels depends strongly on the vortex lattice geometry. Near the points of intersection of the energy branches (12.25) in the $E-Q$ plane one should take into account the splitting of energy levels resulting from the exponentially small tunneling through the classically forbidden regions (the effect of tunneling becomes substantial only for Q close to the MBZ boundaries). As a result, we obtain a set of narrow bands separated by energy gaps.

Van Hove Singularities. Each energy branch has an extremum as a function of the momentum q_x near the MBZ boundary at a certain $\tilde{\varepsilon}_n$ (we neglect here the additional extrema which appear due to the exponentially small splitting of energy levels near the points of intersection of the branches in the $E-Q$ plane). Due to the 1D nature of the low energy spectrum the divergent contributions to the DOS take the form: $\delta N(\varepsilon) \sim |\varepsilon - \tilde{\varepsilon}_n|^{-1/2}$ ($\varepsilon > \tilde{\varepsilon}_n$ for energy minima and $\varepsilon < \tilde{\varepsilon}_n$ for maxima). The distance between these peaks $\delta\varepsilon \sim \varepsilon^*/(2\sigma\alpha)$ coincides with a characteristic energy scale corresponding to van Hove singularities which occur when the CO intersects MBZ boundaries in the q_y ($q_\parallel$) direction (see Fig. 12.1). For $\varepsilon \gtrsim \varepsilon_c$ the q_y-dependence of energy becomes essential and results in appearance of 2D critical points, i.e. 2D local

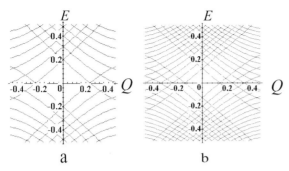

Fig. 12.3. Energy branches for $\pi\sigma\alpha = 50$ (**a**) and $\pi\sigma\alpha = 100$ (**b**)

maxima (or minima) and saddle–points. Thus, instead of the square-root van Hove singularities we obtain a set of discontinuities and logarithmic peculiarities ($\delta N(\varepsilon) \sim -\ln|1-\varepsilon/\tilde{\varepsilon}_n|$) in the DOS, respectively. Obviously, these 2D singularities are more sensitive to temperature and finite lifetime effects and, consequently, the suppression of the corresponding oscillatory structure in the DOS should be stronger in the high energy regime. The above analysis can be generalized for gap nodes at $\boldsymbol{k} = \pm k_F \boldsymbol{y}_0$: the corresponding energy scales take the form $\delta\varepsilon \sim 0.5\varepsilon^*/\alpha$, $\varepsilon_c \sim 0.5\varepsilon^*/\sigma$.

Even in the low energy regime the DOS oscillations with the energy scale $\delta\varepsilon$ are surely smeared due to a finite scattering rate Γ and temperature, and can be observed only for a moderate Dirac cone anisotropy and rather large magnetic fields. Comparing our results with the numerical solution [44] of the BdG equations for $\sigma = 1$ and $\alpha = 5/2$ we find that the above mechanism gives a good estimate of the energy scale of a double-peak structure in the tunneling conductance at the core center at $H/H_{c2} = 0.3$ ($\delta\varepsilon \sim 0.1\Delta_0$) and can explain the absence of this structure at low fields $H/H_{c2} = 0.05$ due to temperature broadening ($T = 0.1T_c > \delta\varepsilon \sim 0.05\Delta_0$). In principle, van Hove singularities may account for the peaks with a large energy gap $\sim \Delta_0/4$ observed experimentally at the vortex centers in YBaCuO [17] at $H \simeq 6$ T, provided we assume $\alpha \sim 1$. Unfortunately, the latter assumption is not consistent with the results of thermal conductivity measurements [64] ($\alpha \sim 14$), and, thus, the nature of the experimentally observed peaks is still unclear. It is also necessary to stress here that the existence of the critical points in the DOS is a direct consequence of perfect periodicity, and the introduction of a rather strong disorder surely removes these singularities.

12.4.3 Modified Semiclassical Approach for QP States in a Vortex Lattice

Hereafter we neglect the DOS oscillations discussed above and consider the modified semiclassical model proposed in [39–41]. According to this approach, the Doppler shift of the QP energy, which plays an important role for $\varepsilon \lesssim$

$\Delta_0\sqrt{H/H_{c2}}$, appears to be averaged in the nodal direction due to an extremely large size of a semiclassical wave packet in this energy interval. We also include in our consideration the finite lifetime effects which can strongly influence the behavior of the DOS [25,39–41,65–67]. According to [40] a diagonal (retarded) Green's function can be written in the form:

$$G^R(\mathbf{k}, \varepsilon, \mathbf{r}) = \frac{\varepsilon + i\Gamma - \hbar \mathbf{k}_F \mathbf{V}_{av} + \epsilon_{\mathbf{k}}}{(\varepsilon + i\Gamma - \hbar \mathbf{k}_F \mathbf{V}_{av})^2 - \Delta_{\mathbf{k}}^2 - \epsilon_{\mathbf{k}}^2}, \qquad (12.26)$$

where $\epsilon_{\mathbf{k}}$ is the normal state electron dispersion, $\mathbf{V}_{av} = \langle \mathbf{V}_s \rangle_x + \langle \mathbf{V}_s \rangle_y$. The scattering rate Γ should be determined self-consistently: $\Gamma = N(\Gamma, \varepsilon)/(2N_F\tau)$ (Born limit), $\Gamma = N_F \Gamma_u / N(\Gamma, \varepsilon)$ (unitary limit), where 2τ and N_F are the relaxation time and DOS at the Fermi level in the normal state, $\Gamma_u = n_{imp}/(\pi N_F)$, n_{imp} is the concentration of point potential scatterers, and $N(\Gamma, \varepsilon) = -\mathrm{Im} \int G^R d^2k/(2\pi^3)$ is the local DOS. The behavior of the DOS within this model differs qualitatively from the one expected on the basis of the conventional approach [11] (which takes account of the Doppler shift of the QP energy by the local $\mathbf{V}_s$ field). The expression for the residual local DOS reads:

$$N(\Gamma, \varepsilon = 0) = \frac{\Gamma N_F}{4\pi \Delta_0}\left(4\ln\frac{\Delta_0}{\Gamma} + \sum_{i=x,y} f_i\right) \qquad (12.27)$$

$$f_i = \frac{\varepsilon_i^* |\Phi_i|}{\Gamma}\tan^{-1}\frac{\varepsilon_i^* |\Phi_i|}{2\Gamma} - \ln\left(1 + \frac{(\varepsilon_i^* \Phi_i)^2}{4\Gamma^2}\right), \qquad (12.28)$$

where $\Phi_x = \Phi(x/R_x)$, $\Phi_y = \Phi(y/R_y)$, $\varepsilon_x^* = \pi\hbar V_F H R_x/\phi_0$, and $\varepsilon_y^* = \pi\hbar V_F \sigma/R_y$. Obviously, Born scatterers result only in a moderate change of the DOS (see [65]) since the corresponding Γ value for $\Delta_0 \tau \gg 1$ is very small compared to $\varepsilon_{x,y}^*$ and the local DOS is given by the expression

$$N(\Gamma \to 0, \varepsilon = 0) \simeq \frac{N_F}{8}\sqrt{\frac{\pi \sigma H}{2H_{c2}}} F_1(x,y), \qquad (12.29)$$

where $F_1(x,y) = |\Phi_x|(R_x/R_y) + |\Phi_y|$. If we assume the parameter σ to be field independent, then the magnetic field dependence of the residual spatially averaged DOS (for $\Gamma = 0$) follows the square root behavior which was first predicted in [11]. The deviations from this behavior may appear if the vortex lattice structure and, hence, the σ value change with an increasing magnetic field (see [50,60,61]). The characteristic energy ε of excitations coming into play at finite temperatures is of the order $\sim T$ and, consequently, for thermodynamic and transport quantities one can expect the existence of two crossover parameters $TR_x/(\Delta_0\xi)$ and $TR_y/(\Delta_0\xi)$ which separate the superflow dominating regimes from the temperature dominating regimes for QP contributions of different nodes. Thus, the scaling relations in the variable $T/\sqrt{H}$ for these quantities can be violated due to a possible change of the vortex lattice structure with an increase in a magnetic field.

In the unitary limit expression (12.29) is valid only in the clean case $\Gamma_u \ll \Gamma^*_{x,y} \sim 0.1\varepsilon^2_{x,y}/\Delta_0$ (for a square lattice $\Gamma^*_{x,y} \sim 0.1\Delta_0 H/H_{c2}$). In the dirty limit $\Gamma_u \gg \Gamma^*_{x,y}$ we obtain:

$$N \simeq N(H=0)\left(1 + \frac{\Delta_0 H \sigma}{64\Gamma_u H_{c2}} F_2(x,y)\right), \tag{12.30}$$

where $N(H=0) \simeq 0.5 N_F \sqrt{\Gamma_u/\Delta_0}$, and $F_2(x,y) = \Phi^2_x(R_x/R_y)^2 + \Phi^2_y$. In Fig. 12.4 we display the contour plots of the functions $F_1(x,y)$, $F_2(x,y)$ for a square lattice of type I (which is close to the one observed experimentally in YBaCuO [17,62]). In the vicinity of each vortex center the local DOS exhibits a fourfold symmetry with the maxima along the nodal directions, which agrees well with numerical calculations based on the Eilenberger theory [38]. Note that according to (12.30) the spatially averaged DOS in the dirty limit varies as H rather than $H \ln H$ (the latter dependence has been predicted in [25,65] within the semiclassical approach taking account of the local $\boldsymbol{V}_s$ value).

The peculiarities of the local DOS can be detected, e.g., by a scanning tunneling microscope (STM). In particular, the residual local DOS discussed above determines the value of the zero-bias tunneling conductance at $T=0$: $g = g_N N(\Gamma, \varepsilon = 0)/N_F$, where g_N is the normal state conductance. Note that for magnetic fields $H \sim 6\,\mathrm{T}$ (which is typically the field of STM experiment [17,18]) the finite temperature effects appear to be small for $T < T^* \sim \Delta_0 \sqrt{H/H_{c2}} \sim 20\,\mathrm{K}$ [40]. One can see that there are two important consequences of an increase in Γ_u: (i) first, the spatial dimensions of the peaks in the local tunneling conductance become rather small compared to the intervortex distance (see Fig. 12.4b); (ii) second, in the dirty limit the amplitude of the peaks appears to be strongly suppressed. For $H = 6\,\mathrm{T}$ these finite lifetime effects can be substantial if we assume $\Gamma_u \gtrsim 10^{-2}\Delta_0$. Thus, our approach allows us to explain rather narrow conductance peaks observed near the vortex centers in YBaCuO [17], even without taking account of

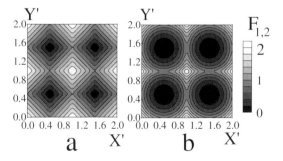

Fig. 12.4. Contour plots of the functions F_1 (**a**) and F_2 (**b**) which determine the spatial variation of the local DOS for a square lattice ($x' = x/a$, $y' = y/a$)

the nontrivial structure of the tunneling matrix element, discussed in [66]. With a further increase of the Γ_u value the amplitude of the peaks at the vortex centers vanishes: $\delta g \sim g_N \sqrt{\Delta_0/\Gamma_u}(H/H_{c2})$. Such a high sensitivity of the $\delta \tilde{g}$ value to finite lifetime effects can probably explain the difficulties in observation of these peaks in the mixed state of BiSrCaCuO [18].

12.5 Conclusions

To sum up, we have discussed an extremely important role of quantum mechanical effects in the behavior of low energy QP excitations in the mixed state of superconductors with gap nodes. Within the BdG theory linearized in gradient terms we have considered the peculiarities of extended QP states for isolated flux lines and regular vortex arrays. Such a model allowed us to develop a 2D quantum mechanical description of these states, taking account of both the long-range magnetic field effects and the periodic V_s potential. The important energy scales in the band spectrum are shown to be determined by the parameter V_F/V_Δ (which characterizes the Dirac cone anisotropy) and the vortex lattice geometry. It is hoped that the physical picture considered in this review can provide a starting point for the analysis of static and dynamic properties of the mixed state in various d-wave systems, including, probably, high-T_c copper oxides.

I am pleased to acknowledge useful discussions with I.D. Tokman, D.A. Ryndyk, A.A. Andronov, N.B. Kopnin, Yu.S. Barash, M. Franz, and Z. Tešanović. This work was supported, in part, by the Russian Foundation for Fundamental Research (Grant No. 99–02–16188).

References

1. C. Caroli, P.G. de Gennes, J. Matricon: Phys. Lett. **9**, 307 (1964)
2. E. Canel: Phys. Lett. 16, 101 (1965)
3. U. Klein: Phys. Rev. B **40**, 6601 (1989)
4. F. Gygi and M. Schluter: Phys. Rev. Lett. **65**, 1820 (1990)
5. B. Pöttinger and U. Klein: Phys. Rev. Lett. **70**, 2806 (1993)
6. N. Hayashi, M. Ichioka, and K. Machida: Phys. Rev. Lett. **77**, 4074 (1996)
7. N. Hayashi, M. Ichioka, and K. Machida: Phys. Rev. B **55**, 6565 (1997)
8. R.P. Huebener, O.M. Stoll, A. Wehner, and M. Naito: Physica C **332**, 187 (2000)
9. K.A. Moler et al.: Phys. Rev. Lett. **73**, 2744 (1994)
10. K.A. Moler et al.: Phys. Rev. B **55**, 3954 (1997)
11. B. Revaz et al.: Phys. Rev. Lett. **80**, 3364 (1998)
12. N.E. Phillips et al.: J. of Supercond. **12**, 105 (1999)
13. D.A. Wright et al.: Phys. Rev. B **82**, 1550 (1999)
14. R.A. Fisher et al.: Phys. Rev. B **61**, 1473 (2000)
15. K. Krishana et al.: Science **277**, 83 (1997)
16. H. Aubin et al.: Phys. Rev. Lett. **82**, 624 (1999)

17. I. Maggio-Aprile, Ch. Renner, A. Erb et al.: Phys. Rev. Lett. **75**, 2754 (1995)
18. Ch. Renner, B. Revaz, K. Kadowaki et al.: Phys. Rev. Lett. **80**, 3606 (1998)
19. B.W. Hoogenboom et al.: Phys. Rev. B **62**, 9179 (2000)
20. S.H. Pan et al.: Phys. Rev. Lett. **85**, 1536 (2000)
21. G.E. Volovik: Pis'ma Zh. Eksp. Teor. Fiz. **58**, 457 (1993) [JETP Lett. **58**, 469 (1993)]
22. N.B. Kopnin and G.E. Volovik: Pis'ma Zh. Eksp. Teor. Fiz. **64**, 641 (1996) [JETP Lett. **64**, 690 (1996)]
23. G.E. Volovik: Pis'ma Zh. Eksp. Teor. Fiz. **65**, 465 (1997) [JETP Lett. **65**, 491 (1997)]
24. S.H. Simon and P.A. Lee: Phys. Rev. Lett. **78**, 1548 (1997)
25. C. Kübert and P.J. Hirschfeld: Solid St. Commun. **105**, 459 (1998)
26. C. Kübert and P.J. Hirschfeld: Phys. Rev. Lett. **80**, 4963 (1998)
27. I. Vekhter, J.P. Carbotte, and E.J. Nikol: Phys. Rev. B **59**, 1417 (1999)
28. L.P. Gor'kov and J.R. Schrieffer: Phys. Rev. Lett. **80**, 3360 (1998)
29. P.I. Soininen, C. Kallin, and A.J. Berlinsky: Phys. Rev. B **50**, 13883 (1994)
30. Y. Wang and A.H. MacDonald: Phys. Rev. B **52**, R3876 (1995)
31. Y. Morita, M. Kohmoto, and K. Maki: Europhys. Lett. **40**, 207 (1997)
32. M. Franz and Z. Tešanović: Phys. Rev. Lett. **80**, 4763 (1998)
33. N.B. Kopnin and G.E. Volovik: Phys. Rev. Lett. **79**, 1377 (1997)
34. N.B. Kopnin: Phys. Rev. B **57**, 11775 (1998)
35. Yu.G. Makhlin: Phys. Rev. B **56**, 11872 (1997)
36. N. Schopohl and K. Maki: Phys. Rev. B **52**, 490 (1995)
37. M. Ichioka, N. Hayashi, N. Enomoto, and K. Machida: Phys. Rev. B **53**, 15316 (1996)
38. M. Ichioka, A. Hasegawa, and K. Machida: Phys. Rev. B **59**, 184 (1999)
39. A.S. Mel'nikov: Report No. cond-mat/9806188; J. Phys. Condens. Matter **11**, 4219 (1999)
40. A.S. Mel'nikov: Pis'ma Zh. Eksp. Teor. Fiz. **71**, 472 (2000) [JETP Lett. **71**, 327 (2000)]
41. A.S. Mel'nikov: Physica B **284-288**, 781 (2000)
42. P.W. Anderson: Report No. cond-mat/9812063
43. M. Franz and Z. Tešanović: Phys. Rev. Lett. **84**, 554 (2000)
44. K. Yasui and T. Kita: Phys. Rev. Lett. **83**, 4168 (1999)
45. B. Jankó: Phys. Rev. Lett. **82**, 4703 (1999)
46. L. Marinelli, B.I. Halperin, and S.H. Simon: Phys. Rev. B **62**, 3488 (2000)
47. N.B. Kopnin and V.M. Vinokur: Phys. Rev. B **62**, 9770 (2000)
48. G.E. Volovik and N.B. Kopnin: Phys. Rev. Lett. **78**, 5028 (1997)
49. S.H. Simon and P.A. Lee: Phys. Rev. Lett. **78**, 5029 (1997)
50. M. Franz, I. Affleck, and M.H.S. Amin: Phys. Rev. Lett. **79**, 1555 (1997)
51. J. Ye: Phys.Rev.Lett. **86**, 316 (2001)
52. Y. Aharonov and D. Bohm: Phys. Rev. **115**, 485 (1959)
53. R.M. Cleary: Phys. Rev. **175**, 587 (1968)
54. E.B. Sonin: Phys. Rev. B **55**, 485 (1997)
55. M.G. Alford and F. Wilczek: Phys. Rev. Lett. **62**, 1071 (1989)
56. Ph. de Sousa Gerbert: Phys. Rev. D **40**, 1346 (1989)
57. C.R. Hagen: Phys. Rev. Lett. **64**, 503 (1990)
58. A.S. Mel'nikov: Report No. cond-mat/0007156; Phys. Rev. Lett., accepted for publication

59. H. Won and K. Maki: Phys. Rev. B **53**, 5927 (1996)
60. M.H.S. Amin, I. Affleck, and M. Franz: Phys. Rev B **58**, 5848 (1998)
61. J. Shiraishi, M. Kohmoto, and K. Maki: Phys. Rev B **59**, 4497 (1999)
62. B. Keimer et al.: Phys. Rev. Lett. **73**, 3459 (1994)
63. E.O. Kane and E.I. Blount: In: *Tunneling Phenomena in Solids*, ed. by E. Burstein and S. Lundqvist (Plenum Press, New York, 1969)
64. M. Chiao, R.W. Hill, C. Lupien et al.: Phys. Rev. Lett. **82**, 2943 (1999)
65. Yu.S. Barash, V.P. Mineev, and A.A. Svidzinskii: Pis'ma Zh. Eksp. Teor. Fiz. **65**, 606 (1997) [JETP Lett. **65**, 638 (1997)]
66. M. Franz and Z. Tešanović: Phys. Rev. B **60**, 3581 (1999)
67. E. Schachinger and J.P. Carbotte: Phys. Rev. B **62**, 592 (2000)

13 Magnetic Field Dependence of the Vortex Structure Based on the Microscopic Theory

Masanori Ichioka, Mitsuaki Takigawa, and Kazushige Machida

Summary. We investigate the vortex structure in the vortex lattice state by the self-consistent numerical calculation based on the microscopic theory, and focus on the magnetic field dependence. First, we study the vortex structure comparatively for the s-wave and the $d_{x^2-y^2}$-wave pairing cases by the quasi-classical Eilenberger theory. We show the spatial variation of the local density of states (LDOS), and discuss the field dependence of the spatially averaged density of states, vortex core radius, and internal field distribution. Second, we propose the site-selective NMR relaxation time measurement as a new method to observe the quasiparticle state around the vortex core. We explain the relation between the LDOS and the temperature dependence of the site-selective T_1 based on the Bogoliubov–de Gennes theory, and discuss the field dependence of the T_1-behavior.

13.1 Introduction

Much attention has been focused on a vortex state in high-T_c superconductors. Many researchers try to detect the $d_{x^2-y^2}$-wave nature of the superconductivity in the vortex structure of the mixed state. In the $d_{x^2-y^2}$-wave superconductors, low-energy quasiparticle excitations around the vortex core are completely different from the conventional s-wave superconductors. In the s-wave pairing, low-energy quasiparticles are bounded within the vortex core, and the energy levels are quantized with the energy gap of the order Δ_0^2/E_F (Δ_0 is the superconducting gap and E_F is the Fermi energy) [1–4]. While Δ_0/E_F is negligibly small in conventional superconductors, high-T_c superconductors have large Δ_0/E_F [5]. In the $d_{x^2-y^2}$-wave pairing, low-energy quasiparticles around the vortex core extend far from the vortex core due to the line node of the $d_{x^2-y^2}$-wave superconductivity [6–9]. Then, theoretical studies [10,11] suggest that the zero-energy peak appears in the local density of states (LDOS) at the vortex core instead of the gap Δ_0^2/E_F. Since most of the experiments for thermodynamics or transports detect the spatially averaged density of states (DOS), they cannot directly observe the $d_{x^2-y^2}$-wave nature in low-energy quasiparticle excitations around the core. But, if we consider the field H dependence, we extract the $d_{x^2-y^2}$-wave nature, i.e., the contributions of low-energy quasiparticles extending away from the vortex core. As for the zero-energy DOS $N(0)$, Volovik [6] suggested that

$N(0) \propto \sqrt{H}$ in the $d_{x^2-y^2}$-wave pairing, while $N(0) \propto H$ in the s-wave pairing.

Our purpose is to numerically calculate the vortex structure based on the microscopic theory, and to check the field dependence of the vortex structure [12,13]. Then, since the vortex structure is a self-consistently obtained result within the theory, we exclude artificial assumptions about the vortex structure. In Sect. 13.2, we report results of the self-consistent calculation by the quasi-classical Eilenberger theory. By comparing the s-wave and the $d_{x^2-y^2}$-wave pairing cases, we discuss the $d_{x^2-y^2}$-wave nature about the LDOS, the spatially averaged DOS and the internal field distribution. We also discuss the field dependence of the vortex core radius and the effect of quasiparticle transfer between neighboring vortices. As for the pair potential and the internal field distributions, the vortex core structure is deformed to square-like shape in the $d_{x^2-y^2}$-wave pairing, while it is circular-shape in the s-wave pairing [8]. Since the square-shape structure is restricted only near the vortex core, its contribution becomes eminent when the core region shares large area of the vortex lattice at high fields.

The scanning tunneling microscopy (STM) is a powerful method for the direct observation of quasiparticle states around the vortex [14]. In high-T_c superconductors, some groups observed the vortex image by STM [5,15,16]. But, they did not see the eminent zero-energy states at the vortex core. There appear only small mysterious shoulders or isolated peaks at higher energy. It may reflect the exotic character of the strongly correlated electron system in high-T_c superconductors [17]. However, we can not exclude the possibility that the STM spectrum is a result largely affected by the surface effect or the c-axis transfer effect [18]. Then, another experimental method is expected to confirm the spatial profile of quasiparticle states around the vortex. Then, we propose the site-selective nuclear magnetic resonance (NMR) measurement [19]. As the resonance spectrum of NMR reflects the internal field distribution of the vortex lattice, we can specify the spatial position in the vortex lattice by tuning the resonance field. If we observe the nuclear spin relaxation time T_1 at the selected position, we can obtain information of the quasiparticle state at that position. By studying the position dependence, we can clarify the spatial structure of quasiparticles around the vortex. At the vortex core, fast relaxation is expected due to the low-energy quasiparticle states. In Sect. 13.3, we perform a model calculation of the site-selective T_1 by self-consistently solving the Bogoliubov–de Gennes (BdG) equation for the extended Hubbard model. The temperature dependence of T_1 reflects the LDOS around the vortex. We also study the field dependence of T_1. The last section is devoted to concluding remarks.

13.2 Magnetic Field Dependence of the Vortex Structure

Our calculation is performed following the quasiclassical method of [20]. We consider the case of the clean limit and a cylindrical Fermi surface. These are appropriate to high-T_c superconductors and low-dimensional organic superconductors. After we explain the formulation, we show our results about the field dependence of the vortex structure.

13.2.1 Quasiclassical Eilenberger Theory

First, we obtain the pair potential and vector potential self-consistently by solving the Eilenberger equation in the Matsubara frequency $\omega_n = (2n + 1)\pi T$. We consider the quasiclassical Green's functions $g(i\omega_n, \theta, \boldsymbol{r})$, $f(i\omega_n, \theta, \boldsymbol{r})$ and $f^\dagger(i\omega_n, \theta, \boldsymbol{r})$, where $\boldsymbol{r}$ is the center of mass coordinate of a Cooper pair. The direction of relative momentum of the Cooper pair, $\hat{\boldsymbol{k}} = \boldsymbol{k}/|\boldsymbol{k}|$, is denoted by an angle θ measured from the x axis. The Eilenberger equation is given by [21]

$$\left\{\omega_n + \frac{i}{2}v_F \cdot \left(\frac{\nabla}{i} + \frac{2\pi}{\phi_0}\boldsymbol{A}(\boldsymbol{r})\right)\right\}f(i\omega_n, \theta, \boldsymbol{r}) = \Delta(\theta, \boldsymbol{r})g(i\omega_n, \theta, \boldsymbol{r}),$$

$$\left\{\omega_n - \frac{i}{2}v_F \cdot \left(\frac{\nabla}{i} - \frac{2\pi}{\phi_0}\boldsymbol{A}(\boldsymbol{r})\right)\right\}f^\dagger(i\omega_n, \theta, \boldsymbol{r}) = \Delta^*(\theta, \boldsymbol{r})g(i\omega_n, \theta, \boldsymbol{r}), \quad (13.1)$$

where $g(i\omega_n, \theta, \boldsymbol{r}) = [1 - f(i\omega_n, \theta, \boldsymbol{r})f^\dagger(i\omega_n, \theta, \boldsymbol{r})]^{1/2}$, $\mathrm{Re}\,g(i\omega_n, \theta, \boldsymbol{r}) > 0$ and $\boldsymbol{v}_F$ ($= v_F\hat{\boldsymbol{k}}$) is the Fermi velocity. The vector potential is written as $\boldsymbol{A}(\boldsymbol{r}) = \frac{1}{2}\boldsymbol{H} \times \boldsymbol{r} + \boldsymbol{a}(\boldsymbol{r})$ in the symmetric gauge, where $\boldsymbol{H} = (0, 0, H)$ is a uniform field and $\boldsymbol{a}(\boldsymbol{r})$ is related to the internal field $\boldsymbol{h}(\boldsymbol{r}) = (0, 0, h(\boldsymbol{r}))$ as $\boldsymbol{h}(\boldsymbol{r}) = \nabla \times \boldsymbol{a}(\boldsymbol{r})$. As for the pair potential $\Delta(\theta, \boldsymbol{r}) = \Delta(\boldsymbol{r})\phi(\theta)$ and the pairing interaction $V(\theta', \theta) = \bar{V}\phi(\theta')\phi(\theta)$, we set $\phi(\theta) = \sqrt{2}\cos 2(\theta - \theta_0)$ for the $d_{x^2-y^2}$-wave pairing and $\phi(\theta) = 1$ for the s-wave pairing. Here, θ_0 is the angle between the x-axis and the a-axis of the crystal coordinate. By solving (13.1) in the so-called explosion method [13,20] under $\Delta(\boldsymbol{r})$ and $\boldsymbol{a}(\boldsymbol{r})$ of the vortex lattice case, we estimate the quasiclassical Green's functions. The self-consistent condition is given as

$$\Delta(\theta, \boldsymbol{r}) = N_0 2\pi T \sum_{\omega_n > 0} \int_0^{2\pi} \frac{D\theta'}{2\pi} V(\theta', \theta) f(i\omega_n, \theta', \boldsymbol{r}), \quad (13.2)$$

$$\nabla \times \nabla \times \boldsymbol{a}(\boldsymbol{r}) = -\frac{\pi\phi_0}{\kappa^2 \Delta_0 \xi_0^3} 2\pi T \sum_{\omega_n > 0} \int_0^{2\pi} \frac{D\theta}{2\pi} \frac{\hat{\boldsymbol{k}}}{i} g(i\omega_n, \theta, \boldsymbol{r}). \quad (13.3)$$

We obtain new $\Delta(\boldsymbol{r})$ and $\boldsymbol{a}(\boldsymbol{r})$ from the quasiclassical Green's functions, and use them at the next step calculation of (13.1). This iteration procedure is repeated until a sufficiently self-consistent solution is obtained. We use the material parameters appropriate to YBCO, i.e., the BCS coherence length $\xi_{\rm BCS} = 16$ Å and the GL parameter $\kappa = 100$. To study the field dependence, calculations are done for various fields at the fixed temperature $T/T_{\rm c} = 0.5$. In the following, energies and lengths are measured in units of Δ_0 and $\xi_0 = v_{\rm F}/\Delta_0 = \pi\xi_{\rm BCS}$ (Δ_0 is the uniform gap at $T=0$), respectively. The shape of the vortex lattice is assumed to be a square lattice tilted by 45° from the a axis. This vortex lattice configuration is suggested at a higher field for $d_{x^2-y^2}$-wave superconductors, or s-wave superconductors with a four-fold symmetric Fermi surface [22–25]. We confirm that the square lattice has a lower free energy than the conventional 60° triangular lattice at a higher field in the $d_{x^2-y^2}$-wave pairing by our framework[12].

Next, we solve (13.1) for $\eta - iE$ instead of ω_n using the self-consistently obtained $\Delta(\bm{r})$ and $\bm{a}(\bm{r})$ to obtain $g(i\omega_n \to E + i\eta, \theta, \bm{r})$. The LDOS for energy E is calculated as

$$N(E,\bm{r}) = N_0 \int_0^{2\pi} \frac{D\theta}{2\pi} {\rm Re}\, g(i\omega_n \to E + i\eta, \theta, \bm{r}) , \qquad (13.4)$$

where N_0 is the DOS at the Fermi surface. Typically, we choose $\eta = 0.03$.

13.2.2 Local Density of States

The LDOS $N(E=0,\bm{r})$ for the s-wave pairing is shown in Fig. 13.1. At low fields, the LDOS for low energy excitations distributes in the restricted circular region around the vortex core, as in the single vortex case [7,26]. For finite E, the LDOS is distributed on a circle around the vortex center. The radius of the circle increases from 0 upon raising E. With increasing field, low energy excitations around the vortex overlap each other between nearest neighbor (NN) vortices, as shown in Fig. 13.1b. Then, the vortex lattice effect appears. There, $N(E=0,\bm{r})$ is suppressed on the line connecting NN vortex centers. The small suppression also occurs along the line between next nearest neighbor (NNN) vortex centers. They reflect the quasiparticle transfer between vortex cores. The LDOS has finite distributions all over the unit cell at the high field. It may be related to the fact that the dHvA oscillation is possible even in the superconducting state.

The $d_{x^2-y^2}$-wave case is shown in Fig. 13.2. As in the single vortex calculation (see Fig. 8(d) of [8]), the LDOS for low energy excitations around a vortex has the eminent tails extending along the 45° direction, reflecting the node structure of the superconducting gap. But, even at a low field $H/H_{c2} = 0.02$ [Fig. 13.2a], the tail structure of $N(E=0,\bm{r})$ along the 45° direction of the single vortex case is modified due to the vortex lattice effect. Each tail along the 45° direction splits into two ridges by the suppression of the vortex lattice effect along the line of the NN vortex direction. Then, the tails extend in rather different directions than the exact 45° direction. The split tail structure from the low field means that the quasiparticle transfer between vortices is large due to the tail structure of the LDOS in the $d_{x^2-y^2}$-wave pairing, compared with the s-wave pairing case. It may be related to the difference in the quasiparticle band between the s-wave and the $d_{x^2-y^2}$-wave pairings, as reported in [27]. There, the quasiparticle band is dispersive in the $d_{x^2-y^2}$-wave

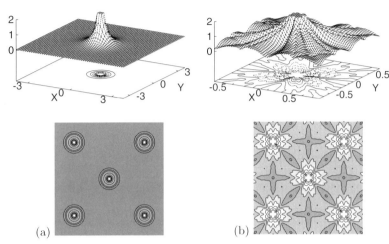

Fig. 13.1. Spatial variation of the LDOS $N(E,\boldsymbol{r})/N_0$ for the s-wave pairing. Upper panels show the region of a Wigner–Seitz cell of the square vortex lattice. The x-axis and the y-axis are along the nearest neighbor vortex directions, which is $45°$ from the a-axis of the crystal coordinate. The peak height is truncated at $N(E,\boldsymbol{r})/N_0 = 2$. Lower panels show the contour lines in wider region. The a-axis and b-axis are along the horizontal and vertical directions. There is a vortex center at each white area. (**a**) At a low field $H/H_{c2}= 0.021$. (**b**) At a high field $H/H_{c2}= 0.54$.

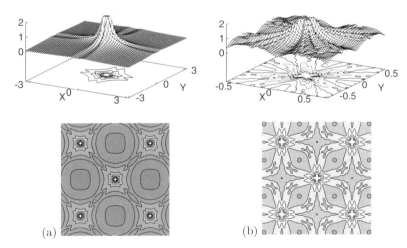

Fig. 13.2. (**a,b**) The same as Fig. 13.1, but for the $d_{x^2-y^2}$-wave pairing case.

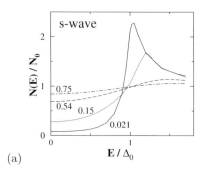 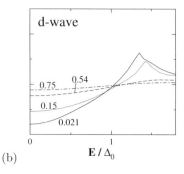

Fig. 13.3. Spectrum of the spatially averaged DOS, $N(E)/N_0$, at $H/H_{c2}=0.021$, 0.15, 0.54 and 0.75. (**a**) For the s-wave pairing. (**b**) For the $d_{x^2-y^2}$-wave pairing.

pairing, reflecting large quasiparticle transfer between vortex cores. With increasing field, the vortex lattice effect increases and the splitting of the tail structure in the 45° direction becomes prominent. The LDOS for finite E is reported in [12] and [13].

The spatially averaged DOS $N(E)$ is presented in Fig. 13.3 for the s-wave and the $d_{x^2-y^2}$-wave pairings. We note that the energy gap in the $d_{x^2-y^2}$-wave pairing is $\sqrt{2}\Delta_0$ in Fig. 13.3b. The energy gap structure at zero field is gradually buried by low energy excitations of the vortex as a field increases. While the local spectrum $N(E, \mathbf{r})$ has several peaks depending on the pairing symmetry and the position[8], there is no peak in $N(E)$ below the energy gap. The peak structure is smeared when the spatial average is taken.

We consider the field dependence of the zero-energy DOS $N(0)$, and analyze it in view of the above structure of low-energy excitations around the core. According to Volovik [6], the contribution to $N(0)$ mainly comes from the tail structure along the node direction in $N(E=0, \mathbf{r})$ for the $d_{x^2-y^2}$-wave pairing. The length of the tail is of the order of $H^{-1/2}$ (lattice constant of the vortex lattice). As the vortex density is proportional to H, $N(0)$ is roughly estimated as $N(0) \sim H^{-1/2} \cdot H = \sqrt{H}$. This estimate is now modified, since the tail structure along the node directions is affected by the vortex lattice effect as shown in Fig. 13.2. We present our results for the field dependence of $N(0)$ in Fig. 13.4a. While the difference between the $d_{x^2-y^2}$-wave and the s-wave clearly appears, the dependence in the $d_{x^2-y^2}$-wave pairing deviates from exact $\sqrt{H}$ behaviour (dotted line). The best fit is obtained by $N(0)/N_0 = (H/H_{c2})^{0.41}$ (solid line). Its exponent 0.41 is slightly smaller than 0.5 of the Volovik theory. The deviation from $\sqrt{H}$ behaviour was also reported by the calculation based on the BdG theory[10]. Experimentally, $N(0)$ is obtained from the coefficient of the T-linear term in the specific heat $C(T)$, i.e., $N(0) \propto \gamma(H) = \lim_{T \to 0} C(T)/T$. So far, the $\sqrt{H}$ behaviour of

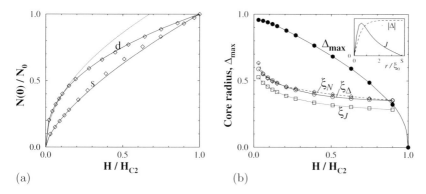

Fig. 13.4. (a) Field dependence of $N(0)/N_0$ in the $d_{x^2-y^2}$-wave and the s-wave pairings (*points* ◇). We also show fitting lines (*solid line*) and a $\sqrt{H}$ curve (*dotted line*) (b) Field dependence of core radius ξ_Δ, ξ_J and ξ_N for the s-wave pairing. Maximum amplitude of the order parameter $\Delta_{\max}$ is also plotted. The radius and $\Delta_{\max}$ are, respectively, scaled by ξ_0 and Δ_0. Lines are guides for the eye. The inset shows profiles of the pair potential $\Delta(r)$ and the current amplitude $J(r)$ from the vortex center to the midpoint S of the nearest neighbor vortices at $H/H_{c2}= 0.021$ in the s-wave pairing.

$\gamma(H)$ was examined within the low field region [28,29]. The deviation from $\sqrt{H}$ is expected when $\gamma(H)$ is measured in higher field regions.

As for the s-wave pairing case, we expect that $N(0) \propto H$, since $N(0)$ is proportional to the vortex density. However, our data in Fig. 13.4a deviate from the H-linear relation. The best fit to our data is obtained by $N(0)/N_0 = (H/H_{c2})^{0.67}$ (solid line). To understand the origin of the deviation, we show the field dependence of the vortex core radius in Fig. 13.4b. The radius ξ_Δ is defined from the initial slope of the pair potential by fitting as $|\Delta(r)| = \Delta_{\max} r/\xi_\Delta$ at the vortex center. There, $\Delta_{\max}$ is defined as the maximum of $|\Delta(r)|$ along the NN vortex direction. The radius ξ_J is the maximum point of the screening current around a vortex. As H increases, both ξ_Δ and ξ_J decrease similarly. In the calculation of a single vortex [30], zero-energy DOS per vortex $N(0)/H$ is proportional to an area of the vortex core $\pi\xi_N^2$, and the radius ξ_N corresponds to our ξ_Δ. If ξ_N is independent of H, we obtain the naively expected relation $N(0) \propto H$. However, it is not the case. In Fig. 13.4b, we also plot the core radius ξ_N estimated from $N(0)$, where $\xi_N = 0.35[(N(0)/N_0)/(H/H_{c2})]^{1/2}$ with a fitting parameter 0.35. The radius ξ_N decreases similarly as ξ_Δ with increasing H. This means that the deviation from H-linear in $N(0)$ reflects the field dependence of the core radius. The deviation from the H-linear in s-wave superconductors was reported experimentally for V_3Si [31] and $CeRu_2$ [32]. The shrinkage of the core radius was also reported by STM[33] and μSR[34] experiments. A review of the vortex state through μSR studies is given by [35].

The origin of the shrinkage is the compression from neighboring vortices. It is understood when we see the supercurrent profile $J(r)$ around the vortex core. After the peak at $r = \xi_J$, $J(r)$ is reduced to 0 at the midpoint S of the NN vortices. With increasing field, the midpoint approaches the vortex center, and consequently the peak at $r = \xi_J$ also approaches the vortex center. The inset in Fig. 13.4b shows the profile at a low field $H/H_{c2} = 0.021$. In this case, the midpoint is at $r = 3.8\xi_0 \sim 12\xi_{\rm BCS}$ (the order of $\sqrt{H_{c2}/H}$). Then, the midpoint S affects well to the shrinkage of the vortex core, since it is narrower than the penetration depth ($\sim \kappa\xi_{\rm BCS}$).

The core radius (ξ_Δ and ξ_J) shows almost the same field dependence also in the $d_{x^2-y^2}$-wave pairing. However, ξ_N shows the different behaviour from the core radius. Then, the $\sqrt{H}$-like behaviour of $N(0)$ in the $d_{x^2-y^2}$-wave pairing is not understood simply by the H-dependence of the core radius. It includes the effect of the quasiparticle distribution extending outside of the vortex core.

13.2.3 Pair Potential and Internal Field Distribution

The $d_{x^2-y^2}$-wave nature appears in the shape of the contour lines of $|\Delta(\boldsymbol{r})|$ and $h(\boldsymbol{r})$ around the vortex core. While the contour lines are circular around each vortex center in the s-wave pairing, they become four-fold symmetric square-like shape in the $d_{x^2-y^2}$-wave pairing. There, $|\Delta(\boldsymbol{r})|$ is suppressed along the a- and b-axis directions around the vortex core. And $h(\boldsymbol{r})$ extends toward the a- and b-axis directions. This square-like structure is restricted within the vortex core region. When leaving the core region, $h(\boldsymbol{r})$ immediately reduces to the circular structure [8,36].

When the contribution of the core region increases at high fields, the $d_{x^2-y^2}$-wave nature becomes eminent. In Figs. 13.5a and b, we show the contour lines of $h(\boldsymbol{r})$ at $H/H_{c2}=0.54$ for the s-wave and the $d_{x^2-y^2}$-wave pairings, respectively. In the s-wave pairing, $h(\boldsymbol{r})$ has minimum at the boundary of the NNN direction. Also, there is a saddle point of $h(\boldsymbol{r})$ at the boundary of each NN direction. In the $d_{x^2-y^2}$-wave pairing, by the effect of the square-like structure around the vortex core, $h(\boldsymbol{r})$ is enhanced at the minimum point of the NNN direction, and suppressed at the saddle point of the NN direction. Then, with increasing field, the minimum of $h(\boldsymbol{r})$ increases and approaches the logarithmic peak field of the saddle points, as shown in Fig. 13.5c. It may be observed as the resonance line shape in the μSR or NMR experiments. The neutron scattering can also detect $h(\boldsymbol{r})$ through the form factor $h_{m,n}$ of the internal field. While the $d_{x^2-y^2}$-wave nature does not so clearly appear in the field dependence of the dominant factor $h_{1,0}$, it affects on higher factors $h_{1,1}$ and $h_{2,0}$ [12].

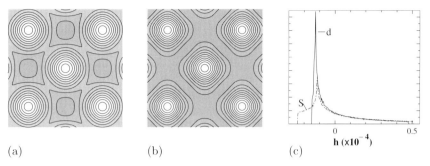

Fig. 13.5. Spatial variation of the internal magnetic field $h(\mathbf{r})$ at a high field $H/H_{c2}=0.54$. Contour lines of $h(\mathbf{r})$ are plotted. The white area shows large $h(\mathbf{r})$ around the vortex core. (**a**) For the s-wave pairing. (**b**) For the $d_{x^2-y^2}$-wave pairing. (**c**) Magnetic field distribution function $\rho(h)$ for (a) and (b). h is normalized by ϕ_0/ξ_0^2.

13.3 Site-Selective Nuclear Spin Relaxation Time

As discussed in the previous section, there are drastic differences between the s-wave and the $d_{x^2-y^2}$-wave pairings in the low-energy quasiparticle structure around the vortex core. In the s-wave superconductors, low-energy quasiparticles are bound within the vortex core. Then, if we consider the quantum effect by the BdG theory, the effect of the quantized energy level will appear in the quasiparticle state [1–4]. In the $d_{x^2-y^2}$-wave pairing, the low energy state around the vortex core extends outside the core due to the node of the superconducting gap, and theoretically we expect the zero-energy peak instead of the quantized energy level [8,10,11]. However, the STM experiments suggested quantized energy levels with a large gap in YBCO [5]. In BSCCO, there is no eminent low energy quasiparticle state in the vortex core [15], or the shoulder structure appears within the superconducting gap [16]. Then, it is an open question whether the STM spectrum reflects the exotic electronic structure in high-T_c superconductors [17], or the spectrum is a result largely affected by the intrinsic problem associated with the STM experiment, such as the surface effect or the c-axis transfer effect [18]. To clarify this point, we need another experimental method to check the quasiparticle structure around the vortex observed by STM.

Here we propose a novel spatially resolved method, that is, vortex imaging to see electronic excitations associated with a vortex core by using NMR, and demonstrate how the T-dependence of the nuclear spin relaxation time T_1 is site-sensitive [19]. A similar idea is also proposed in [37] theoretically. Through this analysis, we are able to produce a spatial image of the low-lying excitation spectrum around the core. Among various experimental methods, NMR experiment is unique because it provides us with two kinds of information. Namely, the nuclear resonance spectrum reflects the magnetic field distribution and T_1 probes electronic excitations through its T-dependence. The resonance spectrum reflects information of internal magnetic field distribution of the vortex lattice [38,39]. Thus, by choosing the resonance field, we can specify the spatial position to detect the NMR signal. The signal at the maximum (minimum) cutoff comes from the vortex center (the furthest) site. The T_1-measurement is a powerful method to discover the low energy DOS structure. For example, the power law $T_1^{-1} \propto T^3$ (T^5) behaviour is taken as definitive evidence for a line (point) node in the gap structure of unconventional superconductors. This conclusion comes from a simple power counting of the DOS: $N(E) \propto E$ (E^3) for a line (point) node. It can observe the full gap structure or the remained zero-energy DOS. By studying the position dependence of T_1 around vortices through the resonance frequency dependence, we can clarify the detail of the spatial structure of the electronic state around the vortex core. Since NMR observes the bulk properties, it is free from the surface problem.

The NMR imaging is actually tested experimentally in high T_c superconductors by some groups [40–42]. They report that T_1 becomes faster when approaching the vortex core. It is qualitatively consistent with our calculation.

13.3.1 Bogoliubov–de Gennes Theory for Extended Hubbard Model

In order to analyze the expected NMR data and propose suitable NMR experiments, we perform a model calculation of T_1. The position dependence of the NMR signal was theoretically studied in the s-wave pairing under some approximations [2,43]. Here, we calculate it microscopically from wave functions obtained by self-consistently solving the BdG equation for the extended Hubbard model. We report mainly the $d_{x^2-y^2}$-wave pairing case. You will also find the s-wave pairing case in [19].

We consider the extended Hubbard model on a two dimensional square lattice of atomic sites. First, by following the method of [10], we obtain the eigen-energy E_α and the wave functions $u_\alpha(\boldsymbol{r}_i)$, $v_\alpha(\boldsymbol{r}_i)$ at i-site by the BdG theory, where α is the index of the eigen-state. The BdG equation for the extended Hubbard model is given by

$$\sum_j \begin{pmatrix} K_{i,j} & D_{i,j} \\ D_{i,j}^\dagger & -K_{i,j}^* \end{pmatrix} \begin{pmatrix} u_\alpha(\boldsymbol{r}_j) \\ v_\alpha(\boldsymbol{r}_j) \end{pmatrix} = E_\alpha \begin{pmatrix} u_\alpha(\boldsymbol{r}_i) \\ v_\alpha(\boldsymbol{r}_i) \end{pmatrix} , \tag{13.5}$$

where $K_{i,j} = -t_{i,j} \exp\left[i\frac{\pi}{\phi_0} \int_{\boldsymbol{r}_i}^{\boldsymbol{r}_j} \boldsymbol{A}(\boldsymbol{r}) \cdot D\boldsymbol{r} \right] - \delta_{i,j}\mu$ with the transfer integral $t_{i,j}$, the chemical potential μ and the flux quantum ϕ_0. $D_{i,j} = \delta_{i,j} U \Delta_{i,i} + \frac{1}{2} V_{i,j} \Delta_{i,j}$ with the on-site interaction U and the NN interaction $V_{i,j}$. For the NN site pair $\boldsymbol{r}_i$ and $\boldsymbol{r}_j$, $t_{i,j} = t$ and $V_{i,j} = V$, and $t_{i,j} = V_{i,j} = 0$

otherwise. The vector potential $\mathbf{A}(\mathbf{r}) = \frac{1}{2}\mathbf{H} \times \mathbf{r}$ in the symmetric gauge. The self-consistent condition for the pair potential is

$$\Delta_{i,j} = -\frac{1}{2}\sum_\alpha u_\alpha(\mathbf{r}_i)v_\alpha^*(\mathbf{r}_j)\tanh\frac{E_\alpha}{2T}. \quad (13.6)$$

We consider the square vortex lattice case where NN vortex is located at the 45° direction from the a-axis, as in Sect. 13.2. The unit cell in our calculation is the square area of N_r^2 sites where two vortices are accommodated. Then, the magnetic field is determined as $H = 2\phi_0/(cN_r)^2$, where c is the atomic lattice constant. We specify the strength of H as H_{N_r} by the unitcell size N_r. By introducing the quasi-momentum of the magnetic Bloch state, we consider the system where many unit cells are included. The periodic boundary condition is imposed by the translational symmetry under magnetic field, where the gauge factor appears. To understand the behaviour of the position-dependent $T_1(\mathbf{r})$, we also consider the LDOS given by

$$N(E,\mathbf{r}) = \sum_\alpha\{|u_\alpha(\mathbf{r})|^2\delta(E-E_\alpha)+|v_\alpha(\mathbf{r})|^2\delta(E+E_\alpha)\}. \quad (13.7)$$

For finite temperatures, the δ-functions in (13.7) are replaced by the derivative $-f'(E)$ of the Fermi distribution function $f(E)$. This finite temperature LDOS corresponds to the differential tunnelling conductance of STM experiments.

We construct the Green's functions from the eigen-energy and the wave functions, and evaluate the spin-spin correlation function $\chi_{-,+}(\mathbf{r},\mathbf{r}',i\Omega_n)$ [43]. Then, we obtain the nuclear spin relaxation rate,

$$R(\mathbf{r},\mathbf{r}') = \mathrm{Im}\chi_{-,+}(\mathbf{r},\mathbf{r}',i\Omega_n \to \Omega+i\eta)/(\Omega/T)|_{\Omega\to 0}$$
$$= -\sum_{\alpha,\beta} u_\alpha(\mathbf{r})u_\beta^*(\mathbf{r})[u_\alpha(\mathbf{r}')u_\beta^*(\mathbf{r}')$$
$$+v_\alpha(\mathbf{r}')v_\beta^*(\mathbf{r}')]\pi T f'(E_\alpha)\delta(E_\alpha-E_\beta). \quad (13.8)$$

We consider the case $\mathbf{r} = \mathbf{r}'$, assuming that the nuclear relaxation event occurs locally such as in Cu-site of high T_c cuprates. Then, the $\mathbf{r}$-dependent relaxation time is given by $T_1(\mathbf{r}) = 1/R(\mathbf{r},\mathbf{r})$. In (13.8), we use $\delta(x) = \pi^{-1}\mathrm{Im}(x-i\eta)^{-1}$ to consider the discrete energy level of the finite size calculation. We typically use $\eta = 0.01t$.

In our calculation, the average electron density per site ~ 0.9 by adjusting appropriately the chemical potential μ. For the $d_{x^2-y^2}$-wave pairing, $U = 0$ and $V = -4.2t$. The resulting order parameter $\Delta_0/t = 1.0$ at $T=0$ and $H=0$. For the s-wave pairing, for example, we choose $U = -2.32t$ and $V = 0$. The pair potential $\Delta(\mathbf{r})$ is suppressed at the vortex center, and $|\Delta(\mathbf{r})|$ exhibits the Friedel oscillations around the core with a period $\sim 1/k_\mathrm{F}$ (k_F is the Fermi wave number) for both the s-wave and the $d_{x^2-y^2}$-wave pairings. The amplitude of this quantum oscillation increases as the attractive interactions $|U|$

and $|V|$ become large because the quantum effects are enhanced when Δ_0/E_F increases. These characteristics coincide with those of the s-wave pairing in our previous study [4]. As for the temperature dependence of the pair potential, the vortex core radius shrinks with decreasing T by the Kramer–Pesch effect [8,44]. We confirm this for the $d_{x^2-y^2}$-wave pairing too. The shrinkage is saturated at a low temperature in both the s-wave and the $d_{x^2-y^2}$-wave pairings. There, the structure of $\Delta(\boldsymbol{r})$ is almost independent of T. This is a quantum-limit effect which occurs for $T/T_c < \Delta_0/E_F$ [4]. We calculate the low temperature behaviour of $T_1(\boldsymbol{r})$ from the saturated pair potential. At higher temperatures, we calculate $T_1(\boldsymbol{r})$ from the self-consistently obtained pair potential at each T.

13.3.2 Field Distribution and Site-Selective NMR

In order to determine the one-to-one correspondence between the site and the magnetic field, we calculate the supercurrent around the vortex from the wave functions, and obtain the magnetic field distributions similar to Fig. 13.5. The magnetic field distribution function is depicted in Fig. 13.6a. This corresponds to the resonance frequency distribution in NMR experiments. We identify each site (V, A, B, C and S) in the vortex lattice to the distribution. The one-to-one correspondence between the site position and resonance frequency can be used as a guide for the site-selective NMR experiment. Namely, as shown in Fig. 13.6a, the NMR signal at the maximum cutoff of the resonance spectrum as a function of applied field or probe frequency comes from the

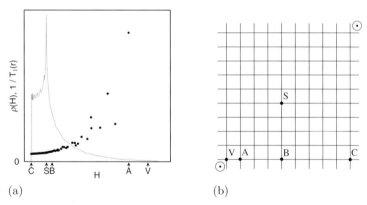

Fig. 13.6. (a) Magnetic field distribution function $\rho(H)$ (*dotted line*) and the site-dependent $T_1^{-1}(\boldsymbol{r})$ (*points* •) as a function of $H(\boldsymbol{r})$. The former corresponds to the NMR line shape. The letters V, A, B, C and S in the figure denote the sites defined in (b). (b) Positions of the sites V, A, B, C and S. The figure shows a quarter region of the vortex unit cell which contains two vortices. The vortex center is shown by ⊙. The nearest neighbour vortex is located at 45° to the a-axis. The solid lines show the atomic lattice.

vortex core at the V-site. When going away from the center (V→A→B→C), the resonance field decreases. The signal at the minimum cutoff comes from the C-site far from the vortex. The logarithmic singularity of the resonance field comes from the saddle point of the field at the S-site. Thus, it is possible to perform the site-selective $T_1(r)$ measurement by tuning the resonance frequency. In Fig. 13.6a, we also show the calculated $T_1^{-1}(r)$ corresponding each $H(r)$. Approaching the vortex core, $T_1(r)$ becomes fast, reflecting the low-energy excitations around the vortex core. It is qualitatively consistent with the experimental data [40–42].

13.3.3 Temperature Dependence of T_1

We show the LDOS $N(E, r)$ and the T-dependence of the site-dependent $T_1(\mathbf{r})$ in the $d_{x^2-y^2}$-wave pairing at H_{20}, and discuss their relation. The LDOS around the vortex is shown in Fig. 13.7a. In $N(E, \mathbf{r})$ at the vortex center (the V-site), the gap edge at Δ_0 in the zero-field case (dotted line U) is smeared, and low-energy states appear around the vortex core. In the $d_{x^2-y^2}$-wave pairing, the core state shows a broad zero-energy peak instead of the split peaks in the s-wave pairing [10]. There is no small gap. The weight of the low-energy states decrease when going away from the vortex centre (V→A→B→C). Far from the vortex, $N(E, \mathbf{r})$ is reduced to the DOS of the zero-field case. But, a small weight remains of the low-energy states extending from the vortex core.

As for the $d_{x^2-y^2}$-wave pairing, we plot $T_1(\mathbf{r})^{-1}$ vs. T in Fig. 13.7b, and re-plot it as a log-log plot in Fig. 13.7c. At zero field (line U), we see the expected power law relation $T_1^{-1} \sim T^3$ of the $d_{x^2-y^2}$-wave pairing below $T/T_c \simeq 0.1$. In the presence of vortices, $T_1(\mathbf{r})^{-1}$ deviates from the T^3-relation, and follows $T_1(\mathbf{r})^{-1} \sim T$ at low temperatures. This deviation was reported in

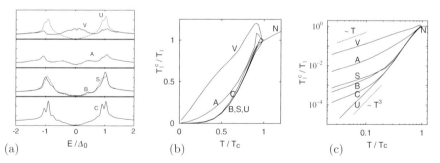

Fig. 13.7. (a) The LDOS $N(\mathbf{r}, E)$ at sites V, A, B, C and S for H_{20} and the $d_{x^2-y^2}$-wave pairing. The DOS at zero-field is also presented (*dotted line* U) with the DOS at the V-site (*solid line*). (b) Temperature dependence of $T_1(\mathbf{r})$ in the $d_{x^2-y^2}$-wave pairing at the sites V, A, B, C and S. $T_1(T_c)/T_1(T)$ is plotted as a function of T/T_c. Line U shows the zero field case. The line N is for the normal state at $T > T_c$. (c) The log-log plot of (b).

the experiments on high-T_c superconductors [45]. The origin of the T-linear behaviour is the low-energy state around the vortex. As seen in Fig. 13.7a, the superconducting gap is buried by the low-energy state around the vortex. When approaching the vortex center, the temperature region of the T-linear behaviour is enlarged and it appears from higher temperatures. The coherence peak below T_c is taken as a manifestation of the s-wave symmetry. In the $d_{x^2-y^2}$-wave pairing, the coherence peak is absent. But in the vortex core region, T_1^{-1} has a peak below T_c even in the $d_{x^2-y^2}$-wave pairing. We should be careful not to mistake this peak due to the vortex core relaxation for the usual coherence peak in the NMR experiment when identifying the gap symmetry.

Here we investigate the relationship between $T_1^{-1}(\boldsymbol{r})$ at low temperatures and the low energy excitations $N(E=0,\boldsymbol{r})$. Figure 13.8a shows the LDOS $N(E=0,\boldsymbol{r})/N(0)$. $N(0)$ is the zero-energy DOS in the normal state. It indicates that the LDOS around the core exceeds the normal state value. In the low temperature region, T_1^{-1} shows T-linear behaviour. We plot the spatial distribution of $T_1^{-1}(\boldsymbol{r}, T \sim 0)$ at H_{32} in Fig. 13.8b. It is normalized by the normal state value T_1^N. This shows that $T_1^{-1}(\boldsymbol{r}, T \sim 0)$ increases when approaching the vortex core, and exceeds its normal state value at the

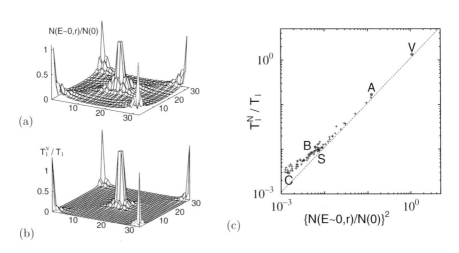

Fig. 13.8. (a) The zero-energy local density of states. (b) Site-dependent relaxation time. We plot topographic view of $N(E=0,\boldsymbol{r})/N(0)$ and $T_1^N/T_1(\boldsymbol{r})$ in a unit cell (32×32 atomic sites) for the $d_{x^2-y^2}$-wave pairing at H_{32}. They are normalized by the normal state value. The vortices are located at the center and corners of the figure. (c) The linear relationship between $T_1^N/T_1(\boldsymbol{r})$ and $[N(E=0,\boldsymbol{r})/N(0)]^2$. We plot the 32 × 32 lattice point values of the normalized $T_1^N/T_1(\boldsymbol{r})$ shown in b and the square of the normalized LDOS $N(E=0,\boldsymbol{r})/N(0)$ shown in (a). The letters in the figure denote the sites defined in Fig. 13.6b. The dotted line shows the linear relation $T_1^N/T_1(\boldsymbol{r}) \propto [N(E=0,\boldsymbol{r})/N(0)]^2$.

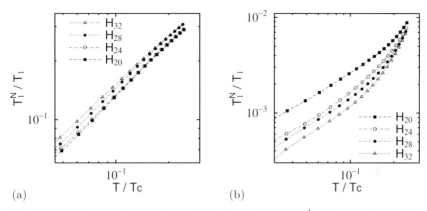

Fig. 13.9. The low temperature behaviour of $T_1(T)^{-1}$. The log–log plots of $T_1(T_c)/T_1(T)$ vs. T/T_c are presented for the $d_{x^2-y^2}$-wave pairing. Applied field increases as $H_{32} \to H_{28} \to H_{24} \to H_{20}$. T_c and $T_1(T_c)$ are defined by each H. (**a**) At the V-site. (**b**) At the S-site

vortex center. This topography of $T_1^{-1}(\mathbf{r}, T \sim 0)$ in Fig. 13.8b looks similar to that of the LDOS $N(E = 0, \mathbf{r})$ in Fig. 13.8a. We plot $T_1^{-1}(\mathbf{r}, T \sim 0)$ vs. $N^2(E = 0, \mathbf{r})$ in Fig. 13.8c. It shows the linear relationship between them, i.e., $T_1^{-1}(\mathbf{r}, T \sim 0)$ at low T is governed by the low energy excitations at each site. Then, the site-selective $T_1^{-1}(\mathbf{r}, T \sim 0)$ is a good measure of the local core excitations. The LDOS around the vortex core can be estimated quantitatively from the slope of $T_1^{-1}(\mathbf{r})$ at low T.

Next, we see how the T-dependence varies with field. We show the results for the V-site and the S-site (other sites exhibit a similar behaviour to the S-site, thus not shown here). They reflect the field dependence of the LDOS at the vortex core and outside of the core[10]. The log-log plots of T_1^{-1} vs. T are presented in Fig. 13.9 to show the low temperature behaviour. At the V-site shown in Fig. 13.9a, the T-linear coefficient in $T_1^{-1}(T)$ is depressed with increasing H at low temperatures. This is because the inter-vortex quasi-particle transfer smears the low energy quasi-particle peak of the LDOS at the vortex core. At the S-site, T_1^{-1} below T_c is suppressed as H increases. It reflects that the low energy state around the vortex smears the $d_{x^2-y^2}$-wave superconducting gap-edge. But, $T_1^{-1}(T)$ grows with increasing H at low temperatures as shown in Fig. 13.9b. Because the vortex contribution is increased and the amplitude of the low energy state extending outside the vortex core becomes large, T_1 at low temperatures becomes short with increasing external magnetic field. This tendency coincides qualitatively with the observation of high-T_c cuprates [45] or an organic superconductor κ-(ET)$_2$Cu[N(CN)$_2$]Br [46].

13.4 Concluding Remarks

In Sect. 13.2, we have studied the vortex structure and its field dependence in the $d_{x^2-y^2}$-wave and the s-wave superconductors by self-consistent calculations of the quasiclassical Eilenberger theory for the vortex lattice case. We have shown the $d_{x^2-y^2}$-wave nature in the LDOS, and the field dependence of the DOS and the internal magnetic field distribution. Here, the effect of neighboring vortices appears and the vortex core radius shrinks with increasing field. If we carefully analyze the experimental data and compare with the theoretical calculations, we may extract the interesting properties of the vortex state.

In Sect. 13.3, we have calculated the nuclear relaxation time T_1 in the mixed state based on the microscopic framework of the BdG theory, and demonstrated that the site-selective T_1 is a good probe to extract detailed information on low lying vortex core excitations. We have investigated the site-dependence, T-dependence and field-dependence of T_1.

It is noted that in the clean limit the vortex core region is not a simple core filled by normal state electrons. Reflecting the rich wave function structure of the low energy state around the vortex core, the LDOS observed by STM shows the characteristic structure. The characteristic T-dependence is expected in $T_1^{-1}(\boldsymbol{r})$ near the vortex core. We expect that the NMR imaging study just explained here will provide vital information for the vortex core state in high-T_c superconductors. If the fast relaxation is observed at the vortex core, it shows that a zero-energy peak exists in the LDOS, as suggested in the theoretical study. If the relaxation is slow even at the vortex core, or it shows the gap-like behaviour, it endicates that there are no eminent low energy quasiparticles as suggested in the STM experiments and reflects the exotic electronic structure in high-T_c superconductors.

Finally, we stress that, to perform a desired site-selective NMR measurement, we need a clear NMR resonance line shape of the vortex lattice as shown in Fig. 13.6a or as obtained by a beautiful experiment on vanadium [38,39]. It is needed to identify the position in the vortex lattice. However, even in the case when the resonance line shape of the vortex lattice is not clear, if we analyze the fast and slow relaxation processes separately, the fast relaxation process includes information of the vortex core contributions [47]. We hope that the site-selective NMR experiments confirm the relation of $T_1(\boldsymbol{r})$ and $N(E,\boldsymbol{r})$. If this experimental method is established, it can be a powerful method to investigate the exotic mechanism of the unconventional superconductors by spatially imaging the low energy quasi-particle excitations around the vortex cores.

References

1. C. Caroli, P.-G. de Gennes, J. Matricon: Phys. Lett. **9**, 307 (1964)
2. C. Caroli, J. Matricon: Phys. Kondens. Mater. **3**, 380 (1965)
3. F. Gygi, M. Schlüter: Phys. Rev. Lett. **65**, 1820 (1990); Phys. Rev. B **41**, 822 (1990)
4. N. Hayashi, T. Isoshima, M. Ichioka, K. Machida: Phys. Rev. Lett. **80**, 2921 (1998)
5. I. Maggio-Aprile *et al.*: Phys. Rev. Lett. **75**, 2754 (1995); J. Low Temp. Phys. **105**, 1129 (1996)
6. G.E. Volovik: JETP Lett. **58**, 469 (1993)
7. N. Schopohl, K. Maki: Phys. Rev. B **52**, 490 (1995)
8. M. Ichioka, N. Hayashi, N. Enomoto, K. Machida: Phys. Rev. B **53**, 15316 (1996)
9. K. Machida, M. Ichioka, N. Hayashi, T. Isoshima: 'Vortex Structure in Clean Type II Superconductors with s-wave and d-wave Pair Symmetries'. In: *The Superconducting State in Magnetic Fields*, ed. by C.A.R. Sá de Melo (World Scientific, Singapore, 1998) Chap. 13
10. Y. Wang, A.H. MacDonald: Phys Rev. B **52**, 3876 (1995)
11. M. Franz, Z. Tešanović: Phys. Rev. Lett. **80**, 4763 (1998)
12. M. Ichioka, A. Hasegawa, K. Machida: Phys. Rev. B **59**, 184 and 8902 (1999); J. Superconductivity **12**, 571 (1999)
13. M. Ichioka, N. Hayashi, K. Machida: Phys. Rev. B **55**, 6565 (1997)
14. H.F. Hess *et al.*: Phys. Rev. Lett. **62**, 214 (1989); **64**, 2711 (1990); Physica B **169**, 422 (1991)
15. Ch. Renner *et al.*: Phys. Rev. Lett. **80**, 3606 (1998)
16. E.W. Hudson *et al.*: Science **285**, 88 (1999) S.H. Pan *et al.*: Phys. Rev. Lett. **85**, 1536 (2000)
17. For example, M. Ogata: Int. J. of Mod. Phys. B **13**, 3560 (1999) A. Himeda, M. Ogata, Y. Tanaka, K. Kashiwaya: J. Phys. Soc. Jpn, **66**, 3367 (1997) J.H. Han, D.H. Lee: Phys. Rev. Lett. **85**, 2754 (2000)
18. C. Wu, T. Xiang, Z. Su: Phys. Rev. B **62**, 14427 (2000)
19. M. Takigawa, M. Ichioka, K. Machida: Phys. Rev. Lett. **83**, 3057 (1999); J. Phys. Soc. Jpn, **69**, 3943 (2000)
20. U. Klein: J. Low Temp. Phys. **69**, 1 (1987); Phys. Rev. B **40**, 6601 (1989) B. Pöttinger, U. Klein: Phys. Rev. Lett. **70**, 2806 (1993)
21. G. Eilenberger: Z. Phys. **214**, 195 (1968)
22. M. Ichioka, N. Enomoto, K. Machida: J. Phys. Soc. Jpn. **66**, 3928 (1997)
23. Y. De Wilde *et al.*: 'Vortex Lattice Structure in $LuNi_2B_2C$'. In: *The Superconducting State in Magnetic Fields*, ed. by C.A.R. Sá de Melo, (World Scientific, Singapore, 1998), Chap. 7
24. V.G. Kogan, P. Miranović, D. M^{c}K. Paul: 'Vortex Lattice Transitions'. In: *The Superconducting State in Magnetic Fields*, ed. by C.A.R. Sá de Melo, (World Scientific, Singapore, 1998), Chap. 8
25. H. Won, K. Maki: Phys. Rev. B **53**, 5927 (1996)
26. S. Ullah, A.T. Dorsey, L.J. Buchholtz: Phys. Rev. B **42**, 9950 (1990)
27. K. Yasui, T. Kita: Phys. Rev. Lett. **83**, 4168 (1999)
28. K.A. Moler *et al.*: Phys. Rev. Lett. **73**, 2744 (1994)
29. R.A. Fisher *et al.*: Physica C **252**, 237 (1995)

30. A.L. Fetter, P.C. Hohenberg : 'Theory of Type II Superconductors'. In: *Superconductivity*, ed. by R.D. Parks (Marcel Dekker, New York, 1969), p.891
31. A.P. Ramirez: Phys. Lett. A **211**, 59 (1996)
32. M. Hedo *et al.*: J. Phys. Soc. Jpn. **67**, 33 (1998)
33. A.A. Golubov, U. Hartmann: Phys. Rev. Lett. **72**, 3602 (1994)
34. J. Sonier *et al.*: Phys. Rev. Lett. **79**, 1742 (1997)
35. J. Sonier *et al.*: Rev. Mod. Phys. **72**, 769 (2000)
36. N. Enomoto, M. Ichioka, K. Machida: J. Phys. Soc. Jpn. **66**, 204 (1997)
37. R. Wortis, A.J. Berlinsky, C. Kallin: Phys. Rev. B **61**, 12342 (2000) D.K. Morr, R. Wortis: Phys. Rev. B **61**, 882 (2000) D.K. Morr: cond-mat/0007393
38. W. Fite II, A.G. Redfield: Phys. Rev. Lett. **17**, 381 (1966)
39. A. Kung: Phys. Rev. Lett. **25**, 1006 (1970)
40. N.J. Curro, C.P. Slichter, W.C. Lee, D.M. Ginsberg: Phys. Rev. B **62**, 3473 (2000)
41. B.I. Halperin: private communication
42. K. Kumagai, Y. Matsuda: private communication
43. R. Leadon, H. Suhl: Phys. Rev. **165**, 596 (1968)
44. L. Kramer, W. Pesch: Z. Phys. **269**, 59 (1974) W. Pesch, L. Kramer: J. Low Temp. Phys. **15**, 367 (1974)
45. K. Ishida *et al.*: Solid State Commun. **90**, 563 (1994); J. Phys. Soc. Jpn. **63**, 1104 (1994)
46. H. Mayaffre *et al.*: Phys. Rev. Lett. **75**, 4122 (1995)
47. Y. Matsuda: Chap. 17 of this book

14 Quasiparticle Spectrum in the Vortex State in Nodal Superconductors

H. Won and K. Maki

14.1 Introduction

As is well known, most of novel superconductors, like heavy fermion superconductors [1] charge conjugated organic superconductors [2], high T_c cuprate superconductors and recently discovered Sr_2RuO_4 [3] are unconventional or nodal superconductors [4]. For example, d-wave symmetry of both the hole-doped and the electron doped high T_c cuprate superconductor has been established rather recently [5,6].

In addition, organic superconductors $(TMTSF)_2X$ with $X=PF_6$, ClO_4 etc. are most likely of p-wave [7,8] while κ-(BEDT-TTF) salts are most likely of d-wave [9-12] though we are not sure if it is $d_{x^2-y^2}$ or d_{xy}. Also the nodal lines of the triplet superconductivity in Sr_2RuO_4 is still unknown. But 2D f-wave superconductor with $\Delta(\boldsymbol{k}) \sim \cos(ck_3)e^{\pm i\phi}$ is the most likely candidate [13,15]. In Fig. 14.1 we show a few examples of $\Delta(\boldsymbol{k})$ of nodal superconductors.

In the past 10 years we have learned that the mean field theory as embodied in Landau's Fermi liquid theory [16] and the BSC theory [17] for unconventional superconductors are most successful in elucidating the new fascinating properties of nodal superconductors.

For example, the BCS theory of nodal superconductors provides an excellent description of both the thermodynamics and transport properties of high temperature superconductors, Sr_2RuO_4, UPt_3 etc. Of course the fluctuation corrections to the mean field theory are not negligible in general, as encountered in the pseudogap phenomenon in high T_c cuprate superconductors in the underdoped regime [18–20].

As first noted by Volovik [21], the presence of nodes and/or nodal lines in $\Delta(\boldsymbol{k})$ is most readily seen in the $\sqrt{H}$ dependence of the specific heat in the vortex state. Indeed such $\sqrt{H}$ dependence of the specific heat has been seen in YBCO [22,23], LSCO [24], κ-(ET)$_2$ salts [9] and Sr_2RuO_4 [25,26]. Later, the quasi-classical approximation used by Volovik was extended for all thermodynamic properties [26] and thermal conductivity [13,27–29].

In particular, the angular dependence of the specific heat in d-wave superconductors in the presence of a planar magnetic field is considered in [29] and Vehkter et al. have predicted a rather large angular dependence. We believe their circular Fermi surface is unrealistic. Rather the Fermi surface should be

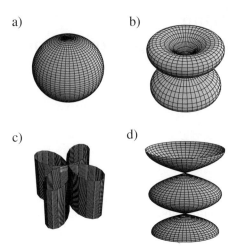

Fig. 14.1. $|\Delta(\mathbf{k})|$'s for s-wave and some unconventional superconductors are shown: (a) s-wave superconductor, (b) f-wave one in UPt$_3$, (c) d-wave one in high T_c cuprates, (d) 2D f-wave one as one of the candidates for Sr$_2$RuO$_4$

more like a cylinder, which gives much smaller angular dependence [13,30]. Also, an improper spatial averaging appeared to have been used in [28,29].

As far as the extended states are concerned, we think the local information is lost at the level of the quasiparticle density of states when we are dealing with the system in the superclean limit (i.e. with the quasiparticle mean free path of the order of a few microns). In such a situation, the spatial average has to be taken before the momentum average over the Fermi surface is taken [13].

The object of the present chapter is to review our work on the thermal conductivity tensor in the vortex state in nodal superconductors. We focus our attention mostly on three nodal superconductors $d_{x^2-y^2}(\Delta(\mathbf{k}) \sim \cos(2\phi))$, $d_{xy}(\Delta(\mathbf{k}) \sim \sin(2\phi))$ and 2D $f(\Delta(\mathbf{k}) \sim \cos(ck_3)e^{\pm i\phi})$. Other 2Df-wave superconductors like $\cos(2\phi)e^{\pm i\phi}$ and $\sin(2\phi)e^{\pm i\phi}$ have the same angular depedence of the thermal conductivity tensor as $d_{x^2-y^2}$ and d_{xy} respectively [31]. The result for $T \ll \epsilon, (\equiv \frac{1}{2}\tilde{v}\sqrt{eH}, \tilde{v} = \sqrt{vv_c}$ where v and v_c are the Fermi velocities in the $a-b$ plane and parallel to the c axis, respectively) have appeared in [13].

More recently the earlier result is extended for $T \gg \epsilon$ [32]. Indeed, all experiments on the thermal conductivity tensor in YBCO, until now, appear to have done in the regime $T \gg \epsilon$ [33–35]. Indeed, the present result for $T \gg \epsilon$ appear to describe the observed angular dependence of the thermal conductivity tensor κ_{xx} and κ_{xy} in YBCO reasonably well.

On the other hand, the recent works [14,15] on Sr$_2$RuO$_4$ cover both $T \ll \epsilon$ and $T \gg \epsilon$. In particular, for $T \ll \epsilon$ κ_{xx} appears to increase linearly in H as predicted in [13]. Therefore, single crystals of YBCO, Bi2212 and

Sr$_2$RuO$_4$ will provide ideal laboratory to test the new concept in the nodal superconductors.

14.2 The Volovik Effect

Perhaps it is useful to summarize the Volovik effect in the simplest case. Let us consider a $d_{x^2-y^2}$ superconductor in a magnetic field $\boldsymbol{H} \parallel \boldsymbol{c}$. In the vortex state the quasiparticle density of states is given by

$$N(E, \epsilon)/N_0 = \text{Re}\left\langle \frac{|E - \boldsymbol{v} \cdot \boldsymbol{q}|}{\sqrt{(E - \boldsymbol{v} \cdot \boldsymbol{q})^2 - \Delta^2 \cos^2(2\phi)}} \right\rangle \simeq \langle |E - \boldsymbol{v} \cdot \boldsymbol{q}| \rangle / \Delta \quad (14.1)$$

for $|E|, |\boldsymbol{v} \cdot \boldsymbol{q}| \ll \Delta$. Here the average $\langle \ldots \rangle$ means both the spatial average over the unit cell in the square vortex lattice [7,36,37] and the average over the nodal directions. Here $\boldsymbol{v} \cdot \boldsymbol{q}$ is the Doppler shift [38] due to the pair momentum $2\boldsymbol{q}$ and $\boldsymbol{v}$ is the quasi-particle velocity. The residual density of states is obtained within the Wigner–Seitz approximation as

$$N(0, \epsilon) = \frac{4}{\pi d^2 \Delta} \int_0^d r\, dr \int_0^2 d\alpha \cos\alpha \frac{v}{2r} = \frac{2}{\pi} \frac{v\sqrt{eH}}{\Delta} = \frac{4}{\pi} \frac{\epsilon}{\Delta}. \quad (14.2)$$

Here, use is made of the relation $d = (eH)^{-\frac{1}{2}}$ for a square lattice. From (1) it follows that the specific heat C_s, the spin susceptibility, the superfluid density and nuclear spin lattice relaxation rate T_1^{-1} for $T \ll \epsilon$ are given by [26]

$$C_s/\gamma_N T = \frac{2}{\pi} \frac{v\sqrt{eH}}{\Delta} \quad (14.3)$$

$$\chi/\chi_N = 1 - \rho_s(0, H)/\rho_s(0, 0) = \frac{2}{\pi} \frac{v\sqrt{eH}}{\Delta} \quad (14.4)$$

$$T_1^{-1}/T_{1N}^{-1} = \left(\frac{2}{\pi}\right)^2 \frac{v^2(eH)}{\Delta^2}. \quad (14.5)$$

As already mentioned (2) has been verified semi-quantitatively in YBCO, LSCO, $\kappa - (ET)_2$ salts and Sr$_2$RuO$_4$. Further, the $\sqrt{H}$ dependence of χ and the H linear dependence of T_1^{-1} have been seen in slightly underdoped YBCO [39]. Within the same approximation it is not difficult to incorporate the E dependence. Then we obtain the scaling behavior of C_s, χ and T_1^{-1}, which depends on ϵ and T through functions in ϵ/T [26]. Therefore the quasi-classical approximation allows us to find the scaling functions discussed by Simon and Lee [40].

Also, we expect that the scaling relations hold for $\epsilon, T \ll \Delta(T)$. Very recently the scaling behavior of the specific heat in YBCO is analyzed experimentally [41].

When a magnetic field $\boldsymbol{H}$ is applied in the a–b plane, the analysis becomes somewhat more involved. In the present situation the quasiparticle density of states depends on θ the angle $\boldsymbol{H}$ makes from the a direction [13].

Now we have

$$C_{\rm s}/\gamma_N T = \frac{2}{\pi}\frac{\tilde{v}\sqrt{eH}}{\Delta}I(\theta) \tag{14.6}$$

$$\chi/\chi_N = 1 - \rho_{sxx}(0,H,\theta)/\rho_{sxx}(0,0) = \frac{2}{\pi}\frac{\tilde{v}\sqrt{eH}}{\Delta}I(\theta)$$

$$\rho_{sxy}(0,H,\theta)/\rho_{sxx}(0,0) \cong 0.29\sin(2\theta)\frac{2}{\pi}\frac{\tilde{v}\sqrt{eH}}{\Delta} \tag{14.7}$$

$$T_1^{-1}/T_{1N}^{-1} = \left(\frac{2}{\pi}\right)^2 \frac{\tilde{v}^2(eH)}{\Delta^2} \tag{14.8}$$

where $\tilde{v} = (vv_c)^{\frac{1}{2}}$ and

$$I(\theta) = \frac{1}{\pi}\left(\sqrt{\frac{3+s}{2}}E\left(\sqrt{\frac{2}{3+s}}\right) + \sqrt{\frac{3-s}{2}}E\left(\sqrt{\frac{2}{3-s}}\right)\right)$$
$$\simeq 0.955 + 0.0285\cos(4\theta) \tag{14.9}$$

and $\theta = \sin(2\theta)$ and $E(k)$ is the complete elliptic integral of the second kind. Therefore c_s, χ and ρ_{sxx} exhibit the fourfold symmetry. Perhaps the most remarkable is the Hall superfluid density $\rho_{sxy}(0,H,\theta)$. We expect this term will play an important role in the vortex dynamics in d-wave superconductors. We note also that the θ dependence disappears from T_1^{-1}.

14.3 Thermal Conductivity for $T \ll \epsilon$

The thermal conductivity in the vortex state in single crystals in YBCO and Bi2212 is one of the rather controversial subjects. We do not want to enter the controversy, but rather address to two issues which are clearly defined. a) The low temperature thermal conductivity in YBCO and Bi2212 when $\boldsymbol{H} \parallel \boldsymbol{c}$. It appears that if we limit ourselves to the experimental data below $T = 1K$, it is possible to separate the phonon contribution from the electronic contribution. Then, it appears that the quasi-classical approximation we are going to discuss is adequate [26,28,42]. b) The angular dependence of the thermal conductivity tensor in single crystals of YBCO in a planar magnetic field [33,34,35], where no satisfactory theory is found until now.

In the following we assume that the quasiparticle life time is mostly due to impurities. Also, the scatterer gives the scattering in the unitarity limit like substituted Zn in the CuO_2 plane [43]. Further, for simplicity we limit ourselves to the superclean limit ($\Gamma \ll \epsilon \ll \Delta$, where Γ is the quasiparticle scattering rate due to impurities). In [28] the same limit is called the clean limit. But we reserve the word "the clean limit" for $\epsilon \ll \Gamma \ll \Delta$, since we think that the system with $\Gamma \ll \Delta$ should not be called "the dirty limit".

Indeed in these systems the quasiparticle mean free path is the order of a few microns [13]. Also we consider here the vortex state when $\boldsymbol{H}$ is applied within the $a-b$ plane. Then the quasiparticle damping at $E=0$ is given by [27]

$$C_0 = \frac{\Gamma}{\Delta}\left(\langle x\rangle + \frac{2}{\pi}C_0\left(\langle \ln\frac{4}{x}\rangle - 1\right)\right)^{-1} \tag{14.10}$$

where

$$C_0 = \lim_{E\to 0} Im\tilde{E}/\Delta \tag{14.11}$$

and

$$\langle x\rangle = \frac{1}{\Delta}\langle|\boldsymbol{v},\boldsymbol{q}|\rangle = \frac{4}{\pi}\frac{\epsilon}{\Delta}I(\theta) \tag{14.12}$$

where $\tilde{E}$ is the renormalized frequency and $\epsilon = \frac{1}{2}(vv_c eH)^{\frac{1}{2}}$ and $I(\theta)$ has been defined in (8). Then the thermal conductivity tensor is given by [44] for $\Delta(\boldsymbol{k}) \sim \cos(2\phi)$:

$$\kappa_{xx}/\kappa_n = \frac{2}{\pi^2}\frac{vv_c(eH)}{\Delta^2}I^2(\theta) \tag{14.13}$$

$$\kappa_{xy}/\kappa_n = -\frac{2}{\pi^2}\frac{vv_c(eH)}{\Delta^2}I(\theta)L(\theta) \simeq -0.29\sin(2\theta)\times\frac{2}{\pi^2}\frac{vv_c(eH)}{\Delta^2} \tag{14.14}$$

where κ_n is the thermal conductivity in the normal state and

$$L(\theta) = \frac{1}{\pi}\left(\sqrt{\frac{3+s}{2}}E\left(\sqrt{\frac{2}{3+s}}\right) - \sqrt{\frac{3-s}{2}}E\left(\sqrt{\frac{2}{3-s}}\right)\right)$$
$$\simeq 0.29\sin(2\theta)\ . \tag{14.15}$$

First of all, κ_n increases linearly in H. This linear dependence has not been found in earlier works [28,29]. This difference arises from how we treat the spatial average in studying the quasiparticle density of states as we discussed earlier. Second, κ_{xx} has the fourfold term $\sim \cos(4\theta)$. Finally, $\kappa_{xy} \sim -\sin(2\theta)$.

On the other hand, comparing with the existing experimental data from single crystals of YBCO [33–35], there are surprising contrasts. First, the experiment shows that κ_{xx} decreases almost linearly in H with increasing H. Second, the fourfold term in κ_{xx} has the opposite sign from the observation. Finally, κ_{xy} has also the opposite sign. However, the result in (12) and (13) are readily deduced if the quasiparticle transport is dominated by the quasiparticles generated by the Doppler shift. Indeed we shall see that when $T \gg \epsilon$ the situation changes dramatically [27,32].

For $\Delta(\boldsymbol{k}) \sim \sin(2\theta)$ we obtain similarly

$$\kappa_{xx}/\kappa_n = \frac{2}{\pi^2}\frac{vv_c(eH)}{\Delta^2}I\left(\theta - \frac{\pi}{4}\right)J(\theta) \tag{14.16}$$

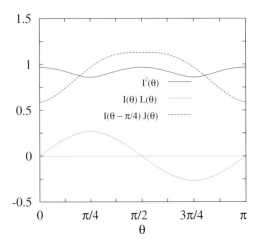

Fig. 14.2. Angular variations of $I^2(\theta)$, $I(\theta)L(\theta)$, and $F(\theta)(=I(\theta-\pi/4)J(\theta))$ are shown.

and $\kappa_{xy} = 0$. Here

$$J(\theta) = \frac{2}{\pi}\sqrt{\frac{3-c}{2}}E\left(\sqrt{\frac{2}{3-c}}\right) \simeq 0.955 - 0.245\cos(2\theta) \tag{14.17}$$

So in the present case κ_{xx} exhibits the twofold symmetry. The functions $I^2(\theta)$, $I(\theta)L(\theta)$ and $I(\theta-\pi/4)J(\theta)$ versus θ are shown in Fig. 14.2. Also, the present result will provide a simple means to distinguish $d_{x^2-y^2}$ from d_{xy}-wave by thermal conductivity measurement.

For a possible candidate state $\Delta(\kappa) \sim \cos(ck_3)e^{\pm i\phi}$ for Sr_2RuO_4, we obtain

$$\frac{\kappa_{xx}}{\kappa_n} = \frac{2}{\pi^2}\frac{vv_c(eH)}{\Delta^2}\left(1.478 - 0.1253\cos(2\theta)\right) \tag{14.18}$$

$$\frac{\kappa_{xy}}{\kappa_n} = -0.1253\sin(2\theta)\frac{2}{\pi^2}\frac{vv_c(eH)}{\Delta^2} \tag{14.19}$$

So in the present case κ_{xx} has twofold term of about 8%. The linear H dependence of the thermal conductivity has been seen in Sr_2RuO_4 for $T \simeq 0.35K$ [14,15]. Further, κ_{xx} exhibits a rather weak twofold term (about 3%) [14,15]. In any case, $\cos(ck_3)e^{\pm i\phi}$ is most consistent with the experiment, compared with the two other candidates.

14.4 Thermal Conductivity Tensor for $T \gg \epsilon$

For $T \gg \epsilon$ we have to solve the renormalized $\tilde{E}$ for $E \gg \epsilon$. Following [27] we find

$$\frac{\tilde{E}}{\Delta} = y + i\frac{\Gamma}{\Delta}\left\langle \frac{1}{y-x} \frac{1 - \frac{2}{\pi}i\ln\left(\frac{4}{y-x}\right)}{1 + \left(\frac{2}{\pi}\right)^2 \ln^2\left(\frac{4}{y-x}\right)} \right\rangle \tag{14.20}$$

where $y = E/\Delta$.

Then, for $\Delta(\mathbf{k}) \sim \cos(2\theta)$, we obtain

$$\kappa_{xx}/\kappa_n = \frac{7\pi^2}{10}\left(\frac{T}{\Delta}\right)^2\left(1 + \left(\frac{2}{\pi}\right)^2 \ln^2\left(\frac{2\Delta}{1.75T}\right)\right)$$

$$-\frac{1}{2\pi^2}\ln\left(\frac{2\Delta}{1.75T}\right)\frac{\tilde{v}^2 eH}{\Delta^2}\left[\ln\left(\frac{4\Delta}{\tilde{v}\sqrt{eH}}\right) - \frac{1}{16}(1 - \cos(4\theta))\right] \tag{14.21}$$

$$\kappa_{xy}/\kappa_n = \frac{1}{(2\pi)^2}\sin(2\theta)\frac{\tilde{v}^2(eH)}{\Delta^2}\ln\left(\frac{2\Delta}{1.75T}\right)\ln\left(\frac{4\Delta}{\tilde{v}\sqrt{eH}}\right) \tag{14.22}$$

where $\tilde{v}^2 = vv_c$.

First of, all κ_{xx} now decreases sublinearly with H. This is due to the fact that the heat transported is dominated by the thermal quasiparticles. For this situation the quasiparticles due to the Doppler shift act as the extra scattering centers. Also, the fourfold term in κ_{xx} has the opposite sign compared with the one for $T \ll \epsilon$. On the other hand, now both the H-dependence and the fourfold term in κ_{xx} is very consistent with observation [33,34].

Finally, the Hall thermal conductivity has also the correct sign and is fully consistent with the recent data from the single crystal of $Y(124.)$ [35]. This suggests indeed the Doppler shift is the most crucial ingredient in the thermal conductivity as long as $T \ll \Delta$.

For $\Delta(\mathbf{k}) \sim \sin(2\phi)$ we obtain

$$\frac{\kappa_{xx}}{\kappa_n} = \frac{7\pi^2}{10}\left(\frac{T}{\Delta}\right)^2\left(1 + \left(\frac{2}{\pi}\right)^2 \ln^2\left(\frac{2\Delta}{1.75T}\right)\right)$$

$$-\frac{1}{2\pi^2}\ln\left(\frac{2\Delta}{1.75T}\right)\frac{\tilde{v}^2(eH)}{\Delta^2}\ln\left(\frac{4\Delta}{\tilde{v}\sqrt{eH}}\right)\left(1 - \frac{1}{2}\cos(2\theta)\right) \tag{14.23}$$

and $\kappa_{xy} = 0$.

Therefore, in the present case, we have a significant twofold term.

Finally, for the candidate for Sr_2RuO_4 ($\Delta(\mathbf{k}) \sim \cos(ck_3)e^{\pm i\phi}$) we obtain

$$\frac{\kappa_{xx}}{\kappa_n} = \frac{7\pi^2}{10}\left(\frac{T}{\Delta}\right)^2\left(1 + \left(\frac{2}{\pi}\right)^2 \ln^2\left(\frac{2\Delta}{1.75T}\right)\right)$$

$$-\frac{1}{2\pi^2}\left(\frac{3}{2} - \frac{1}{4}\cos(2\theta)\right)\ln\left(\frac{2\Delta}{1.75T}\right)\frac{\tilde{v}^2(eH)}{\Delta^2}\ln\left(\frac{4\Delta}{\tilde{v}\sqrt{eH}}\right) \tag{14.24}$$

$$\frac{\kappa_{xy}}{\kappa_n} = \frac{1}{8\pi^2}\sin(2\theta)\ln\left(\frac{2\Delta}{1.75T}\right)\frac{\tilde{v}^2(eH)}{\Delta^2}\ln\left(\frac{4\Delta}{\tilde{v}\sqrt{eH}}\right). \qquad (14.25)$$

So again in this limit, κ_{xx} exhibits a small twofold term, which is consistent with recent data from Sr_2RuO_4 [14]. Further, $\kappa_{xx}(0) > \kappa_{xx}(\pi/2)$ is consistent with the experiment [14].

14.5 Concluding Remarks

We have reviewed our recent work on the thermal conductivity in the vortex state in nodal superconductors. Unlike in conventional s-wave superconductors the low temperature quasiparticle transport is dominated by nodal excitations or extended states. Then, it appears that the quasi-classical approach provides a simple, reliable and accurate means to explore the vortex state as initiated by Volovik.

Now we are confident that the present method is very reliable at least for high T_c cuprate superconductors. Making use of physical parameters discussed in [42], we think the experiments in YBCO and Bi2212 done for $T > 5K$ and $H \sim 10 \sim 20$ Tesla can be considered in the limit $T \gg \epsilon$. Therefore, it is rather gratifying that we can understand the low temperature properties of these vortex states mainly in terms of the Doppler shift for $T \ll \Delta$. On the other hand, we know that the Andreev scattering plays the dominant role near $H = H_{c2}$ or in the high temperature region [45,46,47].

Also, the present result shows that the thermal conductivity in the vortex state provides a unique means to access the nodal structure in $\Delta(\mathbf{k})$. Indeed, very recently Izawa et al. [48] measured the magnetothermal conductivity κ_{xx} of the newly discovered heavy fermion superconductor $CeCoIn_5$ [49] with $T_c = 6,5K$ and found the fourfold symmetry very similar to the one in the single crystal of YBCO. Since a possible f-wave superconductivity can be eliminated from the Knight shift, this appears to provide a clear signature of $d_{x^2-y^2}$ wave superconductor.

Acknowledgements. In the course of the present work we benefitted from discussions with Thomas Dahm, Pablo Esquinazi, Stephan Haas, Koichi Izawa and Yuji Matsuda. Also unpublished data of κ_{xx} and κ_{xy} from single crystals of Y(124) provided by Ocaña and Esquinazi have been crucial to the success of this work.

One of us (KM) acknowledges gratefully the support of the Max-Planck-Institut für Physik komplexer Systeme, where a part of this work is done. HW acknowledges the support from the Korean Science and Engineering Foundation (KOSEF) through the Grant No. 1999-2-114-005-5. We thank Mrs. Regine Adler for her help in preparing the manuscript.

References

1. M. Sigrist and K. Ueda, Rev. Mod. Phys. **63**, 239 (1991)
2. T. Ishiguro, K. Yamaji and G. Saito, "Organic Superconductors", Springer, Berlin (1998)
3. Y. Maeno, T.M. Rice and S. Sigrist, Physics Today **54**, 42 (2001)
4. H. Won and K. Maki, in: "Symmetry and Pairing in Superconductors", ed. by M. Ausloos and S. Kruchinin, Kluwer Academic, Dordrecht (1999), pp. 3–9; K. Maki and H. Won, in "Fluctuating Paths and Fields", World Scientific, Singapore (2001)
5. D.J. van Harlingen, Rev. Mod. Phys. **67**, 515 (1995)
6. C.C. Tsuei and J.R. Kirtley, Rev. Mod. Phys. **72**, 969 (2000)
7. K. Maki, H. Won, M. Kohmoto, J. Shiraishi, Y. Morita, G.-F. Wang, Physica C **317–318**, 353 (1999)
8. K. Maki, E. Puchkaryov and H. Won, Synth. Metals **703**, 1933 (1999)
9. Y. Nakazawa and K. Kanoda, Phys. Rev. B **55**, R8670 (1997)
10. A. Carrington et al., Phys. Rev. Lett. **83**, 4172 (1999)
11. M. Printerić et al., Phys. Rev. B **61**, 7033 (2000)
12. K. Ichimura et al., Synth. Metals **103**, 1812 (1999)
13. H. Won and K. Maki, in: "Rare Earth Transition Metal Borocarbides: Superconducting, Magnetic and Normal State Properties", ed. by K.H. Müller and V. Narozhnyi, Kluwer Academic, Dordrecht 2001, pp. 379–392
14. K. Izawa et al., Phys. Rev. Lett. **86**, 2653 (2001)
15. M.A. Tanatar et al., Phys. Rev. Lett. **86**, 2649 (2001)
16. L.D. Landau, Sov.Phys. JETP **3**, 920 (1957); ibid **5**, 100 (1957)
17. J. Bardeen, L.N. Cooper and J.R. Schrieffer, Phys. Rev. **108**, 1175 (1957)
18. T. Timusk and B. Statt, Rep. Prog. Phys. **62**, 61 (1999)
19. A.A. Varlamov, G. Balestrino, E. Milani and D.V. Livanov, Adv. Phys. **48**, 655 (1999)
20. H. Won and K. Maki, Physica C **341–348**, 891 (2000)
21. G.E. Volovik, JETP Lett. **58**, 496 (1993)
22. K.A. Moler et al., Phys. Rev. Lett. **73**, 2744 (1994); K.A. Moler et al., Phys. Rev. B **55**, 3954 (1997)
23. B. Revas et al., Phys. Rev. Lett. **80**, 3364 (1998)
24. S.J. Chen et al., Phys. Rev. B **58**, R14753 (1998)
25. S. Nishizaki, Y. Maeno and Z. Mao, J. Phys. Soc. Japan **69**, 573 (2000)
26. H. Won and K. Maki, Europhys. Lett. **54**, 248 (2001)
27. Yu.S. Barash, V.P. Mineev, and A.A. Svidzinskii, Sov. Phys. JETP **65**, 638 (1997); Yu.S. Barash and A.A. Svidzinskii, Phys. Rev. B **58**, 476 (1998)
28. C. Kübert and P.J. Hirschfeld, Solid State Commun. **105**, 439 (1998); Phys. Rev. Lett. **80**, 4963 (1998)
29. I. Vehkter, J.P. Carbotte and E.J. Nicol, Phys. Rev. B **59**, 7123 (1999); Phys. Rev. B **59**, R9023 (1999)
30. H. Won and K. Maki, unpublished
31. T. Dahm, K. Maki and H. Won, cond-mat/0006301
32. H. Won and K. Maki, unpublished
33. M.B. Salamon, F. Yu and V.N. Kopylov, J. Superconductivity **8**, 449 (1995); F. Yu et al., Phys. Rev. Lett. **74**, 5136 (1995)
34. H. Aubin et al., Phys. Rev. Lett. **78**, 2624 (1997)

35. R. Ocaña and P. Esquinazi, Phys. Rev. Lett. (submitted)
36. H. Won and K. Maki, Europhys. Lett. **30**, 421 (1995); Phys. Rev. B **53**, 5927 (1996)
37. J. Shiraishi, M. Kohmoto and K. Maki, Phys. Rev. B **59**, 4497 (1999)
38. K. Maki and T. Tsuneto, Prog. Theor. Phys. **27**, 228 (1960)
39. G.Q. Zheng et al., Phys. Rev. B **60**, R9947 (1999)
40. S.H. Simon and P.A. Lee, Phys. Rev. Lett. **78**, 1548 (1997)
41. Y. Wang, B. Revaz, A. Erb and A. Junod, Phys. Rev. B **63**, 94568 (2001)
42. M. Chiao et al., Phys. Rev. Lett. **82**, 2943 (1999); Phys. Rev. B **62**, 3554 (2000)
43. Y. Sun and K. Maki, Europhys. Lett. **32**, 355 (1995)
44. H. Won and K. Maki, cond-mat/0004105
45. K. Maki, Phys. Rev. B **44**, 2861 (1991)
46. I. Vehkter and A. Houghton, Phys. Rev. Lett. **83**, 4626 (1999)
47. K. Maki, G. Yang and H. Won, Physica C **341–348**, 1547 (2000)
48. K. Izawa et al., cond-mat/0104225
49. C. Petrović et al., Europhys. Lett. **53**, 354 (2001); C. Petrović et al., cond-mat/0103168

15 Random-Matrix Ensembles in p-Wave Vortices

Dmitri A. Ivanov

Summary. In disordered vortices in p-wave superconductors the two new random–matrix ensembles may be realized: B and DIII-odd (of $so(2N+1)$ and $so(4N+2)/u(2N+1)$ matrices respectively). We predict these ensembles from an explicit analysis of the symmetries of Bogoliubov–de Gennes equations in three examples of vortices with different p-wave order parameters. A characteristic feature of the novel symmetry classes is a quasiparticle level at zero energy. Class B is realized when the time-reversal symmetry is broken, and class DIII-odd when the time-reversal symmetry is preserved. We also suggest that the main contribution to disordering the vortex spectrum comes from the distortion of the order parameter around impurities.

Since Wigner's modeling Hamiltonians of complex nuclei by random matrices [1], the random–matrix theory (RMT) has played an important role in studying mesoscopic systems. In many cases, a chaotic (non-integrable) mesoscopic system may be accurately described by RMT. The supersymmetric technique by Efetov [2] provides a microscopic explanation of the RMT approximation for disordered systems.

According to the RMT approximation, the only characteristics of the system affecting the eigenvalue correlations at small energy scales are its symmetries. Therefore, a problem arises of classifying symmetries of random–matrix ensembles. It has been suggested by several authors [3–5] that random–matrix ensembles may be classified as corresponding symmetric spaces. The symmetric spaces may be divided into twelve infinite series reviewed in Table 15.1 (we split class DIII in two subclasses: DIII-even and DIII-odd) [7]. Each of the symmetry classes may occur in one of the three forms: positive-curvature, negative-curvature and flat [5]. The corresponding Jacobians in the matrix space are expressed in terms of trigonometric, hyperbolic, and polynomial functions respectively. In the present chapter we shall only discuss the RMT for Hamiltonians forming a linear space and therefore described by zero-curvature (flat) versions of RMT.

The simplest examples of the RMT symmetry classes are the three Wigner–Dyson classes: unitary, orthogonal and symplectic (A, AI, and AII, respectively, in Cartan's notation) [8]. In these classes, the energy level correlations are invariant under translations in energies (at energy scales much smaller

Table 15.1. Symmetric spaces and universality classes of random–matrix ensembles: (**a**) the three Wigner–Dyson classes A (unitary), AI (orthogonal), AII (symplectic) [8]; (**b**) chiral classes $AIII$ (unitary), BDI (orthogonal), CII (symplectic), with dimensions $p \leq q$ have $q-p$ zero modes [9]; (**c**) superconducting classes C, D, CI, $DIII$-even [3]; (**d**) p-wave vortex classes $DIII$-odd, B have one zero mode

Cartan class	Symmetric space	Dimension	Rank	β	α
A [GUE]	$SU(N)$	N^2-1	$N-1$	2	–
AI [GOE]	$SU(N)/SO(N)$	$(N-1)(N+2)/2$	$N-1$	1	–
AII [GSE]	$SU(2N)/Sp(N)$	$(N-1)(2N+1)$	$N-1$	4	–
$AIII$ [chGUE]	$SU(p+q)/S(U(p) \times U(q))$	$2pq$	p	2	$1+2(q-p)$
BDI [chGOE]	$SO(p+q)/SO(p) \times SO(q)$	pq	p	1	$q-p$
CII [chGSE]	$Sp(p+q)/Sp(p) \times Sp(q)$	$4pq$	p	4	$3+4(q-p)$
C	$Sp(N)$	$N(2N+1)$	N	2	2
D	$SO(2N)$	$N(2N-1)$	N	2	0
CI	$Sp(N)/U(N)$	$N(N+1)$	N	1	1
$DIII$-even	$SO(4N)/U(2N)$	$2N(2N-1)$	N	4	1
$DIII$-odd	$SO(4N+2)/U(2N+1)$	$2N(2N+1)$	N	4	5
B	$SO(2N+1)$	$N(2N+1)$	N	2	2

than the spectrum width), and the joint probability distribution of the energy levels ω_i is

$$\frac{DP\{\omega_i\}}{\prod_i D\omega_i} \propto \prod_{i<j} |\omega_i - \omega_j|^\beta \ . \tag{15.1}$$

The parameter β determines the strength of level repulsion and takes values 2, 1, and 4 for the unitary, orthogonal, and symplectic ensembles, respectively.

Other symmetry classes appear when there exists an additional symmetry relating energies E and $-E$. In the corresponding random–matrix ensembles, the levels ω_i repel not only each other, but also their mirror images $-\omega_i$, and the Jacobian has the form

$$\frac{DP\{\omega_i\}}{\prod_i D\omega_i} \propto \prod_{i<j} |\omega_i^2 - \omega_j^2|^\beta \prod_i \omega_i^\alpha \ . \tag{15.2}$$

It is now characterized by the two parameters β and α (the latter responsible for suppressing the density of states near $E=0$).

The three chiral classes ($AIII$, BDI, and CII) appear in systems with a chiral symmetry anticommuting with the Hamiltonian [9]. Four more classes (C, D, CI, and $DIII$) describe mesoscopic superconducting systems (with or without spin-rotational and time-reversal symmetries) [3]. This chapter is devoted to the last two lines in Table 15.1: classes *DIII*-odd and *B*. We shall demonstrate that these two classes correspond to the symmetries of a *p*-wave vortex with or without time-reversal symmetry, respectively (the symmetry class *B* in *p*-wave vortices has also been predicted in [10]; a partial account of the present work appeared as a preprint [11]).

Below we consider three particular *p*-wave vortices and describe their symmetries: the single-quantum vortex in the *A*-phase with a fixed spin orientation of the order parameter, the half-quantum vortex in the *A*-phase with rotating orientation of the order parameter, and the spin (fluxless) vortex in the *B*-phase. We explicitly demonstrate that in the first two examples the symmetries of the Hamiltonian are of type *B*, and, in the last example, of type *DIII*-odd. Then we briefly discuss microscopic requirements for the RMT limit and show that the main contribution to the level mixing comes from inhomogeneous suppression of the order parameter by impurities. Our estimates suggest that the RMT limit may be achieved in a moderately clean superconductor.

15.1 Single–Quantum Vortex

In this section we consider a vortex in the order parameter similar to the *A* phase of ^{3}He. Namely, we assume that the condensate wave function has the form

$$\Psi_\pm = e^{i\varphi}\left[d_x\Big(|\!\uparrow\uparrow\rangle + |\!\downarrow\downarrow\rangle\Big) + id_y\Big(|\!\uparrow\uparrow\rangle - |\!\downarrow\downarrow\rangle\Big)\right.$$
$$\left. +d_z\Big(|\!\uparrow\downarrow\rangle + |\!\downarrow\uparrow\rangle\Big)\right](k_x \pm ik_y)\,, \tag{15.3}$$

and that the vector ***d*** defining the orientation of the triplet is *fixed*. Without loss of generality, we take it to be in the ***z*** direction. We also assume a fixed chirality of the order parameter and take positive sign in place of $\pm$ in (15.3). Then the only allowed vortices are those involving a rotation of the phase of the order parameter, and the magnetic flux in such vortices is quantized in conventional superconducting flux quanta $\Phi_0 = hc/2e$.

Below we analyze the low-lying energy levels obtained as solutions to Bogoliubov–deGennes equations for the mean-field Hamiltonian

$$H = \sum_\alpha \Psi_\alpha^\dagger \left[\frac{(\mathbf{p}-e\mathbf{A})^2}{2m} + V(\mathbf{r}) - \varepsilon_\mathrm{F}\right]\Psi_\alpha$$
$$+\Psi_\uparrow^\dagger\left(\Delta_x * \frac{p_x}{k_\mathrm{F}} + \Delta_y * \frac{p_y}{k_\mathrm{F}}\right)\Psi_\downarrow^\dagger + \mathrm{h.c.}\,, \tag{15.4}$$

where Ψ_α are the electron operators (α is the spin index), $V(\mathbf{r})$ is the external potential of impurities, $\mathbf{A}(\mathbf{r})$ is the electromagnetic vector potential, and $\Delta_x(\boldsymbol{r})$ and $\Delta_y(\boldsymbol{r})$ are the coordinate-dependent components of the superconducting gap. [In the bulk, the preferred superconducting order is one of the two chiral components $\eta_\pm = \Delta_x \pm i\Delta_y$, but in inhomogeneous systems, such as a vortex core, an admixture of the opposite component may be self-consistently generated [12]. We account for this effect by allowing the two independent order parameters Δ_x and Δ_y.] Star (*) denotes the symmetrized ordering of the gradients p_μ and the order parameters Δ_μ [definition: $A * B \equiv (AB + BA)/2$]. At infinity, the order parameter imposes the vortex boundary conditions:

$$\Delta_x(r \to \infty, \phi) = \Delta_0 e^{\pm i\phi}, \qquad \Delta_y(r \to \infty, \phi) = i\Delta_0 e^{\pm i\phi}, \tag{15.5}$$

where r and ϕ are polar coordinates. Plus or minus signs in the exponent correspond to a positive or a negative single-quantum vortex. For an axially-symmetric vortex with the chirality of the order parameter non-self-consistently fixed ($\Delta_y \equiv i\Delta_x$), without the vector-potential $\boldsymbol{A}(\boldsymbol{r})$ and without disorder $V(\boldsymbol{r})$, the low-lying eigenstates of the Hamiltonian (15.4) have been found by Kopnin and Salomaa [13]. The spectrum is

$$E_n = n\omega_0, \quad (p-\text{wave}), \quad n = 0, \pm 1, \pm 2, \ldots \tag{15.6}$$

with $\omega_0 \sim \Delta^2/\varepsilon_F$. This result should be compared with the spectrum of the vortex core in a s-wave superconductor [14]:

$$E_n = \left(n + \frac{1}{2}\right)\omega_0, \quad (s-\text{wave}), \quad n = 0, \pm 1, \pm 2, \ldots \tag{15.7}$$

The common feature of the spectra in the s-wave and p-wave cases is the symmetry with respect to zero energy. If we interpret holes in the negative-energy levels as excitations with positive energies (and with opposite spin), then this symmetry implies that the excitations are doubly degenerate in spin: to each spin-up excitation there corresponds a spin-down excitation at the same energy. For the s-wave vortex, this degeneracy is due to the full spin-rotation $SU(2)$ symmetry. The p-wave Hamiltonian (15.4) has a reduced spin symmetry. Namely, it has the symmetry group $O(2)$ generated by rotations about the z-axis ($\Psi_\uparrow \mapsto e^{i\alpha}\Psi_\uparrow, \Psi_\downarrow \mapsto e^{-i\alpha}\Psi_\downarrow$) and by the spin flip $\Psi_\uparrow \mapsto \Psi_\downarrow, \Psi_\downarrow \mapsto \Psi_\uparrow$. This non-abelian group causes the two-fold degeneracy of all levels, except for the zero-energy level(s) where the symmetry $O(2)$ may mix the creation and annihilation operators for the same state. This symmetry is crucial for our discussion. Note that we have not included in the Hamiltonian neither the spin-orbit term ($\boldsymbol{U}_{\text{SO}} \cdot [\boldsymbol{\sigma} \times \boldsymbol{p}]$), nor the Zeeman splitting $\boldsymbol{H}(\boldsymbol{r}) \cdot \boldsymbol{\sigma}$. Either of these terms would break the spin symmetry $O(2)$, which would eventually result in a different universality class of the disordered system (type D with non-degenerate levels), if these terms are sufficiently strong.

The difference between the s- and p-wave vortices is the zero-energy level in the p-wave case. It has been shown by Volovik that this level has a topological nature [15]. Indeed, suppose we gradually increase disorder in the

Hamiltonian (15.4). The levels shift and mix, but the degeneracy of the levels remains the same as long as the symmetry $O(2)$ is preserved. The total number of levels remains *odd*, and therefore the zero-energy level *cannot shift* if the final Hamiltonian is a continuous deformation of the original one (without disorder), i.e. if the topological class of the boundary conditions (15.5) remains the same.

Now we can identify the symmetry class of the Bogoliubov–deGennes Hamiltonian. The only symmetry of the mean-field Hamiltonian (15.4) is the spin symmetry $O(2)$. The time-reversal symmetry is already broken by the vortex and by the pairing, and therefore neither the vector potential $\mathbf{A}(x)$ nor local deformations of Δ_μ can reduce the symmetry of the Hamiltonian. When projected onto spin-up excitations $\gamma_\uparrow^\dagger = \int [u(\mathbf{r})\Psi_\uparrow^\dagger(\mathbf{r}) + v(\mathbf{r})\Psi_\downarrow(\mathbf{r})]d^2\mathbf{r}$, the Bogoliubov–deGennes Hamiltonian for the two-component vector (u,v) takes the form:

$$H_{\mathrm{BdG}} = \begin{pmatrix} \left[\frac{(-\mathrm{i}\nabla - e\mathbf{A})^2}{2m} + V(\mathbf{r}) - \varepsilon_\mathrm{F}\right] & \left[\frac{\Delta_x}{k_\mathrm{F}} * (-\mathrm{i}\nabla_x) + \frac{\Delta_y}{k_\mathrm{F}} * (-\mathrm{i}\nabla_y)\right] \\ \left[\frac{\Delta_x^*}{k_\mathrm{F}} * (-\mathrm{i}\nabla_x) + \frac{\Delta_y^*}{k_\mathrm{F}} * (-\mathrm{i}\nabla_y)\right] & -\left[\frac{(-\mathrm{i}\nabla + e\mathbf{A})^2}{2m} + V(\mathbf{r}) - \varepsilon_\mathrm{F}\right] \end{pmatrix}.$$
(15.8)

In an arbitrary orthonormal basis of electronic states, this Hamiltonian may be written as a matrix

$$H_{\mathrm{BdG}} = \begin{pmatrix} h & \Delta \\ \Delta^\dagger & -h^* \end{pmatrix}. \tag{15.9}$$

From the hermiticity of the Hamiltonian, it follows that $h^\dagger = h$. From the explicit form of the p-wave pairing, $\Delta = -\Delta^T$ (it is here that the p-wave structure of the pairing is important; for s-wave pairing we would have $\Delta = \Delta^T$ instead). These are the only restrictions on the Hamiltonian (15.9). If we define

$$U_0 = \frac{1}{\sqrt{2}} \begin{pmatrix} 1 & 1 \\ \mathrm{i} & -\mathrm{i} \end{pmatrix}, \tag{15.10}$$

the restrictions on the Hamiltonian (15.9) are equivalent to the condition that the rotated matrix $\mathrm{i}U_0 H_{\mathrm{BdG}} U_0^{-1}$ is real antisymmetric, i.e. it belongs to the Lie algebra $so(M)$, where M is the dimension of the Hilbert space (the same rotation of the Hamiltonian was used in [3] to identify the D universality class).

The last step in our argument is to note that, under the vortex boundary conditions, the dimension of the Hamiltonian (15.9) is odd, not even (this may be difficult to visualize from the particle-hole representation (15.9), but easier from the rotated Hamiltonian $U_0 H_{\mathrm{BdG}} U_0^{-1}$). Thus, for a single-quantum vortex, we identify the space of the Hamiltonians as $so(2N+1)$ (class B in Cartan's notation, see Table 15.1).

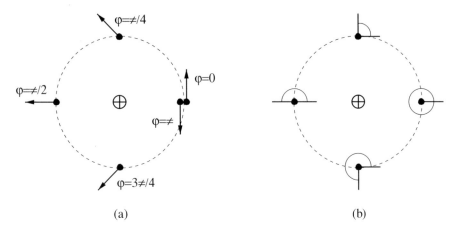

Fig. 15.1. (a) Half-quantum vortex. Arrows denote the direction of vector $\mathbf{d}$. (b) Spin vortex. The angle θ is shown

15.2 Half–Quantum Vortex

Now let us consider a slightly different order parameter: suppose that the condensate wave function is described by the same expression (15.3), but now the vector $\mathbf{d}$ is able to rotate (either in plane or in all three dimensions). The order parameter does not change under simultaneous change of sign of the vector $\mathbf{d}$ and phase shift by π of the phase φ: $(\varphi, \mathbf{d}) \sim (\varphi + \pi, -\mathbf{d})$. This makes possible a half-quantum vortex with magnetic flux $\Phi_0/2$ [16]. On going around such a vortex, both the vector $\mathbf{d}$ and the phase φ rotate by π (Fig. 15.1a) (see also the discussion of such vortices in [17,18]).

First consider first the case when the vector $\mathbf{d}$ rotates in a plane (without loss of generality, let $\mathbf{d}$ rotate in the x-y plane). Then the condensate wave function (15.3) is the sum of two decoupled components containing $|\uparrow\uparrow\rangle$ and $|\downarrow\downarrow\rangle$ pairings:

$$\Psi(r,\theta) = \left[\Delta_+^{(1)}(r,\theta)\,|\uparrow\uparrow\rangle + \Delta_+^{(2)}(r,\theta)\,|\downarrow\downarrow\rangle\right](k_x + ik_y) \tag{15.11}$$

[here r and θ are the polar coordinates]. In a half-quantum vortex, one of the pairing components $\Delta_+^{(1)}$ and $\Delta_+^{(2)}$ has an odd winding number of the phase around the vortex, while the other has an even winding number of the phase. In the simplest vortex of this type, the even winding number is zero, the corresponding component of the order parameter is nearly uniform and does not produce states near Fermi energy. The component with an odd winding number (without loss of generality, let it be $\Delta_+^{(1)}$, and the winding number be one) has vortex boundary conditions. The mean-field Hamiltonian decouples and the vortex part of the Hamiltonian (spin-up sector) takes the form [we have also assumed an axially symmetric order parameter]:

$$H = \int D^2 r \left[\Psi_\uparrow^\dagger \left(\frac{p^2}{2m} - \varepsilon_F \right) \Psi_\uparrow + e^{i\theta} \Delta(r) \Psi_\uparrow^\dagger \left(\nabla_x + i \nabla_y \right) \Psi_\uparrow^\dagger + \text{h.c.} \right]. \quad (15.12)$$

In this form, the half-quantum vortex is equivalent to a single-quantum vortex in a superconductor of spinless fermions (such a superconductor must have odd pairing). The quasiparticles do not have a definite spin projection, but mix particles and holes in the same spin-up sector: $\gamma^\dagger = \int [u(\mathbf{r})\Psi_\uparrow^\dagger(\mathbf{r}) + v(\mathbf{r})\Psi_\uparrow(\mathbf{r})] D^2 r$. The Bogoliubov–deGennes equations for u and v turn out to be identical to those for a single-quantum vortex (15.8), and the solution of Kopnin and Salomaa [13] is equally applicable to this half-quantum vortex resulting in the same spectrum (15.6). However, because of the relation $\gamma^\dagger(E) = \gamma(-E)$, the number of subgap states in the half-quantum vortex is one half of those in a single-quantum vortex: particles in negative-energy levels are identical to holes in positive-energy levels. The zero-energy state is a Majorana fermion [18,19].

Now add a disorder term to the Hamiltonian of a clean vortex (15.12). Consider the following perturbations:

- potential scattering: $H_V = \Psi_\alpha^\dagger V(\mathbf{r}) \Psi_\alpha$,
- Zeeman field: $H_h = \Psi_\alpha^\dagger (\mathbf{h}(\mathbf{r}) \boldsymbol{\sigma}_{\alpha\beta}) \Psi_\beta$,
- vector potential: $H_a = \Psi_\alpha^\dagger (\mathbf{a}(\mathbf{r}) * \mathbf{p}) \Psi_\alpha$,
- spin-orbit scattering: $H_U = \Psi_\alpha^\dagger (\mathbf{U}_{\text{SO}}(\mathbf{r}) * [\mathbf{p} \times \boldsymbol{\sigma}_{\alpha\beta}]) \Psi_\beta$,
- deformation of order parameter:

$$H_\Delta = \Psi_\alpha^\dagger \left(\Delta_{\alpha\beta}^{(x)}(\mathbf{r}) * p_x + \Delta_{\alpha\beta}^{(y)}(\mathbf{r}) * p_y \right) \Psi_\beta^\dagger + \text{h.c.}$$

We assume that the perturbation may be switched on adiabatically in such a way that subgap levels stay localized. One possible way to fulfill this condition is to require that the perturbation is introduced only in a finite region around the vortex core.

The symmetry classification proceeds slightly differently depending on whether the perturbation preserves the decoupling of the Hamiltonian into spin-up and spin-down sectors.

The Hamiltonian remains decoupled under the perturbations H_V, H_a, H_h with $\mathbf{h} \parallel \hat{\mathbf{z}}$, H_U with $\mathbf{U}_{\text{SO}} \perp \hat{\mathbf{z}}$, and H_Δ with diagonal $\Delta_{\alpha\beta}$. With these perturbations, the "spin-up" sector of the Hamiltonian in the basis (u,v) preserves its form (15.9) with the same symmetries as for a single-flux vortex. The symmetry class of the Hamiltonian is therefore identified as $so(2N+1)$ (class B).

If other perturbations are present, so that the Hamiltonian no longer splits into "spin-up" and "spin-down" parts, the full matrix of the Hamiltonian needs to be considered. Without any spin structure, the Hamiltonian breaking time-reversal and spin symmetries belongs to the $so(M)$ symmetry class, as shown by Altland and Zirnbauer [3]. Their argument also works in our case, with the reservation that the number of levels M is odd (with each level counted twice: once as a particle and the other time as a hole). It follows from the usual continuity consideration (the Majorana fermion at zero energy survives any local perturbation). Therefore the symmetry class is again $so(2N+1)$ (class B). [In the examples considered by Altland and Zirnbauer, the number of levels was even, which lead to the symmetry class D or $so(2N)$].

It is important however, that to reach the RMT limit in the latter case, such a system must have sufficiently strong disorder to bring the quasiparticles from the "spin-down" sector (gapped in the clean vortex) to the Fermi energy and to mix them strongly with the quasiparticles in the "spin-up" sector. In this chapter, we do not discuss conditions for reaching this limit.

15.3 Spin Vortex

In this section we consider a hypothetical phase of a triplet superconductor analogous to the phase B of ^{3}He. Namely, the condensate wave function is assumed to be

$$\Psi = \Delta e^{i\varphi}\left[d_x(\bm{k})\Big(|\uparrow\uparrow\rangle + |\downarrow\downarrow\rangle\Big) + id_y(\bm{k})\Big(|\uparrow\uparrow\rangle - |\downarrow\downarrow\rangle\Big)\right],$$
$$\bm{d}(\bm{k}) = R_\theta(\bm{k}),\qquad(15.13)$$

and R_θ is a rotation in the plane by angle θ (like in the previous sections, we consider a two-dimensional problem, and $\bm{k}$ is a two-dimensional vector). In this phase, the vector $\bm{d}$ rotates a full turn as the vector $\bm{k}$ goes around the circular Fermi surface. The parameter θ denotes the angle between vectors $\bm{k}$ and $\bm{d}$. We neglect spin-orbit interactions and assume that all values of θ are allowed.

It is important that this phase preserves the time-reversal symmetry. Moreover, a vortex structure is possible without time-reversal symmetry breaking. It is realized by rotating the angle θ a full turn of 2π when going around the vortex center and keeping the phase φ constant (Fig. 15.1b). This vortex preserves the time-reversal symmetry and is a topological defect in the spin structure of the triplet pairing, thus we shall call it a "spin vortex".

The Hamiltonian of the clean axially symmetric spin vortex splits into "spin-up" and "spin-down" sectors, similarly to the case of the half-quantum vortex considered in the previous section. In the spin vortex, both the "spin-up" and the "spin-down" components have vortex structure. The "spin-up" Hamiltonian is identical to the Hamiltonian H of (15.12), while the "spin-down" component is its complex conjugate H^* (with spin reversed) – or vice versa.

In a clean spin vortex, the spectrum of each of the two vortices (in the "spin-up" and the "spin down" sectors) are identical to that in the half-quantum vortex of the previous section, and therefore the total spectrum coincides with that of the single-flux vortex of Section I (with the two Majorana fermions combining into a single zero-energy level). In contrast to the single-quantum vortex, it is now not the spin symmetry which is responsible for the double degeneracy of the levels with non-zero energy, but the time-reversal symmetry (the degeneracy is due to Kramers' theorem).

Now suppose that the Hamiltonian is perturbed by a disorder term preserving the time-reversal symmetry. The allowed perturbations include potential scattering H_V, spin-orbit scattering H_U, as well as deformation of the order parameter H_Δ in a time-reversal invariant way. Then the levels with non-zero energy stay doubly degenerate, and the zero-energy level cannot shift (because it is non-degenerate). The symmetry class of the perturbed Hamiltonian may be identified from the argument of Altland and Zirnbauer as $DIII$ (time-reversal symmetry present, spin symmetry broken) [3]. In the p-wave vortex, the number of subgap levels is odd, and therefore the symmetry class is $DIII$-odd, in contrast to class $DIII$-even studied by Altland and Zirnbauer [similarly to distinguishing between classes B and D, we find it useful to distinguish between classes $DIII$-even and $DIII$-odd: they have different parameters α, as a consequence of the zero-energy level in odd dimensions].

15.4 Level Mixing by Disorder

In this section we discuss whether the RMT limit may be realized in disordered vortices. We suppose that the superconductor contains a finite concentration n_{imp} of spinless impurities with strong electron scattering (Born parameter is of order one). In this case the effect of impurities is dual: they directly contribute a potential term H_V in the Hamiltonian and they also suppress the superconducting order parameter, thus inducing an inhomogeneous pairing perturbation H_Δ. We shall see that the latter effect has a much stronger influence on shifting the energy levels of the subgap states.

Consider first the potential term H_V with the impurity potential $V(\mathbf{r}) = \sum_i V_i \delta(\mathbf{r} - \mathbf{r}_i)$. The matrix element of the impurity potential between the two localized states is equal to

$$(H_V)_{mn} = \sum_i V_i \varrho_{mn}(\mathbf{r}_i),$$
$$\varrho_{mn}(\mathbf{r}) = (u_m^*(\mathbf{r})u_n(\mathbf{r}) - v_m^*(\mathbf{r})v_n(\mathbf{r})). \quad (15.14)$$

From explicitly solving Bogoliubov–deGennes equations for a clean p-wave vortex [13], one finds that u_n and v_n are proportional to each other, to the leading order in E/Δ (where $E = n\omega_0$ is the energy of the n-th level). Therefore the matrix element ϱ_{mn} between any two low-lying states nearly vanishes (this was noticed by Volovik in [15]). A more accurate calculation shows that matrix elements of the charge density between states at energy E is of the order of $\xi^{-2}(E/\Delta)$, which is by the factor (E/Δ) smaller than that in the s-wave vortex. As a consequence, even for strong impurities (with

$V_i \sim \varepsilon_F k_F^{-2}$), an extremely high impurity concentration ($n_{\rm imp} \sim k_F^2$) is required to mix low-energy levels – a much higher disorder than that destroying p-wave superconductivity ($n_{\rm imp} \sim k_F \xi^{-1}$) [20,21]. Therefore, for a uniformly disordered p-wave superconductor, the H_V term is insufficient for the RMT regime in the vortex core.

The suppression of the order parameter H_Δ turns out to be a more important effect. In unconventional superconductors, spinless impurities suppress pairing in the region of size ξ which is of the same order as the size of the vortex core. For a homogeneous order parameter (without vortex), the suppression of the order parameter $\delta\Delta(r)$ decreases as R^{-1} (in our two-dimensional problem) at distances up to ξ, and is negligible at distances much larger than ξ [22,23], and the integral suppression may be estimated as $\int \delta\Delta\, D^2r \sim \Delta \xi k_F^{-1}$. Also, in a chiral superconductor, the component of the order parameter of opposite chirality is admixed in the vicinity of impurity [24]. The two effects gives comparable contributions to the matrix elements between subgap states. An accurate calculation of the impurity influence on the states in the vortex core requires a full self-consistent solution for the order parameter, taking into account both the vortex and the impurities. This goes beyond the scope of this chapter, and we only estimate the order of magnitude of the corresponding matrix elements. The deformation of the order parameter H_Δ has the matrix elements between subgap states m and n:

$$(H_\Delta)_{mn} = \int D^2 r \left(u_m^*(\boldsymbol{r}) \left[\delta\Delta^{(\alpha)}(\boldsymbol{r}) * (i\nabla_\alpha)\right] u_n(\boldsymbol{r}) \right.$$
$$\left. + v_m^*(\boldsymbol{r}) \left[\delta\Delta^{(\alpha)*}(\boldsymbol{r}) * (i\nabla_\alpha)\right] v_n(\boldsymbol{r}) \right). \tag{15.15}$$

Unlike the case of the matrix elements of H_V, there is no cancellation in $(H_\Delta)_{mn}$, and we may estimate, for one impurity in the vortex core

$$(H_\Delta)_{mn} \sim \frac{1}{\xi^2} \int \delta\Delta(\boldsymbol{r})\, D^2 r \sim \Delta(k_F \xi)^{-1} \sim \omega_0. \tag{15.16}$$

Thus, a single impurity introduces matrix elements of the order of interlevel spacing. These estimates suggest that the RMT regime may be achieved when the number of impurities in the vortex core is much greater than one (which corresponds to the moderately clean regime $\Delta^{-1} \ll \tau \ll \omega_0^{-1}$).

15.5 Summary and Discussion

We have shown that symmetries of vortex Hamiltonians in p-wave superconductors correspond to one of the two random–matrix ensembles: B or DIII-odd. These two RMT ensembles are distinguished from the rest of ensembles by the presence of zero-energy modes (zero-energy modes are also present in chiral ensembles at pn_cq, see Table 15.1). Zero-energy levels appear in p-wave vortices as a consequence of odd pairing symmetry combined with the vortex (topologically nontrivial) boundary conditions. We have checked the symmetries of the Hamiltonian explicitly taking three vortices as examples: a single-quantum vortex with spin symmetry, a half-quantum vortex, and a flux-less (spin) vortex without time-reversal symmetry breaking. Based on these examples, we may conjecture that classes B and DIII-odd appear in any vortex-like structure with odd pairing, whenever a zero-energy mode is present. In cases with time reversal symmetry, the vortex belongs to the class DIII-odd; when the time-reversal symmetry is broken – to the class B. Of course, this can also be checked explicitly in any particular case.

The two classes B and DIII-odd considered in this chapter are odd-dimensional counterparts of the classes D and DIII-even, respectively (see Table 15.1), with the even-dimensional classes realized in conventional (singlet-pairing) normal-superconducting structures [3]. Far from zero energy, the statistics of energy levels is not affected by the zero mode. In the immediate vicinity of zero energy, repulsion from the zero mode in classes B and DIII-odd increases the exponent α (see Table 15.1), suppressing the average density of states $\langle \varrho(\omega) \rangle$ as $\langle \varrho(\omega) \rangle \propto \omega^2$ for class B, and $\langle \varrho(\omega) \rangle \propto |\omega|^5$ for class DIII-odd.

Because of the zero modes, the classes B and DIII-odd stand somewhat separately from the rest of RMT ensembles. In particular, they do not appear in the table of correspondence between symmetries of the Hamiltonian and those of the transfer matrix [5,6] (for RMT ensembles without zero-energy levels, this table establishes a one-to-one correspondence). Neither of these two ensembles are known to be derived from a supersymmetric sigma-model as other classes are [4]. Understanding the latter fact remains an interesting problem, as well as a possible microscopic derivation of the random–matrix theory for p-wave vortices with the supersymmetric method, analogously to that for the s-wave vortex [25,26].

Several more comments can be made on comparing our results to the level statistics in s-wave vortices. The latter is known to belong to the symmetry class C [25,26], having spin-rotational symmetry and broken time-reversal symmetry [3]. In contrast with the $SU(2)$ spin symmetry in conventional superconductors, the single-quantum vortex considered in this chapter has the $O(2)$ spin symmetry instead, which leads to a somewhat different symmetry classification. The resulting symmetry is $so(N)$, which leads to classes D (in even dimensions, no vortex) or B (odd dimensions, vortex with zero mode).

The joint probability distributions of the energy levels (15.2) for classes B (p-wave vortex) and C (s-wave vortex) coincide (with $\alpha = \beta = 2$), and only the zero-energy level distinguishes the two ensembles. In particular, the average density of states in the class B random–matrix ensemble is

$$\langle \varrho(\omega) \rangle = \frac{1}{\omega_0} - \frac{\sin(2\pi\omega/\omega_0)}{2\pi\omega} + \delta(\omega) \,. \tag{15.17}$$

(for class C it is the same but without the $\delta(\omega)$ term).

In s-wave vortices, Koulakov and Larkin found that in a wide range of impurity concentrations, the level mixing does not lead to a class C, but has more symmetries [27]. These symmetries arise from the fact that the impurities are local scatterers in a chiral quasi-one-dimensional two-channel system, resulting in an ensemble of 2×2 random matrices, instead of $N \times N$ matrices with large N. Such effect seems improbable in p-wave vortices where the main contribution to level mixing comes from the distortion of the order parameter around impurities which is extended in space to distances of the order of the core size. In this chapter we performed only an order-of-magnitude estimate suggesting that the RMT limit may be reached in moderately clean vortices (with the number of impurities per vortex core much greater than one). A more accurate self-consistent treatment of impurities in the vortex core is required for a quantitative study of the effects of disorder.

The author wishes to thank M. V. Feigel'man for suggesting this problem, discussions, and comments on the manuscript. At different stages of work, the author benefited from discussions with G. Blatter, V. Geshkenbein, D. Gorokhov, R. Heeb, L. Ioffe, C. Mudry, M. Skvortsov, G. E. Volovik, and M. Zhitomirsky. The author thanks Swiss National Foundation for financial support.

References

1. E. P. Wigner: Proc. Camb. Phil. Soc. **47**, 790 (1951)
2. K. Efetov: *Supersymmetry in disorder and chaos* (Cambridge University Press, Cambridge 1997)
3. A. Altland, M. R. Zirnbauer: Phys. Rev. B **55**, 1142 (1997) [cond-mat/9602137]
4. M. R. Zirnbauer: J. Math. Phys. **37**, 4986 (1996)
5. M. Caselle: A new classification scheme for random matrix theories, cond-mat/9610017
6. P. W. Brouwer, C. Mudry, A. Furusaki: Phys. Rev. Lett. **84**, 2913 (2000) [cond-mat/9904200];
   P. W. Brouwer, A. Furusaki, I. A. Gruzberg, C. Mudry: Localization and delocalization in dirty superconducting wires, cond-mat/0002016
7. S. Helgason: *Differential geometry, Lie groups, and symmetric spaces* (Academic Press, New York 1978)
8. F. J. Dyson: J. Math. Phys. **3**, 1199 (1962)
9. J. Verbaarschot: Phys. Rev. Lett. **72**, 2531 (1994)
10. M. Bocquet, D. Serban, M. R. Zirnbauer: Disordered 2d quasiparticles in class D: Dirac fermions with random mass, and dirty superconductors, cond-mat/9910480
11. D. A. Ivanov: The energy-level statistics in the core of a vortex in a p-wave superconductor, cond-mat/9911147

12. R. Heeb, D. F. Agterberg: Phys. Rev. B **59**, 7076 (1999) [cond-mat/9811190]
13. N. B. Kopnin, M. M. Salomaa: Phys. Rev. B **44**, 9667 (1991)
14. C. Caroli, P.-G. de Gennes, J. Matricon: Phys. Lett. **9**, 307 (1964)
15. G. E. Volovik: Pis'ma Zh. Exp. Teor. Fiz. **70**, 601 (1999) [JETP Lett. **70**, 609 (1999); cond-mat/9909426]
16. G. E. Volovik, V. P. Mineev: Pis'ma Zh. Exp. Teor. Fiz. **24**, 605 (1976) [JETP Lett. **24**, 561 (1976)]
17. G. E. Volovik: Pis'ma Zh. Exp. Teor. Fiz. **70**, 776 (1999) [JETP Lett. **70**, 792 (1999); cond-mat/9911374]
18. D. A. Ivanov: Non-abelian statistics of half-quantum vortices in p-wave superconductors, cond-mat/0005069
19. N. Read, D. Green: Phys. Rev. B **61**, 10267 (2000) [cond-mat/9906453]
20. R. Balian, N. R. Werthammer: Phys. Rev. **131**, 1553 (1963)
21. K. Ueda, T. M. Rice: 'Heavy electron superconductors – some consequences of the p-wave pairing'. In: *Theory of Heavy Fermions and Valence Fluctuations*, ed. by T. Kasuya, T. Saso (Springer, Berlin 1985)
22. D. Rainer, M. Vuorio: J. Phys. C: Solid State Phys. **10**, 3093 (1977)
23. C. H. Choi, P. Muzikar: Phys. Rev. B **39**, 9664 (1989)
24. Y. Okuno, M. Matsumoto, M. Sigrist: Analysis of impurity-induced circular currents for the chiral superconductor, cond-mat/9906093
25. R. Bundschuh, C. Cassanello, D. Serban, M. R. Zirnbauer: Nucl. Phys. B **532**, 689 (1998) [cond-mat/9806172]
26. M. A. Skvortsov, M. V. Feigel'man, V. E. Kravtsov: Pis'ma Zh. Exp. Teor. Fiz. **68**, 78 (1998) [JETP Lett. **68**, 84 (1998)]
27. A.A. Koulakov, A.I. Larkin: Phys. Rev. B **60**, 14597 (1999) [cond-mat/9810125]

Part IV

Selected Experiments on Vortices in Superconductors

16 Scanning Tunneling Spectroscopy on Vortex Cores in High-T_c Superconductors

B.W. Hoogenboom, C. Renner, I. Maggio-Aprile, and Ø. Fischer

Summary. Scanning tunneling spectroscopy (STS) with its unique capacity for tunneling spectroscopy with sub-nanometer spatial resolution, has opened new ways to look at the flux lines and their distribution in superconductors. In contrast to all other imaging techniques, which are sensitive to the local magnetic field, STM relies on local changes in the density of states near the Fermi level to generate a real space image of the vortex distribution. It is thus sensitive to the vortex cores, which in high temperature superconductors have a size approaching the interatomic distances. The small size of the vortex cores and the anisotropic character of the high temperature superconductors allow pinning to play a large role in determining the vortex core positions. Vortex hopping between different pinning sites, again down to a sub-nanometer scale, has been studied by STM imaging as a function of time. These studies give microscopic indications for quantum tunneling of vortices. Moreover, STM provides new insights into the detailed electronic vortex core structure, revealing localized quasiparticles.

16.1 Introduction

Since the discovery of high temperature superconductors (HTSs) in 1986 [1], research on vortices has intensified and has revealed many aspects that had not been seen before. Macroscopically observable vortex phenomena in HTSs have already shown interesting physics [2], and nowadays many techniques are available to also study vortices at microscopic scales. The latter allows one to verify the interpretation of macroscopic vortex behaviour – creep, pinning, and phase transitions, for example – by directly observing individual vortices. Among the techniques giving direct images of vortices in real space one can cite Bitter decoration [3,4], magnetic force microscopy [5–7], scanning hall-probe microscopy [8], scanning SQUID microscopy [9,10], and Lorentz microscopy [11]. All these methods are characterized by their sensitivity to the local magnetic field, thus showing the magnetic magnitude of the vortices, with a size of the order of the penetration depth λ (of the order of 10^3 Å in HTSs).

The only technique, to our knowledge, that is sensitive to the electronic change due to the presence of a vortex – the vortex core – is Scanning Tunneling Microscopy (STM) [12,13]. Its abilities to image vortex cores were first demonstrated by Hess et al. [14], on NbSe$_2$. In the vortex core the superconducting order parameter is suppressed over a length scale of the order of the coherence length ξ, which in HTSs is typically 2 orders of magnitude smaller than λ. Thus, even near the upper critical field H_{c2} it is possible to observe individual vortex cores by STM. With the STM one can also perform spatially resolved tunneling spectroscopy (scanning tunneling spectroscopy – STS), resulting in information about the electronic structure of vortex cores.

In this review, we will concentrate on STM studies of vortex cores in HTSs. After a description of the experimental method in Sect. 16.2, we will give some details about quasiparticle states in vortex cores in Sect. 16.3. Sect. 16.4 presents results obtained on NbSe$_2$, and serves mainly for comparison of BCS results with those obtained on the HTSs YBa$_2$Cu$_3$O$_{7-\delta}$ (YBCO), in Sect. 16.5, and Bi$_2$Sr$_2$CaCu$_2$O$_{8+\delta}$ (BSCCO), in Sect. 16.6. Finally, Sect. 16.7 gives some conclusions and a brief outlook. For completeness we quote STM vortex studies on LuNi$_2$B$_2$C [15], YNi$_2$B$_2$C [16] and CeRu$_2$ [17], which will not be discussed here.

16.2 Studying Vortex Cores by STM

The principles of an STM experiment [18], and more specific those of vortex imaging, are depicted in Fig. 16.1. The STM tip and the sample surface form both sides of a tunnel junction, separated by a vacuum tunnel barrier. Moving the tip over the surface one can obtain topographic information by keeping either the current constant, while measuring the therefore required z tip motion (this is the mode used to obtain the images of YBCO and BSCCO presented below), or by keeping the tip z position constant, while measuring variations in the tunnel current.

The dependence of the tunnel current I on the bias voltage V_b allows spectroscopic measurements [19,20]. Writing down an Hamiltonian of the

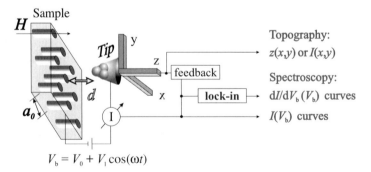

Fig. 16.1. Principles of STM vortex imaging. At the left is the sample. A magnetic field H is applied along the tunneling direction and results in vortex cores with lattice spacing a_0. The position of the STM tip is governed by a feedback mechanism. By varying the bias voltage V_b, $I(V)$ spectra can be obtained. dI/dV spectra are taken using a lock-in amplifier.

system as consisting of two (commuting) parts for both sides of the tunnel junction, and a part transferring electrons from one side to the other, one can derive [21,22] for a metal-superconductor junction (where the metallic DOS is assumed to be constant near the Fermi level) :

$$\frac{dI}{dV} = \int_{-\infty}^{\infty} \frac{d\omega}{2\pi} \left(-\frac{\partial f(\omega - eV)}{\partial eV}\right) \sum_{\mathbf{k}} |T_{\mathbf{k}}|^2 A_S(\mathbf{k},\omega). \tag{16.1}$$

Assuming a constant tunneling matrix element $|T_{\mathbf{k}}|$ and with A_S the (local) spectral function of the superconductor, this results in the quasiparticle local DOS convoluted by the derivative of the Fermi-function (causing temperature broadening). Using planar junctions, Giaever and coworkers thus measured the BCS DOS in led [23]. In general, and specifically in HTSs, for explaining the *detailed* shapes of the dI/dV spectra, a treatment going beyond the approximations of a constant matrix elements and a BCS spectral function (see [24]) is required.

The presence of a vortex core leads to a local suppression of the order parameter, and thus the local DOS in the core and in the superconducting state are expected to be different. By scanning over the surface and measuring the conductance $dI/dV(x,y)$ at an energy eV at which this difference is most pronounced, one can obtain spectroscopic images mapping the vortex core distribution. By recording $dI/dV(V)$ quasiparticle excitation spectra can be measured locally.

16.3 Quasiparticle Excitations in Vortex Cores

As was shown by Caroli, de Gennes and Matricon [25], the suppression of the order parameter in the vortex core in a conventional (BCS) superconductor results in localized quasiparticle states. These states, with energies smaller than the superconducting gap Δ, exist on both sides of the Fermi level E_{F}. The states with lowest energies ($\ll \Delta_0$) are separated by $\sim \Delta^2/E_{\mathrm{F}}$. In conventional superconductors this quantity is small compared to the temperature smearing and experimental resolution in a typical low-temperature STM experiment. The local DOS in a vortex core will thus appear as a broad peak at zero bias, consisting of many localized quasiparticle states.

In the HTSs the situation is quite different. First, the coherence length is extremely short, approaching the interatomic distances [26]. In analogy to conventional superconductors one would therefore expect the energy spacing between the first localized quasiparticle excitations to increase due to the decrease in size of the "quantum-mechanical box" formed by the vortex core. This would give clearly separated peaks in the LDOS. However, this way of reasoning neglects the d-wave character of the order parameter in HTSs. Basically, quasiparticles can escape from the vortex core along the nodes of the gap function, and will thus not be truly localized. In fact, quasiparticle states in the vortex core will scatter to quasiparticle states that are extended

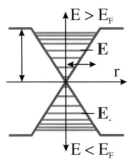

Fig. 16.2. Schematic representation of quasiparticle excitations localized in the vortex core in a BCS superconductor (for a more rigorous picture we refer to numerical calculations of Gygi and Schlüter [28]). The order parameter Ψ goes to zero in the vortex core, over a characteristic distance ξ. Inside the vortex core one can find quasiparticle states, the lowest of which correspond to $E_\mu \sim \mu \Delta^2/E_F$ ($\mu = \pm 1/2, \pm 3/2, \pm 5/2, ...$), smaller than the gap Δ. The continuum of quasiparticle states above the gap is not shown in this picture.

along the nodes of the gap function, leading to a broad peak at zero bias again. It should be clear, however, that the nature of this peak is different from the one expected for conventional superconductors.

Both for conventional superconductors [27,28] and for HTSs [29,30] these intuitive pictures have been confirmed by numerical calculations.

16.4 NbSe$_2$

STS results on NbSe$_2$ perfectly match the behaviour of conventional superconductors described in the previous paragraph [14,31]. The well ordered hexagonal vortex lattice is shown in Fig. 16.3a. In the light region the dI/dV spectra show a superconducting gap, in the centers of the dark spots – the vortex cores – a broad zero-bias peak is observed (Fig. 16.3c). The diameter of the vortex core is about 300 Å. According to a BCS analysis [28], the superconducting spectrum is recovered at a distance of about 6ξ from the center of the core. This leads to a coherence length $\xi \sim 25$ Å. The zero-bias peak is split into two just outside the center of the core, and the r dependence (where r is the distance from the center of the core) is in agreement with the calculations quoted before [27,28]. Furthermore, it has been shown that the localized quasiparticle states are very sensitive to the sample purity, in the dirty limit (Nb$_{0.8}$Ta$_{0.2}$Se$_2$) the zero-bias peak is completely suppressed [31].

For a later comparison with HTSs, Fig. 16.3b shows the result of a scan covering a longer time, where vortex motion is observed [32,33]. Since the image is taken line by line, and each line takes a certain time, there is and intrinsic time axis in the image: in Fig. 16.3c from the lower left corner to the upper right one. Slowly moving vortices thus appear as elongated objects.

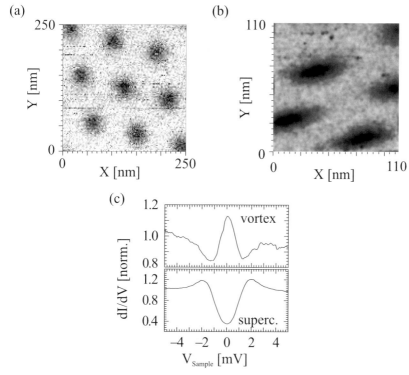

Fig. 16.3. Vortex imaging and spectroscopy on NbSe$_2$ at 1.3 K and 0.3 T. (**a**) Flux line lattice. (**b**) A slow scan ($\sim$ 1 hour) Vortex cores appear elongated due to vortex drift. (**c**) Zero-bias peak in the vortex core (top) and gap in the superconducting state (bottom).

Recently, vortex motion in NbSe$_2$ was also studied in a fast-scanning STM experiment [34].

16.5 YBa$_2$Cu$_3$O$_{7-\delta}$

The first STM vortex images on a HTS were surprisingly obtained on YBCO [35]. YBCO does not have an easy cleavage plane, making it rather difficult to obtain a clean and well-defined surface as required for spectroscopic measurements. The measurements shown here were all taken on as-grown surfaces, having been exposed to air.

The first conclusion that can be drawn from the vortex cores in Fig 16.4a, is that they are at least a factor of 3 smaller than those observed in NbSe$_2$. This corresponds to the smaller size of the ξ in the HTSs. The vortex cores have an elliptic shape due to in-plane $a - b$ anisotropy [36]. On a large scale,

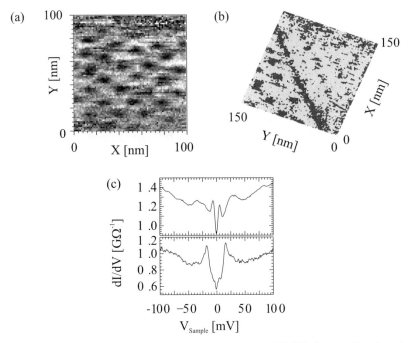

Fig. 16.4. Vortex imaging and spectroscopy on YBCO (optimally doped, $T_c = 91$ K). (a) Vortex distribution at 4.2 K and 6 T. (b) Image of vortex cores near a twin boundary, taken 12 hours after the field was reduced from 3 to 1.5 T (see text). (c) Tunneling spectra in the center of the core (top) and in a superconducting region (bottom).

the vortex distribution is disordered, underlining the importance of pinning. Locally, however, an oblique ordering can be observed.

The most striking result comes from the tunneling spectra. Moving the tip from a superconducting zone (Fig. 16.4c-bottom) into a vortex core (Fig. 16.4c-top), the superconducting coherence peaks disappear, and instead of a broad zero-bias peak, two split and well-pronounced peaks come up at low-energy (± 5.5 meV). This could correspond to an s-wave picture comparable to $NbSe_2$ with a smaller core size, causing most localized quasiparticles to pile up at the top of the low-energy potential well of the vortex core as it has been depicted in Fig. 16.2. Thus, only the first localized state is observed on both sides of the Fermi level.

The observation of quasiparticle states localized in the vortex core, however, is in disagreement with YBCO if seen as a d-wave superconductor. An explanation may be the presence of an additional component in the order parameter in the vortex core, cutting off the nodes of the d-wave gap function. In fact, numerical calculations of the Boguliobov–de Gennes equations, with a $d_{x^2-y^2} + \mathrm{i}d_{xy}$ symmetry of the order parameter, qualitatively account for the spectra observed in the YBCO vortex cores [37].

In Fig. 16.4b a twin boundary is present. After the field is reduced from 3 to 1.5 T, the vortices pile up at one side of the boundary and move away at the other [36]. A detailed study of the spots at the right side of the boundary reveals fractional images of vortex cores. Like the elongated vortices in Fig. 16.3, this can be explained by the presence of an intrinsic time axis in the (line-by-line taken) images. When the first line is scanned, a vortex can be at a certain position. After some lines have been scanned, covering (only) part of the vortex core, the vortex may move to another position. Thus, for the next couple of lines no vortex core is observed, until another vortex appears at the "empty" site, resulting in a vortex signal for the following scanned lines, etc. In contrast to the elongated vortex cores in NbSe$_2$, the vortex cores move by small, quick jumps, staying only short times at each site. Furthermore, the they seem to be restricted to certain positions, probably corresponding to pinning sites.

16.6   Bi$_2$Sr$_2$CaCu$_2$O$_{8+\delta}$

16.6.1   Vortex Core Spectra

In spite of the possibility of *in-situ* cleaving, offering clean and extremely flat surfaces, and the high spatial reproducibility in scanning tunneling spectra [38], it turned out to be difficult to map the vortex cores in this material. Finally, the reason was that the difference between the quasiparticle spectra inside and outside the vortex core is rather small. When proper criteria for mapping the vortex distribution are used, it is indeed possible to observe the vortices [26,39–43]. The image in Fig. 16.5a shows an example with an interesting peculiarity: the structure at the right corresponds to a large defect (where superconductivity is suppressed), which also showed up clearly in the topographic data [40]. This defect allowed for an exact position reference during the whole experimental run, also after having decreased and removed the field. Thus, the relation between the black spots in the images and the vortices can be rigorously proved: the number of vortex cores in this region scales with the applied magnetic field.

A typical spectrum taken in a vortex core is shown in Fig. 16.5b. The striking result here is that the low-bias part of the spectra does not change appreciably when going from outside to inside the vortex core. There is no sign of a zero-bias anomaly, nor of the pronounced split peak as in YBCO. The coherence peak at negative bias is suppressed as one enters the vortex core. However, the peak at positive bias is only weakened and shifts to somewhat higher energy. These are precisely the characteristics of the pseudogap structure observed in zero field above T_c; the pseudogap above T_c and the vortex core spectra are in fact remarkably similar when thermal broadening is taken into account [26,44].

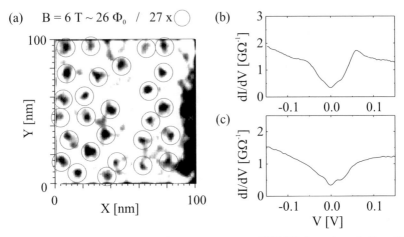

Fig. 16.5. Vortex imaging and spectroscopy on BSCCO (overdoped, $T_c = 77$ K) at 4.2 K and 6 T. (a) Vortex distribution. (b) Tunneling spectra in the center of the core (solid curve) and in the surrounding superconducting region (dotted curve). (c) As in (b), for an other vortex, where the peak at positive bias in the core is less pronounced.

More recent measurements have revealed additional structures in the low-energy part of the vortex core spectra [39,42,43]. They can already be observed in Fig. 16.5b as weak shoulders in the gap. Remarkably, these low-energy structures do show up more clearly when the gap is smoother, and does not show the peak at positive bias, as in Fig. 16.5c. Generally, the suppression of coherence peaks corresponds to a degradation of the (zero-field) superconducting properties of the surface. From these measurements it is not so clear whether the difference in size of the low-energy states in Fig. 16.5b and c is simply because the low-energy states are hidden in the slope of the gap in spectra like Fig. 16.5b, or whether this difference is related to truly intrinsic behaviour. Although the shoulders in the gap in the BSCCO vortex cores are much less pronounced than the peaks observed in YBCO vortex cores (cf. Fig. 16.4), the presence of these signatures of vortex core states suggest similar behaviour in both materials. Finally, one should note the similarity between the low-energy states in the BSCCO vortex cores to the weak shoulders observed in the gap of *zero-field* spectra of YBCO [35].

16.6.2 Vortex Core Images

From Fig. 16.5a it can be deduced that the vortex cores are extremely small. Typically, their size is of the order of 50 Å. If one assumes that a BCS analysis can be applied – which is most probably not strictly valid – this will give a coherence length of about 4 Å, very close to the lattice constant.

A high-resolution image of a vortex core is shown in Fig. 16.6a. The vortex core does not show the square symmetry predicted for *d*-wave superconduc-

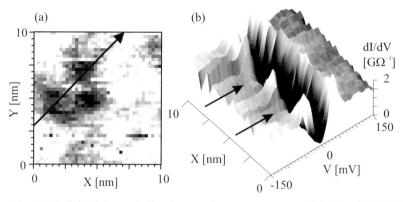

Fig. 16.6. (a) High-resolution image of a vortex core, at 4.2 K and 6 T. The core is small and seems to consist of several elements. (b) Trace of spectra along the arrow indicated in (a): in between two regions where the spectra show vortex core characteristics (the two arrows), the coherence peaks come up again.

tors [29,45,46], and which has been observed around impurity atoms [47]. In fact, the vortex cores generally show irregular shapes, and, more strikingly, often seem to be split into several components. A trace of spectra through the vortex core in Fig. 16.6 reveals that the core is truly split, with the coherence peaks coming up in the middle of the core. This splitting of the core can be interpreted as a vortex moving back and forth between two or more pinning sites with a frequency too high to be observed in these experiments. It should be noted that the smaller elements of split cores cannot correspond to separate vortices: this seems highly improbable because of the vortex-vortex repulsion, but most of all would give a total magnetic flux that is far too high with respect to the applied magnetic field.

With subsequent spectroscopic images one can also study the vortex distribution as a function of time. Generally, measurements are initiated several days after having switched on the field (the measurements presented here were taken on zero-field cooled samples), after which the vortex motion should be negligible [48]. However, though some vortex cores stay on the same positions, many of them slowly move around [40]. Such a moving vortex is shown in Fig. 16.7 (note that the area displayed in this figure is considerably larger than the one in Fig. 16.6). This behaviour is considerably different from the very slow moving vortex cores in $NbSe_2$ (Sect. 16.4) or those moving instantaneously such as those in YBCO (Sect. 16.5). The core moves from the right in Fig. 16.7a to the near-behind in Fig. 16.7c, passing through an intermediate, split state in Fig. 16.7b. As can be seen on the scales next to the images, the vortex core in Fig. 16.7b indeed consist of two parts with lower intensity than the single vortex cores in (a) and (c).

So apparently the vortex cores move around their own positions (resulting in split or irregularly shaped images as in Fig. 16.6a, but are also displaced

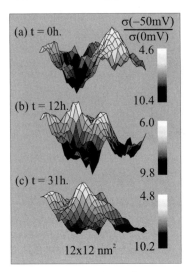

Fig. 16.7. Time-dependent images of a moving vortex core, at 4.2 K and 6 T. Over a square of 12×12 nm^2 the normalized conductivity is mapped in the vertical dimension. A normalized conductance of about 4 corresponds to the presence of a vortex core 1 of > 8 to superconducting spectra. The vortex core is first at the right (**a**), moves finally to the left (**c**), and passes through an intermediate state in (**b**). t is the time at which the image is started. Each image takes about 12 hours.

over larger distances, as in Fig. 16.7. Keeping in mind that the areal density of oxygen vacancies per CuO_2 double layer is such that their average spacing is of the order of 10 Å [49,50], one can relate this motion to hopping between different pinning centers. The fact that this plays a larger role in BSCCO than in YBCO is not so surprising since the larger anisotropy of BSCCO facilitates pinning of the (pancake) vortices, also causing the very irregular distribution of vortices in Fig. 16.5a.

Since these measurements were taken at a temperature close to the crossover temperature from thermal to quantum creep, as inferred from magnetic relaxation measurements [48,51–53], one can wonder about the mechanism of observed vortex motion: thermal hopping or quantum tunneling of vortices. Note that in the case of Fig. 16.7 the long time scale does not have anything to do with the hopping or tunneling time. The overall pinning potential changes very gradually (finally leading to the total displacement of the vortex core in Fig. 16.7), and rapid hopping or tunneling results in the split appearance in these images. In [40] a comparison is made between measurements at different temperatures, showing that an interpretation in terms of motion due to thermal excitations is highly improbable in this context. Other examples of split vortex cores and further details about experimental considerations can be found in this [40] as well. An alternative interpretation of these measurements, as a real splitting of the core which happens for example with the vortex core in superfluid 3He-B [54,55], cannot be excluded, though it seems hard to explain larger scale motion as in Fig. 16.7 by this effect. Therefore, these measurements strongly suggest that the vortex cores can actually occupy a state delocalized between different pinning sites as a result of quantum tunneling, similarly to the case of an ammonia molecule [56].

16.7 Conclusions and Outlook

As can be concluded from the previous sections, vortex cores in HTSs behave differently from those in conventional superconductors. First of all, the vortex cores are extremely small, leading to estimates of the coherence length ξ of the order of the interatomic distances. This is a further evidence of the electron pairs having a size comparable to the distances between the pairs. Therefore the HTSs are in a limit where BCS theory cannot be strictly valid. In this respect, BSCCO appears to be a more extreme case than YBCO (see also temperature dependent tunneling measurements on $Bi_2Sr_2Cu_2O_{6+\delta}$ [57]).

As a consequence of the small size of the vortex cores, both in the in-plane and – because of the anisotropy of the materials – also in the out-of-plane directions, they are very sensitive to pinning. This can be derived, among other ways, from the disorder in the vortex core distributions in STM (high field) measurements. This pinning not only influences the vortex distribution, but also the vortex core shape. Unlike in YBCO, where the vortex cores have an elliptical shape related to the rectangular atomic lattice, in BSCCO they appear with irregular shapes, often even split in separate smaller elements. This splitting implies that the actual vortex cores can be even smaller than expected based on their apparent size. One can interpret the splitting of the vortex cores in terms of hopping between different pinning sites, occurring even at low temperatures, and also over larger distances. One observes vortex cores moving from one site to the other, passing an intermediate split state. Estimates of the strength of the pinning potential wells in temperature dependent measurements strongly suggest that the hopping takes place by means of quantum tunneling rather than by thermal excitations [40]. With ^{3}He STMs it is now possible to reach temperatures of 200 $\sim$ 300 mK [58,59], allowing a definite test to exclude the possibility of thermally induced motion.

The results on the shapes of vortex cores in BSCCO leave open the question of the predicted four-fold symmetry of the cores. One may expect an intrinsic vortex core shape to show up in extremely pure and homogeneous samples, where pinning plays a much smaller role.

Concerning the vortex core electronic DOS, the situation is far from clear. On the one hand, there are apparent core states, with relatively large energy spacing, in both YBCO and BSCCO; on the other hand, the difference in size of these states is considerable, making it difficult to be conclusive about their mutual correspondence. Moreover, the origin of these states is unclear: they cannot be explained for a superconductor with pure $d_{x^2-y^2}$ symmetry, but require additional symmetry components of the order parameter, or another so far unknown mechanism. It is also necessary to obtain spectroscopic data of the vortex cores in other HTS materials in order to establish the common phenomenology (if present) of vortex core states in HTSs. Finally, though so far no large field and temperature dependence of vortex core states has been observed, their full systematic dependence on the applied magnetic field (also possibly implying the formation of bands of vortex core states, see e.g. [30]) and on temperature still has to be established.

References

1. J. G. Bednorz and K. A. Müller: Z. Phys. B **64**, 189 (1986)
2. G. Blatter, M. V. Feigel'man, V. B. Geschkenbein, A. I. Larkin, and V. M. Vinokur: Rev. Mod. Phys. **66**, 1125 (1994), and references therein
3. U. Essmann and H. Träuble: J. Sci. Instrum. **43**, 344 (1966)
4. U. Essmann and H. Träuble: Phys. Lett. **24A**, 526 (1967)
5. P. Rice and J. Moreland: IEEE Trans. Magn. **27**, 5181 (1991)
6. A. Moser, H. J. Hug, I. Parashikov, B. Stiefel, O. Fritz, H. Thomas, A. Baratoff, H.-J. Güntherodt, and P. Chaudari: Phys. Rev. Lett. **74**, 1847 (1995)
7. C. W. Yuan, Z. Zheng, and J. N. Eckstein: J. Vac. Sci. Technol. **14**, 1210 (1996)
8. A. M. Chang, H. L. Kao H. D. Hallen, H. F. Hess, J. Kwo, A. Sudbø, and T. Y. Chang: Europhys. Lett. **20**, 645 (1992)
9. F. P. Rogers: BS/MS thesis (1993)
10. L. N. Vu, M. S. Wistrom, and D. J. van Harlingen: Appl. Phys. Lett. **63**, 1693 (1993)
11. A. Tonomura J. Low Temp. Phys **105**, 1091 (1996), and references therein
12. G. Binnig, H. Rohrer, Ch. Gerber, and E. Weibel: Phys. Rev. Lett. **49**, 57 (1982)
13. G. Binnig and H. Rohrer: Helv. Phys. Acta **55**, 726 (1982)
14. H. F. Hess, R. B. Robinson, R. C. Dynes, J. M. Valles, and J. V. Waszczak: Phys. Rev. Lett. **62**, 214 (1989)
15. Y. DeWilde, M. Iavarone, U. Welp, V. Metlushko, A. E. Koshelev, I. Aranson, G. W. Crabtree, and P. C. Canfield: Phys. Rev. Lett. **78**, 4273 (1997)
16. H. Sakata, M. Oosawa, K. Matsuba, N. Nishida, H. Takeya, and K. Hirata: Phys. Rev. Lett. **84**, 1583 (2000)
17. H. Sakata, N. Nishida, M. Hedo, K. Sakurai, Y. Inada, Y. Onuki, E. Yamamoto, and Y. Haga: preprint
18. H.-J. Güntherodt and R. Wiesendanger: *Scanning Tunneling Microscopy* (Springer-Verlag, Berlin, 1993), Vol. I-III
19. C. B. Duke: *Tunneling in Solids* (Academic Press, New York, 1969)
20. E. L. Wolf: *Principles of Electron Tunneling Spectroscopy*(Oxford University Press, New York, 1985)
21. J. Bardeen: Phys. Rev. Lett. **6**, 57 (1961)
22. G. D. Mahan: *Many-Particle Physics*, 2nd. ed. (Plenum Press, New York, 1983)
23. I. Giaever: Phys. Rev. Lett. **5**, 147 (1960)
24. B. W. Hoogenboom, A. A. Manuel, and Ø. Fischer: preprint

25. C. Caroli, P. G. de Gennes, and J. Matricon: Phys. Lett. **9**, 307 (1964)
26. Ch. Renner, B. Revaz, K. Kadowaki, I. Maggio-Aprile, and Ø. Fischer: Phys. Rev. Lett. **80**, 3606 (1998)
27. F. Gygi and M. Schluter: Phys. Rev. B **41**, 822 (1990)
28. F. Gygi and M. Schluter: Phys. Rev. B **43**, 7609 (1991)
29. M. Franz and Z. Tešanović: Phys. Rev. Lett. **80**, 4763 (1998)
30. K. Yasui and T. Kita: Phys. Rev. Lett. **83**, 4168 (1999)
31. Ch. Renner, A. D. Kent, Ph. Niedermann, Ø. Fischer, and F. Lévy: Phys. Rev. Lett. **67**, 1650 (1991)
32. Ch. Renner: Ph.D. thesis, University of Geneva (1993)
33. Ch. Renner, I. Maggio-Aprile, and Ø. Fischer: in *The superconducting state in magnetic fields, special topics and new trends*, Vol. 13 of *Series on directions in condensed matter physics*, edited by Carlos A. R. Sà de Melo (World Scientific, Singapore, 1998), p. 226
34. A. M. Troyanovski, J. Aarts, and P. H. Kes: Nature **399**, 665 (1999)
35. I. Maggio-Aprile, Ch. Renner, A. Erb, E. Walker, and Ø. Fischer: Phys. Rev. Lett. **75**, 2754 (1995)
36. I. Maggio-Aprile, Ch. Renner, A. Erb, E. Walker, and Ø. Fischer: Nature **390**, 487 (1997)
37. M. Franz and Z. Tešanović: Phys. Rev. B **60**, 3581 (1999)
38. Ch. Renner and Ø. Fischer: Phys. Rev. B **51**, 9208 (1995)
39. B. W. Hoogenboom, M. Kugler, Ch. Renner, B. Revaz, I. Maggio-Aprile, and Ø. Fischer: Physica C **332**, 440 (2000)
40. B. W. Hoogenboom, Ch. Renner, B. Revaz, I. Maggio-Aprile, and Ø. Fischer: accepted for publication in Phys. Rev. B, cond-mat/0002146
41. M. Kugler: Ph.D. thesis, University of Geneva (2000)
42. S. H. Pan, E. W. Hudson, A. J. Gupta, K.-W. Ng, H. Eisaki, and S. Ushida J. C. Davis: cond-mat/0005484
43. N. Nishida: private communication
44. Ch. Renner, B. Revaz, J.-Y. Genoud, K. Kadowaki, and Ø. Fischer: Phys. Rev. Lett. **80**, 149 (1998)
45. A. J. Berlinsky, A. L. Fletter, M. Franz, C. Kallin, and P. I. Soininen: Phys. Rev. Lett. **75**, 2200 (1995)
46. M. I. Salkola, A. V. Balatsky, and D. J. Scalapino: Phys. Rev. Lett. **77**, 1841 (1996)
47. S. H. Pan, E. W. Hudson, K. M. Lang, H. Eisaki, S. Uchida, and J. C. Davis: Nature **403**, 746 (2000)
48. A. J. J. van Dalen, R. Griessen, and M. R. Koblischka: Physica C **257**, 271 (1996)
49. T. W. Li, A. A. Menovsky, J. J. M. Franse, and P. H. Kes: Physica C **257**, 179 (1996)
50. E. W. Hudson, S. H. Pan, A. K. Gupta, and K.-W. Ng. and J. C. Davis: Science **285**, 88 (1999)
51. D. Prost, L. Fruchter, and I. A. Campbell: Phys. Rev. B **47**, 3457 (1993)
52. K. Aupke, T. Teruzzi, P. Visani, A. Amann, A. C. Mota, and V. N. Zavaritsky: Physica C **209**, 255 (1993)
53. D. Monier and L. Fruchter: Phys. Rev. B **58**, 8917 (1998)
54. M. M. Salomaa and G. E. Volovik: Rev. Mod. Phys. **59**, 533 (1987)
55. K. Kondo, J. S. Korhonen, M. Krusius, V. V. Dmitriev, Y. M. Mukharsky, E. B. Sonin, and G. E. Volovik: Phys. Rev. B **54**, 3617 (1996)

56. R. P. Feynman, R. B. Leighton, and M. Sands: *Lectures on Physics* (Addison-Wesley, Reading, Massachusetts, 1965), Vol. III
57. M. Kugler, Ch. Renner, S. Ono, Y. Ando, and Ø. Fischer: preprint
58. S. H. Pan, E. W. Hudson, and J. C. Davis: Rev. Sci. Instrum. **69**, 125 (1998)
59. M. Kugler, Ch. Renner, V. Mikheev, G. Batey, and Ø. Fischer: Rev. Sci. Instrum. **71**, 1475 (2000)

17 Charged Vortices in High-T_c Superconductors

Yuji Matsuda and Ken-ichi Kumagai

Summary. It is well known that a vortex in type II superconductors traps a magnetic flux. Recently the possibility that a vortex can accumulate a finite electric charge as well has come to be realized. The sign and magnitude of the vortex charge not only is closely related to the microscopic electronic structure of the vortex, but also strongly affects the dynamical properties of the vortex. In this chapter we demonstrate that a vortex in high-T_c superconductors (HTSC) indeed traps a finite electronic charge, using the high resolution measurements of the nuclear quadrupole frequencies. We then discuss the vortex Hall anomaly whose relation with the vortex charging effect has recently received considerable attention. We show that the sign of the trapped charge is opposite to the sign predicted by the conventional BCS theory and deviation of the magnitude of the charge from the theory is also significant. We also show that the electronic structure of underlying system is responsible for the Hall sign in the vortex state and again the Hall sign is opposite to the sign predicted by the BCS theory. It appears that these unexpected features observed in both electrostatics and dynamics of the vortex may be attributed to the novel electronic structure of the vortex in HTSC.

17.1 Introduction

Vortices are the circulating current flow, which are familiar and universal phenomena in the fluid dynamics. A vortex in a type II superconductor can be viewed as comprised of two parts; the inner core, characterized by the coherence length ξ, and the circulating superfluid electrons outside, characterized by the magnetic penetration length λ. The superconducting energy gap Δ is suppressed in the core region and goes to zero at the center. Perhaps one of the most fundamental physical properties of the vortex created in type II superconductors is that a vortex line can support a magnetic flux with a flux quantum $\Phi_0 = hc/2e (= 2.07 \times 10^{-7}$ Oe·cm^2) [1]. This fact has been confirmed experimentally more than 4 decades ago. On the other hand, it is only very recently that another prominent feature, namely the possibility that a vortex of the superconductor can accumulate a finite electric charge as well, has been proposed[2–6]. The vortex charge appears as a result of the chemical potential difference between the vortex core and the region away from the core. If electrons are expelled from the vortex core, the core is positively charged and hence a hole-like vortex appears, while if electrons (holes) are trapped inside (expelled from) the core, the core is negatively charged and an electron-like vortex appears (see Fig. 17.1). The sign and magnitude of the vortex charge reflect fundamental natures of the superfluid electrons and the low energy excitation out of the condensate.

The vortex charge is closely related to the microscopic electronic structure of the vortex. Unfortunately, the magnitude of the vortex charge in ordinary superconductors is predicted to be extremely small and hence has never been reported so far. However, as will be discussed later, the extremely short coherence length and the strong electron correlation of high temperature superconductors (HTSC) are expected to enhance the charging effect dramatically. Recent direct measurements of the vortex core structure by scanning tunneling spectroscopy (STS) performed on $YBa_2Cu_3O_7$ and $Bi_2Sr_2CaCu_2O_{8+\delta}$ have revealed that the vortex core structure of HTSC is very different from that of conventional superconductors [7–9]. Although the novel vortex core structure of HTSC has been discussed in terms of several intriguing models [10–22], the issue is still far from being settled. Study of the sign and magnitude of the vortex charge provides complementary and important information on the microscopic electronic structure of the vortex in HTSC.

The vortex charge also plays an important role for the vortex dynamics. When a vortex moves in the superfluid electrons, the core plays a key role in dissipation process [23]. A most striking phenomena related to vortex dynamics is the sign reversal of the flux-flow Hall-effect below T_c, which is observed in most HTSC [24,25]. Here we shall refer to the sign reversal as the vortex Hall anomaly. The vortex Hall anomaly indicates that the vortices move upstream against the superfluid flow. Such unusual motion has never been observed in any other fluid including superfluid ^{4}He and cannot be explained in the framework of the classical hydrodynamic theory [26]. Numerous attempts have been made to explain the vortex Hall anomaly, but the microscopic origin behind it is still not clear. Recently, this phenomena has been discussed in terms of the vortex charge which produces an additional force acting on the vortices [2,27–29]. Thus, the clarification of the issue of the relation between the vortex charge and the vortex Hall-effect serves as a test of the predictions for the vortex dynamics.

In Sect. 17.2, we discuss the electrostatics of the vortex line. We show that the precise measurement of the nuclear quadrupole frequency yields an experimental evidence of the vortex charge. On the basis of the nuclear magnetic resonance (NMR) results, we discuss the sign and magnitude of the vortex charge of $YBa_2Cu_3O_7$ and $YBa_2Cu_4O_8$. In Sect. 17.3, we discuss the dynamical properties of the vortex, focusing our attention on the vortex Hall-effect. We discuss the relation between the vortex core charge and vortex Hall anomaly. In Sect. 17.4, we discuss the unusual features observed both in electrostatics and dynamics of the vortex in HTSC in the light of vortex charging effect.

17.2 Charged Vortices

17.2.1 Electrostatics of a Vortex

In this section, we briefly explain the origin of the vortex charge in conventional s-wave type II superconductors [2]. Within the vortex cores in conventional superconductors, the energy levels of the quasi–particles trapped by the pair potential are discretized due to the Andreev scattering at the boundary of the core. The difference between each discrete energy levels are estimated to be $\sim \Delta^2/\varepsilon_F$, which are merely of the order of a few mK (ε_F is the Fermi energy) [1,30]. Thus, it is sufficient to view the energy levels as forming a continuous spectrum, just like in a normal metallic state. Generally the chemical potential μ changes when the metal goes into the superconducting state in the presence of the electron-hole asymmetry, namely if the density of states (DOS) of the electrons near the Fermi energy is energy dependent. In the superconducting state, the particle density n is determined by the pair occupation probability $v_k^2 = (1 - \xi_k/E_k)/2$, where $\xi_k = \epsilon_k - \mu$ and $E_k = (\xi_k^2 + \Delta^2)^{1/2}$ denote the excitation energies in the normal and superconducting state, respectively. The opening of a gap in the spectrum leads the change in μ which can be expressed as,

$$\mu_s - \mu_n = \Delta\mu \sim \frac{\Delta^2}{N(\varepsilon_F)}\left(\frac{\partial N(\varepsilon)}{\partial \varepsilon}\right)_{\varepsilon=\varepsilon_F}, \qquad (17.1)$$

where μ_s and μ_n are the chemical potential in the superconducting and normal state, respectively, and N is the DOS of the electrons. Assuming, therefore, that the vortex core is a normal metallic region surrounded by the superconducting materials, this difference in μ is expected to arise and should leads to a redistribution of the electrons. In order to maintain the same electrochemical potential ($\mu + e\phi$) on both sides, the charge transfer occurs between the core and the outside.

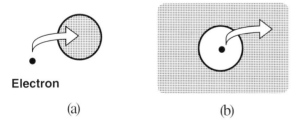

Fig. 17.1. (a) Negatively charged (electron-like) vortex: Electrons are trapped inside the core. (b) Positively charged (hole-like) vortex: Electrons are expelled from the core.

In the framework of the BCS theory, taking into account the metallic screening effect, the charge accumulated within the vortex core Q_ξ per layer normal to the magnetic field is given as [3],

$$Q_\xi \approx \frac{2ek_F s}{\pi^3}\left(\frac{\lambda_{TF}}{\xi}\right)^2\left(\frac{d\ln T_c}{d\ln\mu}\right), \qquad (17.2)$$

where λ_{TF} is the Thomas–Fermi screening length defined as $\lambda_{TF} = \sqrt{8\pi e^2 N(\mu)}$, s is the interlayer distance, and $e(> 0)$ is the electron charge. The factor $(\lambda_{TF}/\xi)^2$ originates from the screening. The sign of the core charge is determined by $d\ln T_c/d\ln\mu$, which measures the electron-hole asymmetry. Outside the core, the charges with opposite sign screen the core charge. The total charge vanishes due to the overall charge neutrality. Far outside the core, the screening charge decreases gradually with the distance from the core like r^{-4} as depicted in Fig. 17.2 and decreases rapidly at still larger distances, $r > \lambda$. In strong fields ($H_{c1} \ll H \ll H_{c2}$) where each vortex overlaps with its neighborhood, the periodic modulation of the charge density appears, similar to the charge density waves in a solid (Fig. 17.3).

We now evaluate $|Q_\xi|$. In ordinary superconductors, $|Q_\xi|$ is estimated to be $\sim 10^{-5} - 10^{-6} e$, using $k_F \sim 1 \text{Å}^{-1}$, $\lambda_{TF} \sim k_F^{-1} \sim 1 \text{Å}$, $\xi \sim 100 \text{Å}$ and $|d\ln T_c/d\ln\mu| \approx \ln(\hbar\omega_D/k_B T_c) \sim 1 - 10$, where ω_D is the Debye frequency. Thus $|Q_\xi|$ is negligibly small and seems to be extremely difficult to observe experimentally. However, we next demonstrate that the vortex in HTSC indeed accumulate a finite charge which is experimentally detectable.

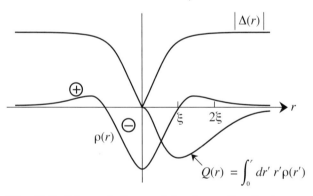

Fig. 17.2. Schematic figure of the charge distribution around a single vortex core when the electrons are trapped within the core (negatively charged core). $\Delta(r)$ is the superconducting energy gap. $\rho(r)$ is the charge density. The charge accumulated inside the core are screened by the charges with opposite sign. $\rho(r)$ decays gradually as r^{-4} well outside the core region. $Q(r)$ is the total charge within the distance r. $Q(r)$ goes to zero as $r \to \infty$ due to the requirement of the overall charge neutrality.

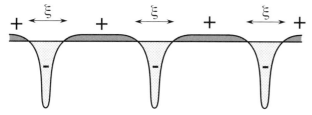

Fig. 17.3. The charge density modulation (negatively charged core) in the strong magnetic field ($H_{c1} \ll H \ll H_{c2}$) where each vortices overlaps.

17.2.2 Vortex Charge Probed by NMR

A straightforward means of identifying the vortex charge is to measure the electron density modulation shown in Fig. 17.3. It is widely accepted that the nuclear quadrupole frequency ν_Q is very sensitive to the local charge density. The nucleus with spin $I > 1/2$ has a quadrupole momentum due to the inhomogeneous distribution of the charged particles. In a solid, an electron distribution with spherical asymmetry, such as unclosed $3d$ shell and non-cubic surrounding ions, induces a local electric field gradient (EFG) in the vicinity of the nuclei. This EFG lifts the degeneracy of the nuclear spin levels, interacting with the nuclear quadrupole moment, and ν_Q corresponds to the nuclear spin energy difference. The relevant information is obtained from the nuclear quadrupole resonance (NQR) in zero magnetic field and nuclear magnetic resonance (NMR) in a finite magnetic field (Fig. 17.4). For the ^{63}Cu nuclei with spin $I=3/2$, the NQR resonance frequency ν_Q^{NQR} is expressed as

$$\nu_Q^{\mathrm{NQR}} = \frac{e^2 Q q_{zz}}{2h}\sqrt{\left(1+\frac{\eta^2}{3}\right)} = \nu_Q\sqrt{\left(1+\frac{\eta^2}{3}\right)} \qquad (17.3)$$

where eq_{zz} is the largest principle (z-axis) component of EFG at the nuclear site, Q ($=-0.211$ barn) is the quadrupole moment of the copper nuclei. The asymmetry parameter η of the EFG defined as $\eta = |(q_{xx} - q_{yy})/q_{zz}|$ is close to zero for the Cu site in the two dimensional CuO$_2$ planes (Cu(2) site). In a strong magnetic field when the Zeeman energy is much larger than the quadrupole energy, each Zeeman level is shifted by the quadrupole interactions and thus, two satellite peaks ($\pm 3/2 \leftrightarrow \pm 1/2$) appear on both sides of the central ($\pm 1/2 \leftrightarrow \pm 1/2$) resonance peak. The frequency difference between the upper and the lower satellites exactly coincides with $2\nu_Q$ for $H \parallel z$. It should be noted that this procedure for obtaining ν_Q is essentially free from the influence of the change of magnetic shift (or Knight shift) [31]. Moreover, the magnetic effect of the asymmetric broadening due to the vortex lattice is exactly cancelled out in the process determining ν_Q. Thus, we obtained the ν_Q values simply from the difference of the peak frequencies of the two satellite lines.

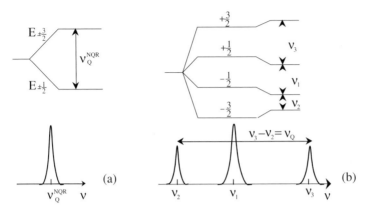

Fig. 17.4. (a) Energy splitting of ^{63}Cu nuclei with spin $I=3/2$ in zero field. EFG lifts the degeneracy of the nuclear spin levels and resonance occurs at $\nu = \nu_Q^{\mathrm{NQR}}$. (b) In a strong magnetic field when the Zeeman energy is much larger than the quadrupole energy, each Zeeman level is shifted by the quadrupole interactions and two satellite peaks ν_3 $(+3/2 \leftrightarrow +1/2)$ and ν_2 $(-3/2 \leftrightarrow -1/2)$ appear on both sides of the central ν_1 $(\pm 1/2 \leftrightarrow \pm 1/2)$ resonance peak. ν_Q is obtained from the difference between the upper and lower resonance frequencies; $\nu_3 - \nu_2 = 2\nu_Q$. In HTSC, ν_Q almost exactly coincides with ν_Q^{NQR} due to very small η.

Generally the EFG originates from two different sources, namely from the on-site distributions $\nu_{\mathrm{on-site}}$ of the electrons and from the surrounding ions ν_{ion}, $\nu_Q = \nu_{\mathrm{on-site}} + \nu_{\mathrm{ion}}$. Recent analysis of ν_Q on the Cu(2) site suggests that $\nu_{\mathrm{on-site}}$ is mainly composed of the Cu 4p and 3d shell terms [32]. In HTSC the holes in the Cu $3d_{x^2-y^2}$ orbital plays the most important role for the onset of superconductivity. Figure 17.5 shows the doping dependence of ν_Q on the Cu(2) site for YBa$_2$Cu$_3$O$_{7-\delta}$ [33], La$_{2-x}$Sr$_x$CuO$_4$ [34] and HgBa$_2$CuO$_{4+\delta}$ [35]. In all materials, ν_Q increases linearly with the number of holes in the planes and can be written as,

$$\nu_Q = A n_{\mathrm{hole}} + C, \tag{17.4}$$

where n_{hole} is the number of holes per Cu(2) atom, and A and C are constants [33–35]. Although C is strongly material dependent, reflecting the difference in ν_{ion}, $A \approx 20 - 30$ MHz per hole per Cu(2) atom is essentially material independent. Thus, the measurement of ν_Q makes possible the accurate determination of the local hole number.

The principle of our experiment is the following [36]. In the measurement, only the resonance of the ^{63}Cu(2) nuclei *outside the vortex core* is detected. This is because the applied field is much less than H_{c2} and hence the core region occupies a smaller area in the sample. If the vortex core traps (expels) a finite amount of electrons, the electron density outside the core should decrease (increase) from that in zero field where the electron distribution is uniform, as shown in Fig. 17.3. We are able to detect the change of carrier

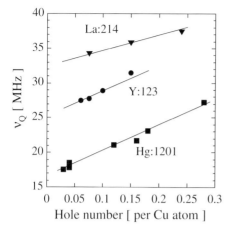

Fig. 17.5. Doping dependence of ν_Q on the Cu(2) site for YBa$_2$Cu$_3$O$_{7-\delta}$ [33], La$_{2-x}$Sr$_x$CuO$_4$[34] and HaBa$_2$CuO$_{4+\delta}$[35]. ν_Q is proportional to the hole number.

density via the shift of the value of ν_Q. In the present measurements, we used slightly overdoped YBa$_2$Cu$_3$O$_7$ and underdoped YBa$_2$Cu$_4$O$_8$ in which the NQR and NMR spectra are very sharp compared to those of other HTSC [37]. Figure 17.6 shows the NQR and NMR spectra for ^{63}Cu(2) site of YBa$_2$Cu$_4$O$_8$. All data were taken by utilizing a phase coherent spin echo spectrometer. The spectra are obtained with superposed method of the FT spectra of the spin echo measured at a certain frequency interval. The NMR experiments were performed in the field cooling condition under a constant field of 9.4T.

We plot, in Fig. 17.7, the difference between ν_Q in zero field and in the vortex state, $\Delta\nu_Q = \nu_Q(0) - \nu_Q(H)$, for YBa$_2Cu_3O_7$ and YBa$_2$Cu$_4$O$_8$. The value of ν_Q in zero field for NQR is obtained after accounting for the correction by the factor of $\sqrt{1+\eta^2/3}$ in (17.1), although this factor is at most 0.03% of ν_Q for $\eta \approx 0.04$ of the present materials. In both materials $\Delta\nu_Q$ is essentially zero above T_c, indicating no modulation of the carrier density. Meanwhile, a nonvanishing $\Delta\nu_Q$ is clearly observed below T_c in both materials. While $\Delta\nu_Q \sim -25$kHz has a negative sign and in YBa$_2$Cu$_3$O$_7$, $\Delta\nu_Q \sim 50$kHz has a positive sign in YBa$_2$Cu$_4$O$_8$ at $T=0$. We point out that the magnetostriction which causes the local lattice distortion cannot be the origin of the nonzero $\Delta\nu_Q$, because such an effect is negligibly small in HTSC for the field cooling condition ($\Delta\ell/\ell < 10^{-8}$ where ℓ is the lattice constant) [38]. Therefore, these nonzero values for $\Delta\nu_Q$ naturally leads us to conclude that the electron density outside the core is different from that in zero field.

We now discuss the issue of the sign and magnitude of the accumulated charge. The negative $\Delta\nu_Q$ in YBa$_2$Cu$_3$O$_7$ indicates the increment of the hole density outside the core. This excess density of holes is nothing but the holes expelled from the core. Therefore, the accumulated charge in the core of YBa$_2$Cu$_3$O$_7$ is negative. By the same reasoning, the positive $\Delta\nu_Q$ in

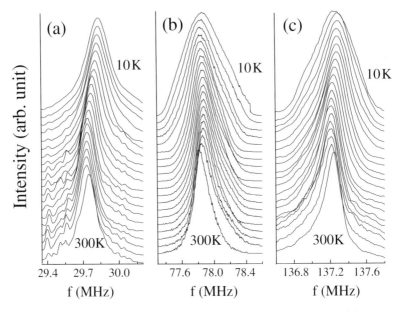

Fig. 17.6. The NQR in zero field (**a**) and lower (**b**) and upper (**c**) NMR-satellite spectra at 9.4 T for ^{63}Cu(2) site of YBa$_2$Cu$_4$O$_8$ for various temperatures (300 K, 220 K, 200 K, 180 K and 160 K to 10 K by a 10 K step.

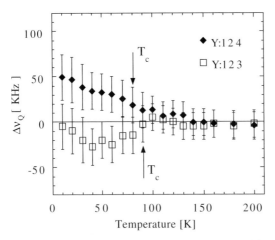

Fig. 17.7. T-dependence of $\Delta\nu_Q = \nu_Q(0) - \nu_Q(H)$, for YBa$_2Cu_3O_7$ and YBa$_2$Cu$_4$O$_8$. In both materials nonzero $\Delta\nu_Q$ is clearly observed below T_c, showing that the electron density outside the core differs from that in zero field.

YBa$_2$Cu$_4$O$_8$ indicates a positive accumulated charge. Meanwhile, since μ decreases monotonically with doping holes, (17.2), $\mathrm{sgn} Q_\xi = \mathrm{sgn}(d\ln T_c/d\ln\mu)$, predicts that $Q_\xi > 0$ in the underdoped regime while $Q_\xi < 0$ in the overdoped regime. This is strikingly in contrast to the sign determined by the present experiment. The deviation of the magnitude of the charge from theory is also noteworthy. The magnitude of charges per pancake vortex, which are roughly estimated using $Q_\xi \approx \Delta\nu_Q H_{c2}/AH$, assuming $H_{c2} \sim 200$ T are $Q_\xi \sim -0.005e$ to $-0.02e$ for YBa$_2$Cu$_3$O$_7$ and $Q_\xi \sim 0.01e$ to $-0.05e$ for YBa$_2$Cu$_4$O$_8$. However, according to (17.1), Q_ξ is estimated to be $\sim 10^{-4} - 10^{-5} e$ where we assumed $\xi \sim 30 Å$. Therefore, $|Q_\xi|$ determined by the present experiments are still one or two order of magnitude larger than expected by (17.1). Thus, *the BCS theory not only predicts the wrong sign of the charge but also underestimates $|Q_\xi|$ seriously.*

17.3 Dynamical Properties of Vortices

17.3.1 Vortex Hall Anomaly

In this section, we shift to the subject of the dynamical properties of the vortex. The vortex motion in the superfluid electrons has presented a persistent problem in the superconducting state of type II superconductors. Here we focus our attention on the vortex Hall-effect. The Hall sign is determined by the topology of the Fermi surface in the normal state, while it is determined by the vortex motion in the superconducting state. In the latter, the vortex motion generates an electric field which results in a longitudinal and Hall resistivities. This electric field is given by the Josephson relation

$$\boldsymbol{E} = \boldsymbol{B} \times \boldsymbol{v}_\mathrm{s} \qquad (17.5)$$

where $\boldsymbol{E}$ is the measured electric field and $\boldsymbol{v}_\mathrm{s}$ is the vortex velocity. From (17.5) it follows that vortex motion perpendicular to the transport current corresponds to a dissipative longitudinal electric field E_x while parallel motion corresponds to a Hall electric field E_y. Knowledge of the Hall-effect enables us to obtain important information on the problem of the energy dissipation process, which in turn reflects the electronic structure of the vortex.

A most puzzling and controversial phenomena is the sign change that has been observed in the Hall-effect in the superconducting state in most HTSC and some conventional superconductors [24,25] (Fig. 17.8). Recent measurements revealed that the second Hall sign reversal occurs at still lower temperature. The occurrence of the sign reversal in one-unit-cell-thick ultrathin YBa$_2$Cu$_3$O$_{7-\delta}$ film demonstrates that the Hall anomaly occurs within a two-dimensional CuO$_2$-plane [39], and is not ascribable to three dimensional interlayer effect. When the vortices move, the Magnus force acts on them [40]. This Hall sign reversal indicates that the vortices move upstream against the superfluid flow. This unusual motion has never been observed in any other fluid and cannot be explained in the framework of the classical hydrodynamic theory[26]. In fact, the classical theories of vortex motion predict that the superconducting and normal states will have the same Hall sign. Thus, in

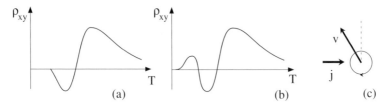

Fig. 17.8. Schematic figures of the vortex Hall anomaly. (**a**) Hall resistivity ρ_{xy} changes the sign below T_c from the normal state. (**b**) Second sign change occurs in very anisotropic Bi- or Tl- basic materials. (**c**) The negative Hall sign indicates that the vortices move upstream the superfluid flow j

spite of the various proposals made to explain the anomaly, e.g., as a vortex pinning induced phenomena [41], collective vortex phenomena [42], the issue is still controversial. Recently, the vortex charging effect has been invoked to account for the vortex Hall anomaly. In this model, the charged vortex produces an additional force which give rise to the vortex motion upstream to v_s [2,27–29]. However, even if the charged vortices were established, the influence of the charging effect on the Hall conductivity is still controversial. For example, the results in [27,28] are opposite in sign to [2,42]. Therefore, it is strongly desirable to have a clarification made on the relation between the vortex charge and the vortex Hall anomaly.

17.3.2 Time Dependent Ginzburg–Landau Equation

Recent theories based on the time dependent Ginzburg–Landau (TDGL) equation have shed light on describing the Hall-effect in the superconducting state [43,44]. According to the TDGL theory, the vortex Hall conductivity $\sigma_{xy}(=\rho_{xy}/(\rho_{xx}^2+\rho_{xy}^2))$ in the flux-flow state can be expressed as a sum of two contributions,

$$\sigma_{xy} = \sigma_{xy}^V + \sigma_{xy}^N, \tag{17.6}$$

where $\sigma_{xy}^V (\propto 1/H)$ is the vortex Hall term arising from the hydrodynamic contribution while $\sigma_{xy}^N \propto H$ is the contribution of the quasiparticle inside the vortex core. Accordingly, the Hall sign reversal occurs at low field when σ_{xy}^V has a sign opposite that of the normal state Hall-effect. Another advantage for analyzing vortex Hall-effect in terms of σ_{xy} is that σ_{xy} is insensitive to disorder, which follows from a general argument of the vortex dynamics [45]. In fact, for HTSC, σ_{xy} stays in the same order of magnitude, even though ρ_{xx} and ρ_{xy} change more than three orders of magnitude within the temperature range of interest. Recently, however, it has been shown that the Hall sign is not pinning independent in the strongly pinned regime [46,47]. We therefore discuss the Hall data in terms of σ_{xy} outside the strongly pinned regime, where the nearly free flux-flow state is realized.

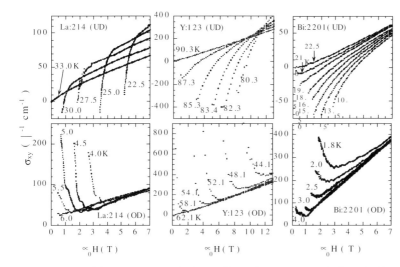

Fig. 17.9. Field dependence of the Hall conductivity σ_{xy} below T_c for some examples. The upper panels ((**a**), (**c**), and (**e**)) display σ_{xy} for the underdoped crystals, and the lower panels ((**b**), (**d**), and (**f**)) display σ_{xy} for the overdoped crystals. (**a**) Underdoped La:214, $La_{2-x}Sr_xCuO_4$, $x=0.15$ ($T_c=33.3K$); (**b**) Overdoped La:214, $La_{2-x}Sr_xCuO_4$, $x=0.28$ ($T_c=9.5K$); (**c**) Underdoped Y:123, $YBa_2Cu_3O_{7-\delta}$, ($T_c=90.3K$); (**d**) Overdoped Y:123, $(Y_{1-x}Ca_x)Ba_2Cu_3O_{7-\delta}$, $x=0.4$ ($T_c=62.1K$); (**e**) Underdoped Bi:2201, $Bi_{1.80}Pb_{0.38}Sr_{2.01}CuO_{6+\delta}$ annealed in a vacuum, ($T_c=21.1K$);(**f**) Overdoped Bi:2201, $Bi_{1.80}Pb_{0.38}Sr_{2.01}CuO_{6+\delta}$, as-grown ($T_c=6.0K$).

Figure 17.9 shows some examples of the field dependence of σ_{xy} for La:214, Y:123 and Bi:2201 [25]. The upper (lower) panels represent a typical case for the underdoped (overdoped) crystals. At high fields, σ_{xy} increases linearly with H in all crystals. With decreasing H, σ_{xy} diverges towards $-\infty$ for underdoped crystals but diverges towards $+\infty$ for overdoped crystals. Here we focus on the sign of σ_{xy}^V, which we determine by the diverging direction of σ_{xy} with decreasing H at low fields.

In Fig. 17.10 we summarize the doping dependence of the Hall sign in the superconducting state of HTSC [25]. In this figure, T_c is normalized by T_c at optimal doping, T_c^{opt}, for the corresponding system. The filled circles depict the electron-like Hall sign (Hall anomaly), and the open circles depict the hole-like Hall sign (no Hall anomaly). Figure 17.10 demonstrates that the Hall anomaly always occurs in the underdoped- and slightly overdoped-regimes. The Hall anomaly is sample dependent near $T_c/T_c^{opt} \sim 0.9$ in the slightly overdoped regime, but no Hall anomaly is observed beyond this regime. This behavior is observed in the crystals with monolayer, double layer, or triple-layer structures. Moreover, the Hall sign depends neither on the magnitude of anisotropy nor on the pinning strength of the materials. We therefore con-

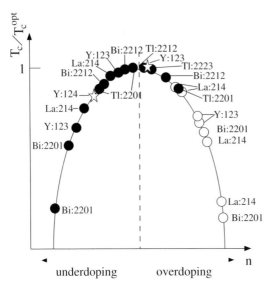

Fig. 17.10. Doping dependence of the Hall anomaly in the superconducting state for various high-T_c cuprates including La-, Y-, Bi-, and Tl-based compounds. Filled circles denote presence of and open circles absence of Hall anomaly. n represents the carrier number, and T_c are normalized by T_c^{opt}. The Hall anomaly is present in the underdoped and slightly overdoped materials, while it is absent in the overdoped materials. The stars show slightly overdoped Y:123 and underdoped Y:124 which are used for the present study. Reference [51] reports the Hall anomaly in Y:124 which contains two one-dimensional conducting chains. However, Hall anomaly in Y:124 occurs only when current is applied parallel to the chains. Moreover H-dependence of ρ_{xy} is very different from that of the other HTSC. Therefore, it is not clear that the vortex Hall anomaly similar to the other HTSC is present in Y:124.

clude that the doping dependence shown in Fig. 17.10 is a universal transport property of HTSC, showing that *the Hall sign in the superconducting state is closely related with the characteristic electronic structure determined by the doping.*

Within the framework within the BCS theory, several authors have calculated σ_{xy}^V and emphasized the importance of the electronic structure of the materials for understanding the Hall-effect. It has been shown that the imaginary part of the complex relaxation time ($\gamma = \gamma_1 + i\gamma_2$) in the TDGL equation gives rise to σ_{xy}^V ($\sigma_{xy}^V \propto -\gamma_2$) [43,44]. Accordingly, the sign of σ_{xy}^V is given by the sign of $\mathrm{sgn}(e)d\ln T_c/d\mu$ [48]. From the effective action for the vortex motion based on the BCS Hamiltonian, σ_{xy}^V has been microscopically derived and it has been pointed out that σ_{xy}^V can be interpreted as the vortex charging effect discussed in Sect. 17.2 [27,49]. Interestingly, the sign of the vortex Hall-effect is opposite to the sign of the accumulated charge inside the core. All these calculations remain valid for *s*-wave weak coupling superconductors regardless of the nature of the interaction.

The present results show that the relation between the sign of the vortex charge and the sign of the vortex Hall-effect is not simple. According to the photoemission experiment, μ decreases monotonically with doping holes [50]. Thus, the BCS theory leads to the conclusion that the Hall sign is positive in the underdoped regime where $\mathrm{sgn(e)}d\ln T_c/d\mu > 0$ and is negative in the overdoped regime where $\mathrm{sgn(e)}d\ln T_c/d\mu < 0$; the sign reversal should occur in the overdoped regime and be absent in the underdoped regime. This is in striking contrast to the results displayed in Fig. 17.10. Moreover, comparing the sign of the vortex charge determined by NMR and the vortex Hall sign, the situation turn out to be more complicated. The positively charged vortex in $YBa_2Cu_4O_8$ probed by NMR appears to be consistent with the negative Hall sign reported in [51], although the Hall anomaly in Y:124 is not clear as discussed in the figure caption of Fig. 17.10. On the other hand, the negatively charged vortex in $YBa_2Cu_3O_7$ probed by NMR leads to positive σ_{xy}^V, namely no Hall anomaly. However, a clear Hall anomaly is observed in $YBa_2Cu_3O_7$.

Therefore, the prediction of the conventional TDGL equation based on the BCS theory fails to evaluate the hydrodynamic force acting on the vortices of HTSC and yields the wrong Hall sign. This discrepancy in Hall sign seems to suggest that the terms which have not been taken into consideration in the conventional TDGL theory give additional contributions to σ_{xy}^V [52]. It has been shown that the term arising from requirement of the charge neutrality, which has not been incorporated into the conventional theory, does not change the sign of σ_{xy}^V from the conventional theory [53]. Quite recently it has been suggested that σ_{xy}^V can be opposite in sign to the conventional result of TDGL theory due to the effect of the local electric field at the vortex core, if the charging and screening effects are taken into account[29]. Thus, summarizing this situation, we conclude that the vortex Hall anomaly is closely related to the electronic structure of the vortex state, but the precise relation between the vortex charge and the vortex Hall-effect still remains to be resolved.

17.4 Enhanced Vortex Charge of High Temperature Superconductors

As discussed in Sect. 17.3, the measurements of the nuclear quadrupole frequency revealed that the sign of the trapped charge is opposite to the sign predicted by the conventional BCS theory and deviation of the magnitude of the charge from the theory is also significant. We discuss here several intriguing possible origins for these discrepancies.

(i) d-wave symmetry; The order parameter of HTSC has a $d_{x^2-y^2}$-wave symmetry, in which the pair potential is no longer uniform, but instead falls to zero along four directions at 45 degree to the crystal lattice vectors. The vortex state of d-wave superconductors is fundamentally different from the s-wave counterparts, in that the extended quasi-particles states around the core play a much more important role [10].

(ii) Quantum limit; Several calculations have been carried out in the framework of semi-classical approximation. However, the coherence length in HTSC is one or two order of magnitude smaller than in the conventional superconductors, and thus $k_F\xi$ may be order of unity, where k_F is the Fermi wave number. In this quantum limit, the description of the quasi-particles in terms of semiclassical wave packets breaks down in contrast to conventional superconductors[14].

(iii) Strong correlation; One of the most striking properties of HTSC is the close proximity between the antiferromagnetic and superconducting phases. There are a number of theories linking the microscopic origin of HTSC to antiferromagnetic correlation. Several groups have pointed out that the AF correlation may be important in the vortex core [17–22,54,55]. In the extreme case, it has been pointed out that the vortex could have an AF insulating core.

These unusual electronic properties appear to promote the charging effect. For example, as demonstrated in [5] in the quantum limit, contribution of the lowest bound state in the vortex core strongly dominates at low temperature. In this situation the state whose spatial distribution is concentrated at the center of the core (angular momentum $\mu=1/2$) does not contribute to the DOS inside the core. Thus, a dramatic change of the quasi-particle distribution inside the core should occur and the charging is enhanced. Recent STM results show that the energy minigap ΔE in HTSC becomes the order of a few tenth K, and hence there are only one or two states localize within the core. Moreover, as suggested by recent theories of the vortex core based on e.g. the $t-J$ or SO(5) models, the antiferromagnetic (AF) state may be energetically preferable to the metallic state in the vortex core of HTSC. If this is indeed so, the AF correlation is expected to enhance the charging effect because it causes a large shift of μ by changing the DOS of the electrons inside the core dramatically. We note here that the present results exclude the possibility of the SO(5) insulating AF core [17,22] in which holes should be expelled from the core and the accumulated charges are always negative; the present result yields the opposite sign for $YBa_2Cu_4O_8$ in the underdoped regime where the AF correlation is important. The situation therefore calls for a microscopic calculation which can evaluate the accumulated charge quantitatively including the correct sign.

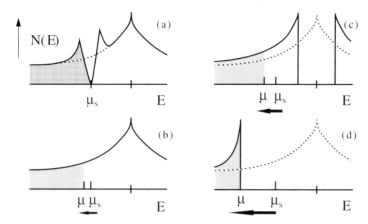

Fig. 17.11. Schematic figures of the DOS of electrons in the superconducting states of d-wave superconductor [56]. In the figure we show the electron states filling up below the van-Hove singularity. (**a**) DOS well outside the vortex core. μ_s is the chemical potential in the superconducting state. (**b**) DOS in the normal metallic core. The chemical potential μ decreases. (**c**) DOS according to the $t-J$ model ($t-J$ core). The antiferromagnetic correlation splits the electron band at the middle. The shift of μ in $T-J$ core is larger than in (**b**), because of the large change of the DOS. (**d**) DOS of the antiferromagnetic insulating core according to SO(5) theory. The shift of μ is largest.

17.5 Summary

In this chapter we demonstrated that a vortex in HTSC indeed traps a finite electric charge, using high resolution measurements of the nuclear quadrupole frequencies. We showed that the sign of the trapped charge is opposite to the sign predicted by the conventional BCS theory and deviation of the magnitude of the charge from the theory is also significant. These unexpected features observed in the electrostatics of the vortex can be attributed to the novel electronic structure of the vortex in HTSC. We also discussed the vortex Hall anomaly. We showed that the sign of the vortex Hall-effect is closely related to the electronic structure of the vortex state and again contradicts the prediction of the BCS theory. Although the vortex Hall anomaly is closely related to the electronic structure of the vortex state, precise relation between the vortex charge and the vortex Hall-effect still remains to be resolved.

Acknowledgements. We are grateful to V.B. Geshkenbein, T. Hanaguri, S. Hikami, N. Ichioka, R. Ikeda, Y. Kato, T. Kita, N.B. Kopnin, W. Lang, T. Mishonov, M. Ogata, N.P. Ong, M. Takigawa, and A. Tanaka, for their helpful discussions. We thank T. Nagaoka and K. Nozaki for their technical assistance and M. Isobe, H. Obara, A. Sawa, S. Shamoto, M. Suzuki, and T. Terashima for preparing the crystals.

References

1. P.G. de Gennes in *Superconductivity of Metal and Alloys*. (Addison–Wesley Publishing Co., Inc., 1989).
2. D.I. Khomskii and A. Freimuth, Phys. Rev. Lett. **75** 1384 (1995).
3. G. Blatter, M. Feigelman, V. Geshkenbein, A. Larkin, A. van Otterlo, Phys. Rev. Lett. **77**, 566 (1996).
4. T. Mishonov, Cond-mat/0004286.
5. N. Hayashi, M. Ichioka, K. Machida, J. Phys. Soc. Jpn, **67** 3368 (1999).
6. J. Kolácek, P. Lapavský, and E.H. Brandt, Phys. Rev. Lett **86**, 312 (2001).
7. I. Maggio-Aprile, Ch. Renner, A. Erb, E. Walker, Ø. Fischer, Phys. Rev. Lett. **75**, 2754 (1995).
8. Ch. Renner, B. Revaz, K. Kadowaki, I. Maggio-Aprile, Ø. Fischer, Phys. Rev. Lett. **80**, 3606 (1998).
9. S.H. Pan, E.W. Hudson, A.K. Gupta, K.-W. Ng, H. Eisaki, S. Uchida, J.C. Davis, Phys. Rev. Lett. **85**, 1536 (2000).
10. N. Schopohl, N. and K. Maki, Phys. Rev. B **52**, 490 (1995), M. Ichioka, N. Hayashi, N. Enomoto, K. Machida, *ibid.* **53**, 15316 (1996).
11. G. E. Volovik, JETP Lett. **58**, 470 (1993).
12. P. I. Soininen, C. Kallin, and A. J. Berlinsky, Phys. Rev. B **50**, 13883 (1994); Yong Ren, Ji-Hai Xu, and C. S. Ting, Phys. Rev. Lett. **74**, 3680 (1995); A.J. Berlinsky, A. L. Fetter, M. Franz, C. Kallin, P.I. Soininen, Phys. Rev. Lett. **75**, 2200 (1995).
13. Y. Wang and MacDonald, Phys. Rev. B **52**, R3876 (1995).
14. Y. Morita, M. Kohmoto, K. Maki, Phys. Rev. Lett. **78**, 4841 (1997).
15. M. Franz, and Z. Tesanovic, Phys Rev. Lett. **80**, 4763 (1998).
16. K. Yasui, and T. Kita, Phys Rev. Lett. **83**, 4168 (1999).
17. D.P. Arovas, A.J. Berlinsky, C. Kallin, S.-C. Zhang, Phys. Rev. Lett. **79**, 2871 (1997).
18. A. Himeda, M. Ogata, Y. Tanaka, S. Kashiwaya, J. Phys Soc. Jpn. **66** 3367(1997).
19. M. Ogata, Int. J. Mod. Phys B **13**, 3560 (1999).
20. J.H. Han, and D,-H. Lee, Phys. Rev. Lett, **85**, 1100 (2000).
21. M. Franz, and Z. Tesanovic, cond-mat/0002137.
22. B. M. Andersen, H. Bruus, P. Hedegård, Phys Rev. B, **61**, 6298 (2000).
23. M. Eschrig, J.A. Sauls, Phys Rev. B **60**, 10447 (1999).
24. S.J. Hagen, A.W. Smith, M. Rajeswari, J.L. Peng, Z.Y. Li, R.L. Greene, S.N. Mao, X.X. Xi, S. Bhattacharya, Qi Li, C.J. Lobb, Phys. Rev. B **47**, 1064 (1993) and references therein.
25. T. Nagaoka, Y. Matsuda, H. Obara, A. Sawa, T. Terashina, I. Chong, M. Takano, M. Suzuki, Phys. Rev. Lett. **80**, 3594 (1998) and references therein.
26. J. Bardeen and M.J. Stephen, Phys. Rev. **140**, A1197 (1965), P. Noziéres and W.F. Vinen, Phylos. Mag. **14**, 667 (1966).
27. A. van Otterlo, M. Feigel'man, V. Geshkenbein, G. Blatter, Phys Rev. Lett., **75**, 3736 (1995).
28. M.V. Feigel'man, V.B. Geshkenbein, A.J. Larkin, V.M. Vinokur, JETP Lett., **62**, 834 (1995).
29. Y. Kato, J. Phys Soc Jpn. **68**, 3798 (1999).
30. F. Gygi, and M. Schlüter, M. Phys. Rev. B **43**, 7609 (1991).

31. G.C. Carter, L.H. Bennett, D.J. Kahan in *Metallic Shifts in NMR* (Pergamon Press 1977).
32. K. Schwarz, Phys. Rev. B, **42**, 2051 (1990), Y. Ohta, W. Koshibae, S. Maekawa, J. Phys. Soc. Jpn. **61**, 2198 (1992), K. Hanzawa, J. Phys Soc. Jpn. **62**, 3302 (1993).
33. G. Zheng, Y. Kitaoka, K. Ishida, K. Asayama, J. Phys Soc. Jpn. **64**, 2524 (1995).
34. H. Yasuoka, in *Spectroscopy of Mott Insulator and Correlated Metals*, edited by A. Fujimori and Y. Tokura p.213 (Springer Series in Solid-State Sciences, Vol.119).
35. A.A. Gippius, E.V. Antipov, W. Hoffmann, K. Lueders, Physca C **276**, 57 (1997).
36. K. Kumagai, K. Nozaki, and Y. Matsuda, Phys. Rev. B **63**, 144502 (2001).
37. K. Muller, M. Mali, J. Roos, O. Brinkmann, Physica C**162–164**, 173 (1989),A. Suter, M. Mali, J. Roos, D. Brinkmann, J. Karpinski, E. Kaldis, Phys. Rev. B**56**, 5542 (1997).
38. H. Ikuta, N. Hirota, Y. Nakayama, K. Kishio, K. Kitazwa, Phys Rev. Lett. **70**, 2166 (1993). T. Hagaguri, private communication.
39. Y. Matsuda and S. Komiyama, Phys. Rev. Lett. **69**, 3228 (1992).
40. For example , E.B. Sonin, Phys Rev. B **55**, 485 (1997).
41. Z.D. Wang, Jinming Dong, C.S. Ting, Phys Rev. Lett. **72**, 3875 (1994).
42. P. Ao , J. Phys. C, **10**, L677 (1998).
43. A.T. Dorsey, Phys. Rev. B **46**, 8376 (1992).
44. N.B. Kopnin, B.J. Ivlev, V.A. Kalatsky, J. Low Temp. Phys. **90**, 1 (1993).
45. V.M. Vinokur, V.B. Geshkenbein, M.V. Feigel'man, G. Blatter, Phys. Rev. Lett. **71**, 1242 (1993).
46. G. D'Anna, V. Berseth, L. Forró, A. Erb, E. Walker, Phys Rev. Lett. **81**, 2530 (1998), W. Göb, W. Liebich, W. Lang, S. Puica, R. Sobolewski, R. Rössler, J.D. Pedarnig, D. Bäuerle, Phys. Rev. B, in press, (2000).
47. N.B. Kopnin, and V.M. Vinokur, Phys. Rev. Lett. **83**, 4864 (1999). R. Ikeda, Physica C **316**, 189 (1999).
48. A. G. Aronov, S. Hikami, and A. I. Larkin, Phys. Rev. B **51**, 3880 (1995).
49. We note that several authors have raised critisism on the derivation of Ref.[27]; E. Simanek, Phys. Lett. A **221**, 277 (1996), M. Stone, cond-mat/9708017, see also G.E. Volovik, JETP Lett. **65**, 676 (1997), A. Tanaka, and M. Machida, Physica C **313**, 141 (1999).
50. A. Ino, T. Mizokawa, A. Fujimori, K. Tamasaku, H. Eisaki, S. Uchida, T. Kimura, T. Sasgawa, K. Kishio, Phys. Rev. Lett. **79**, 2101 (1997). We take μ as the chemical potential of the electron.
51. J. Schoenes, E. Kaldis, J. Karpinski, Phys Rev. B **48**, 16869 (1993).
52. V.B. Geshkenbein, L.B. Ioffe LB, and A.I. Larkin, Phys . Rev. B **55**, 3173 (1997).
53. N.B. Kopnin, and A.V. Lopatin, Phys Rev. B **51**, 15291 (1995).
54. N. Nagaosa and P.A. Lee, Phys Rev. B **45**, 966 (1992).
55. S. Sachidev, Phys Rev. B **45**, 389 (1992).
56. M. Ogata, unpublished.

18 Neutron Scattering from Vortex Lattices in Superconductors

Andrew Huxley

18.1 Introduction

When a magnetic field is applied to a superconductor the magnetic field can penetrate as an array of flux lines. These lines are often ordered to form a lattice, the orientation and geometry of which reflects the anisotropy of the underlying interactions between the lines. Neutron scattering gives an image of the mean magnetic field contrast due to this lattice in the bulk of a crystal, while the weak nature of the scattering means that the charge-neutral neutron does not disturb the system being measured. At the simplest level, small-angle neutron scattering (SANS) provides a direct determination of the geometry and orientation of the flux-line lattice (FLL), which can in some special circumstances provide a sensitive probe of the underlying symmetry of the superconductivity itself. The scattered intensities give further quantitative information about the spatial form of the magnetic field distribution in the sample, usually expressed as the field profile normalised per vortex line. Even with only the most limited data, the field dependence of the intensities of the lowest order Bragg reflections is sufficient to estimate the two fundamental length scales that characterise a superconductor, namely the penetration length (λ) and coherence length (ξ). From the intensities of higher order reflections, the field profile at smaller length scales, and ultimately the vortex cores, can be probed. The measurement of the profiles of the diffracted peaks gives complementary information on the degree to which the flux-line lattice is ordered, a quality that is related to the pinning of the flux-lines. All these possibilities clearly show that SANS offers a valuable tool for the investigation of vortex physics, and have motivated its application to study a growing number of recently discovered unconventional superconductors.

In this chapter we will concentrate on the most directly observable quality of a flux-line lattice, namely its geometry and orientation, touching on other aspects only in so much as they are relevant to this discussion.

18.2 The Measurement Technique

A regular arrangement of the flux lines in two dimensions gives rise to a two-dimensional field distribution in reciprocal or Fourier space (Fig. 18.1).

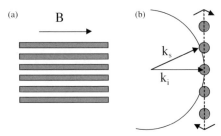

Fig. 18.1. A cut through a flux-line lattice is shown in (**a**) viewed perpendicular to the field direction. The cores of the flux lines are shaded. The Fourier transform of the periodic field distribution gives rise to a two dimensional array of spots in reciprocal space (**b**). These are swept through the Ewald sphere (that defines the locus of points where the elastic scattering condition is fulfilled) as the FLL is rotated. k_i and k_s are the incident and scattered wave vectors.

The experimental measurement is then conceptually very simple. In the most commonly exploited configuration the applied magnetic field is initially aligned with the incident neutron beam. Neutrons, scattered via the interaction of their magnetic moments with spatial modulations of the magnetic field, are detected on a small-angle camera positioned typically 4–40m from the sample. The detected images for several small inclinations (rocking angles) of the sample and field from the incident beam direction are recorded (Fig. 18.2). After subtraction of background scattering (typically measured by heating the sample above T_c) the images give inclined cuts that map directly the Fourier transform of the magnetic field distribution. When the magnetic field is everywhere parallel to the applied field, the scattering produces no spin flip. Transverse fields can however occur when the field is not applied along a symmetry axis and, in this case, spin flip scattering also occurs. The spin flipped neutrons experience a different mean potential ($\mu.B$) and this gives rise to an additional refraction, an effect that has been observed experimentally [1–3]. Most experiments, however, are performed with non-polarised neutrons and with the field aligned to a crystal symmetry direction, for which no transverse fields are expected. Quantitative information is obtained by measuring the total reflected intensity normalised to the incident flux associated with a particular reciprocal lattice position as the sample is rocked through the Bragg condition. After correcting for absorption, the integrated reflected intensity for a sample of thickness t is,

$$R(\bm{q}) = \frac{\gamma^2 \lambda_n^3 t}{16\phi_0^2 \sin(2\theta)} \frac{|h(\bm{q})|^2}{S_0^2}, \tag{18.1}$$

where θ is the Bragg angle, S_0 is the unit cell area of the FLL, $\gamma = 1.91$ (the gyromagnetic ratio of the neutron), and ϕ_0 is the flux quantum. The above quantity is particularly useful as it can be measured in absolute units and is independent of the detailed resolution function of the neutron spectrometer

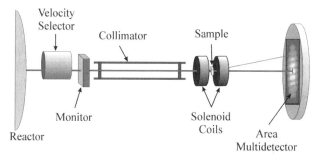

Fig. 18.2. A schematic of the small-angle neutron scattering instrument D22 at the ILL is shown (figure courtesy of S. Lloyd and R. Cubitt). Neutrons of fixed wavelength in the range $\lambda = 6-20$ Å are selected and collimated over a distance of up to 20m. A similar distance separates the sample from the multidetector (128×128 elements of 0.75 cm^2) at the right of the image.

and any small mosaic of the field distribution. The form factor $h(\mathbf{q})$ is the Fourier component of the magnetic field distribution at a given reciprocal lattice vector $\mathbf{q} = \mathbf{G}$. Different values of $\mathbf{q}$ can be measured either by measuring higher order reflections or by changing the magnetic field and, therefore, the lattice spacing. The experiment therefore determines the position and orientation of the $\mathbf{G}$'s and the field profile $h(\mathbf{q})$. The simple geometry of the experiment means that the resolution function of the spectrometer can also be easily calculated [4,5]. The angular width of the rocking curves and the angular width of the diffraction spots on the detector relative to the resolution then allow the sizes of the spots in reciprocal space to be estimated. These dimensions can then be related to the long range structure and degree of order of the flux line lattice. In fact due to the disruptive influence of disorder the FLL is never perfectly ordered and is more accurately described as a Bragg glass [6].

18.3 Field History

The field history prior to measurement can be an important factor in determining the degree of order of the FLL. Historically it is well established that changing the magnetic field creates a flux gradient in the sample, as described by the Bean model. This might suggest that the most ordered lattices would be found when the sample has been cooled in a constant field from above T_c (field cooled). However, in weakly pinned superconductors the FLL is often observed to be better ordered after the lattice has been displaced either by passing a large electric current (well above the critical current to unpin the lattice) or by cycling the field [7–9]. The reason for this is that in the frame of the moving lattice the pinning potentials appear dynamic and the FLL does not have time to deform in response to them. The better ordered lattice is observed to persist even once the flux lattice is again frozen. Careful consideration should be given as to whether the direction of motion during the FLL displacement, either imposed by the specimen geometry or due to channelling along preferred directions, favours a particular FLL orientation or geometry. In the experiments referred to below a field history that gives the most ordered lattice has usually been applied to optimise the sharpness of the Bragg spots and therefore the signal/background measured.

18.4 The Form Factor

For a conventional isotropic superconductor the form factor $h(r)$ has been calculated within various approximations. At large distances, relevant to small q and small fields, the situation is probably the most straightforward since London's theory can be used, which neglects the vortex cores. In the other extreme, very close to the upper critical field, Abrikosov's solution for the flux-line lattice has been employed. This solution can be extended by numerical calculation [10] to cover the whole field range, but has the same limited validity as the Ginzburg–Landau (GL) theory from which it is derived. In particular it overlooks a possible narrowing of the vortex cores predicted at low temperatures for pure materials in more exact treatments [11]. The domain of field and temperature accessible to measurements often lie between the range of applicability of the London and GL theories. The applied field has to be sufficiently large compared with H_{c1} (the lower critical field), to ensure a scattering angle that is sufficient for the diffracted beam to be distinguishable from the transmitted beam. On the other hand, the modulated component of the field becomes very small as H_{c2} (the upper critical field) is approached. More exact numerical calculations that apply only to isolated vortices have been based on the Eilenberger equations [11] valid for $T \gg T_c^2/E_f$ and the De Gennes–Bogoliubov equations.

Fitting this experimental data a modified London model has been the most widely used technique, in which the effect of the vortex core is included by multiplying the London solution by a cut-off function. An appropriate choice of this function is given by a variational solution of the GL equations [12] and reproduces well the exact GL solution at low fields [13]. For large κ this solution can be conveniently approximated by,

$$h(q) = g(q, B/B_{c2})\phi_0/(1 + \lambda^2 q^2) \qquad (18.2)$$

$$g(q, b) = (1 - b^4)\sqrt{2}\xi q . K_1(\sqrt{2}\xi q), \qquad (18.3)$$

where K_1 is a modified Bessel function. At low magnetic fields $g(q)$ only depends on B through the reciprocal lattice vector, q, and the field distribution is just the sum of the profiles of single vortices. Since usually $\lambda q \gg 1$, the 1 may be dropped from the denominator in the first equation, but is kept in the above to preserve the exact result of the London theory, $h(q) = \phi_0/(1 + \lambda^2 q^2)$, at small q. The prefactor g effectively cuts off the London solution at large q of the order of $1/\xi$, and removes the unphysical divergence of $h(\mathbf{r})$ at small $\mathbf{r}$. The field profile in real space from a single vortex thus decreases exponentially with distance at large distances ($r \gg \lambda$), but varies like $\ln(\lambda/r)$ at intermediate distances ($\xi < r < \lambda$) and saturates at the centre of the vortex core at a value larger than H_{c1} (the lower critical field). The flux lines therefore interact substantially with other flux lines within a radius λ.

18.5 Flux Line Lattice Geometries

We will first consider an isotropic material and then consider progressively more complicated cases. Throughout this discussion it is useful to note that the geometry of a 2D primitive lattice in reciprocal space can be obtained by rotating the primitive real space lattice through 90°.

For the GL solution the free energy is minimised by minimising the quantity $\langle \Delta^4 \rangle / \langle \Delta^2 \rangle^2$ where Δ is the superconducting gap, with the result that a two dimensional hexagonal lattice is predicted [14]. For the London theory the free energy is given by a particularly simple expression [15],

$$F = \frac{\phi_0}{8\pi} \sum_{i,j} h_i(\mathbf{r}_j) \tag{18.4}$$

where $h_i(\mathbf{r}_j)$ is the field at the flux line at position r_j due to the line at r_i. For very low fields, since the single vortex field profile decreases exponentially at large distances, only nearest neighbour interactions are important and the energy is minimised by keeping the flux lines as far apart as possible. For a uniform macroscopic flux density this is achieved with a hexagonal lattice. At high fields interactions with more distant neighbours have to be summed. Going over to Fourier components and the modified London field profile introduced above,

$$F = \frac{B^2}{8\pi\phi_0} \sum_{q=G} h_z(\mathbf{q}) \tag{18.5}$$

At very high fields the factor $g(\mathbf{q})$ (18.2) now limits the sum to nearest neighbours in reciprocal space so that again a hexagonal lattice is preferred. Thus it is no surprise that a hexagonal lattice is indeed observed in a large number of superconductors. However, there is no preferred orientation of the lattice in the above model. The simplest way to introduce anisotropy is to examine the case of an anisotropic effective electronic mass.

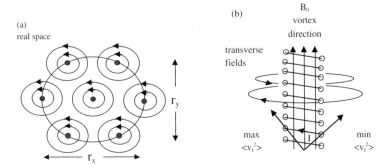

Fig. 18.3. (a) A schematic of the FLL in a material with an anisotropic effective mass. The current distribution and FLL are distorted from the isotropic case by $r_x/r_y = \langle v_{fx}^2 \rangle^{1/2}/\langle v_{fy}^2 \rangle^{1/2} = \langle v_\perp^2 \cos^2(\theta) + v_\parallel^2 \sin^2(\theta) \rangle^{1/2}/\langle v_\perp^2 \rangle^{1/2}$ in real space. When the field is applied along a principal axis no preferred orientation results. (b) When the field is applied away from a principal axis the currents no longer flow in planes perpendicular to the flux lines, but flow preferentially in the direction of the maximum Fermi–velocity. As for a solenoid that has inclined windings, this gives rise to transverse dipole fields directed perpendicular to the axis of rotation. This gives the FLL a preferred orientation, such that some neighbours in direct space lie directly along the axis of the North-South poles. A reciprocal lattice vector then lies along the axis of rotation.

18.6 An Anisotropic Effective Mass

The introduction of an ellipsoidal mass anisotropy can be motivated by considering the energy dispersion relation for electrons in a uniaxial material to be,

$$E = \frac{p_x^2 + p_y^2}{2m_\perp} + \frac{p_z^2}{2m_\parallel}. \tag{18.6}$$

This leads to anisotropic values for the second moments of the Fermi–velocity averaged over the Fermi surface,

$$\frac{\langle v_{f\perp}^2 \rangle}{\langle v_{f\parallel}^2 \rangle} = \frac{m_\parallel}{m_\perp}. \tag{18.7}$$

It is this second definition of the mass anisotropy that is appropriate for the case of more complicated Fermi surfaces. The London equation can then be generalised to,

$$\frac{4\pi}{c}j_i = -\frac{1}{\lambda^2}m_{ij}^{-1}a_j \quad \text{where} \quad a = A + \frac{\phi_0}{2\pi}\nabla\vartheta \tag{18.8}$$

ϑ is the phase of the order parameter, A is the electromagnetic vector potential, and the mass tensor has been normalised so that $\text{Det}(m) = 1$.

From this an expression can be derived for $h(q)$ [16]. For a rotation of the crystal through θ about its y axis with the field applied along z in the unrotated frame,

$$\vec{h}(q) = \frac{\phi_0}{(1 + m_{xx}q_y^2 + m_\perp q_x^2)(1 + m_{zz}q^2) - m_{xz}^2 q_y^2} \begin{Bmatrix} m_{xz}q_y^2 \\ m_{xz}q_xq_y \\ 1 + m_{zz}q^2 \end{Bmatrix} \quad (18.9)$$

with $\begin{Bmatrix} m_{xx} \\ m_{xz} \\ m_{zz} \end{Bmatrix} = \begin{Bmatrix} m_\perp \cos^2(\theta) + m_\| \sin^2(\theta) \\ (m_\perp - m_\|)\cos(\theta)\sin(\theta) \\ m_\perp \sin^2(\theta) + m_\| \cos^2(\theta) \end{Bmatrix}.$ (18.10)

The consequences are;

- The FLL unit cell vectors lie on an ellipse (Fig. 18.3a). The lattice geometry is no longer hexagonal but has been stretched and compressed along the x and y axes. This leads to a gain in energy with respect to a perfect hexagonal lattice that is of the order of $\Delta F \approx |m_\| - m_\perp| B^2 a^2/\lambda^2$ where a is the FLL lattice constant.
- Within the anisotropic London model, when the field is applied along a principal axis ($\theta = 0°$ or $\theta = 90°$), the solution for an isotropic material can be employed with appropriately scaled x and y coordinates. As a consequence the FLL is deformed, but has no preferred orientation [17].
- When the field is applied along a low symmetry direction (i.e. θ is not 0 or 90°) and $H \gg H_{c1}$ the currents are still extended spatially in approximately the same proportion as the lattice distortion, but now a transverse field component has also to be considered. This transverse field (that averages to zero over space) arises because the superconducting screening currents no longer flow in planes perpendicular to the flux lines, but flow in planes that are skewed towards the direction of the largest Fermi–velocity $\langle v_f^2 \rangle$ (Fig. 18.3b). This leads to a preferred orientation of the lattice with a reciprocal unit cell vector parallel to the rotation axis, whenever the mass anisotropy is different from 1. The energy gain for this alignment relative to a deformed lattice, with a reciprocal unit cell vector parallel to the x-axis, is of the order of $\Delta F \approx B^2 |m_\| - m_\perp| (a/\lambda)^4 \sin^2(2\theta)$.

The above arguments can easily be extended to describe the most general mass anisotropy, $m_{xx} \neq m_{yy} \neq m_{zz}$, appropriate to general crystal symmetries and for an orthorhombic symmetry in particular. Further, the result can be generalised to the case of an anisotropic superconducting gap; the mass tensor then becomes $m_{ij} = \langle \Delta^2 v_{fi} v_{fj} \rangle / \text{Det}(m)$.

The above predictions have been nicely demonstrated in measurements on the high-T_c superconductor YBa$_2$Cu$_3$O$_7$ (YBCO) which is orthorhombic (Fig. 18.4), but with a large anisotropy between the c-axis and the perpendicular directions. In YBCO, when the field is applied along the c-axis the

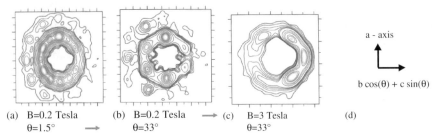

Fig. 18.4. The FLL is shown in YBCO at low temperature for different fields (figures courtesy of Johnson et al. [18]). (**a**) For small inclinations of the field from the c-axis, two orientations of the FLL are seen corresponding to an alignment of the FLL with the two symmetry axis of the crystal. (**b**) For a larger inclination of the field only the lattice favoured by the effective mass anisotropy survives. (**c**) In a larger field however, the other FLL orientation occurs. Further factors must therefore influence the orientation of the FLL. Figure (**d**) shows the orientation of the crystal axis in these figures.

situation is complicated by the fact that the FLL is locked to twin boundaries present in most samples. However Johnson et al. [18] have shown that in weakly twinned samples, as the field is rotated away from the c-direction about the a-axis two intrinsic orientations of the FLL are found. With a further rotation only the orientation predicted from the mass anisotropy model occurs. However, contrary to this, at higher fields the second orientation is preferred (furthermore, there is no preferred orientation at all when the field is rotated about the b-axis, the direction of the CuO chains in this material). The above result indicates that further factors not included in the effective mass model can dominate the effective mass anisotropy in determining the FLL orientation and that these factors become relatively more important as the field is increased.

For conventional superconductors such as Nb [19], V_3Si [20] and more recently in the borocarbides such as YNi_2B_2C [21] a rich variety of flux lattice geometries and orientations have been observed. These structures have been related to the anisotropy of the underlying Fermi surfaces, by extending the London and Ginzburg–Landau theories to include non-local corrections.

18.7 Non-local Corrections and Fermi Surface Anisotropy

Corrections to the London theory stem from the fact that the relationship between the current and vector potential is non-local. The current at a given point depends on the vector potential in a region of extent ξ_0 (the microscopic coherence length) about this point. This can be expressed via,

$$\frac{4\pi}{c} j_i = -\frac{1}{\lambda^2}(m_{ij}^{-1} - \lambda^2 n_{ijlm} k_l k_m + \ldots) a_j \,. \tag{18.11}$$

The first term on the right hand side of the equation is simply the London result with anisotropic masses, while the second term is the lowest order non-local correction. Non-local corrections can also be introduced into the Ginzburg–Landau theory by including higher than quadratic powers of the derivatives of the order parameter in the expression for the free energy. Such a procedure is appropriate to derive the anisotropy of the upper critical field (e.g. [22]). The more technical aspect of the problem is to calculate the coefficients of these terms from microscopic theory. For a one component s-wave superconductor this has been considered by Takanaka [23,24] and more recently by Kogan [25]. For the case of the boro-carbides, these results [26] have been used to describe the rich variety of structures seen. The full analysis requires numerically evaluating equation (18.4) with the field profile found from (18.11). If we consider the leading order non-local correction for a field applied along the tetragonal axis of $LuNi_2B_2C$, we have,

$$h(q) = \phi_0 \frac{g(q)}{1 + \lambda^2 q^2 + \lambda^4 d q_x^2 q_y^2} \quad \text{and}$$

$$F = \frac{B^2}{8\pi} \sum_{q=G} \frac{g(q)}{1 + \lambda^2 q^2 + \lambda^4 d q_x^2 q_y^2} . \tag{18.12}$$

d is determined from the band structure, and is found to have a positive value, since the Fermi surface average $< \langle v_{fx}^4 \rangle \; (= \langle v_{fy}^4 \rangle)$ is relatively large compared to the averages of quartic combinations of other components of v_f, (the coefficient of the isotropic term q^4 turns out to be small and for simplicity has been omitted from (18.12)). The absolute magnitude of d increases slightly with decreasing temperature and is of order ξ_0^2/λ^2, well below T_c, which is small since the borocarbides are extreme type II superconductors.

Since the reciprocal lattice vectors are of order $2\pi\sqrt{B/\sqrt{3/2}\phi_0} < 2\pi/\xi$, and the relevant values of q increase with field, the effect of the non-local corrections becomes more significant at higher field. At very small fields it is better to consider F in real space. Then, only nearest neighbour flux line interactions need to be considered due to the exponential fall-off at large distance of the vortex–vortex interaction. Thus, the predicted FLL is nearly hexagonal (cf. (18.4)), but since the magnetic field profile from a single vortex extends most along the x and y directions, the choice of orientations with nearest neighbours along these directions is avoided. Thus, two equivalent nearly hexagonal lattices with nearest neighbour lattice points aligned at 45° to the crystal axis are expected. As the field is increased, the FLL is deformed to shorten the spacing along these 45° directions and lengthen the lattice spacing along the other lattice vectors. Starting from the opposite extreme of high fields we only need to consider nearest neighbours in reciprocal space, since $g(q)$ in (18.12) strongly reduces the interactions between more distant reciprocal lattice points. The free energy is then minimised by locating nearest neighbour reciprocal lattice points as close as possible to the 45° directions. If the non-local terms are important enough, the gain in energy achieved for this choice of nearest neighbour directions overwhelms the energy cost of having shorter nearest neighbour distances, resulting in a square lattice with nearest neighbours aligned at 45° to the crystal lattice. Since the low field and high field lattices do not continually deform into each other the passage between the two at intermediate fields is rather complicated. A careful calculation [27] reveals that the transition is in fact continuous, with a rapid field evolution through low symmetry lattices, although in practice this might be obscured by disorder or by considering other terms that go beyond the present approximation. The experimental results indeed support the main predictions of the non-local theory very well (Fig. 18.5). For the compound YNi_2B_2C a change in lattice orientations from deformed triangular lattices aligned with crystal [110] directions to deformed lattices aligned with the crystal axes has been observed at a small field h_1. The high field geometries then continuously evolve to yield a square lattice above a higher field h_2. Theoretically the non-local corrections are reduced as the electronic mean free path is decreased which results in higher values of these characteristic fields. This effect has also been confirmed experimentally in a study of doped alloys of another boro-carbide material [28].

18.8 Non-conventional Superconductors

Other factors may also favour an alignment or particular geometry of the FLL. We have already mentioned extended pinning centres, which include sample surfaces. Another possibility relates to magnetic order as in $TmNi_2B_2C$ [29]. We will now look at the role of unconventional superconductivity.

Unconventional superconductivity occurs when the superconducting order breaks additional symmetries besides the gauge symmetry broken in all superconductors. The symmetries in question are the point group symmetry of the crystal (and therefore of the Fermi surface), and the symmetry under time reversal. Since most materials have a centre of symmetry the superconducting state can be conveniently classified to have either an even or odd parity. Since the paired state is built from Fermions, an odd parity state corresponds to spin triplet (parallel spin) pairing, while an even parity state corresponds to a spin singlet pairing (as realised in a conventional superconductor). A further complication arrives if the superconducting state requires two degenerate functions to describe it (this results if the order parameter belongs to a two dimensional representation of the crystal symmetry group). We will return to this special case after considering non-conventional one component order parameters.

For the case of both conventional and unconventional superconductors the superconducting gap can be anisotropic. The main difference is in the degree of the anisotropy. The anisotropy rarely exceeds 10% in a conventional 3 dimensional superconductor, but in a non-conventional superconductor,

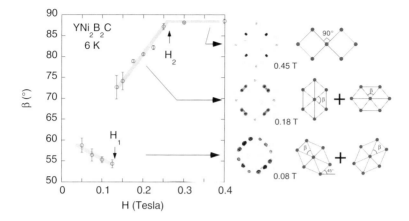

Fig. 18.5. The apex angle and orientation of the FLL in YNi$_2$B$_2$C at low temperature as a function of field applied parallel to the c-axis (courtesy of S. Levett and C. D. Dewhurst). The differently oriented FLL's that make up the complete diffraction images at representative fields are also indicated (the tetragonal crystal axes are horizontal and vertical). A change in the FLL orientation at h_1 and the evolution of the deformation with field agree with the behaviour predicted from non-local corrections. Similar results have been obtained for small inclinations of the field from the c-axis [21], but then only one of the two domains at each field is observed due to the off-axis mass anisotropy.

nodes in the gap are expected to occur if the order parameter changes sign under any of the crystal symmetry operators. For directions along the nodes the coherence length $\xi \approx \hbar v_f / \Delta$ diverges, which suggest that the non-local corrections must be considered even in the case when the Ginzburg–Landau parameter κ, is large. At low temperatures the treatment of the non-local corrections is particularly delicate, since the physics is dominated by excitations close to the nodes [30], and effects such as resonant impurity bound states [31,32] have to be considered. At higher temperatures, however, the non-local picture for dealing with Fermi surface anisotropy can be extended to cover the anisotropy of the superconducting gap [33].

Apart from the high-T_c oxides, one of the most studied non-conventional superconductors is UPt$_3$. For an excellent review of the physics of UPt$_3$ see the recent book by Mineev and Samokhin [34]. A unique quality of UPt$_3$ (for a pure compound) is that it has 3 distinct superconducting phases as a function of field and temperature (Fig. 18.6). There are two transitions in zero field; the lower transition at T_{c-} is particularly noteworthy as it associated with a jump in the specific heat of the same magnitude as at the upper transition, T_{c+}, where bulk superconductivity appears. Most of the theories put forward to explain the phase diagram select a two component order parameter. A slight deviation from hexagonal symmetry, attributed to a weak

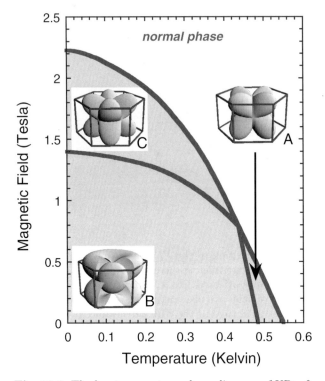

Fig. 18.6. The low temperature phase-diagram of UPt$_3$ for magnetic fields parallel to the hexagonal crystal axis. Second order phase transitions separate three different superconducting phases. The geometric figures show schematically the theoretical superconducting gap symmetries in the 3 phases (A,B and C) for the E_{2u} model [35]. The gap magnitude is represented as the difference between the outer surfaces and inner spheres as a function of angle (a segment of the outer surface has been removed to show the inner sphere in the figure for the B-phase). The frames define the hexagonal crystal symmetry.

symmetry-breaking field, for example due to an observed weak quasi-static anti-ferromagnetic order [36], lifts the degeneracy of the transition temperatures for the two components and produces the double transition. NMR and other experiments [37] indicate the superconductivity a is spin-triplet. In the figure we show the gap symmetries consistent with one model [35] for a spin-triplet state, which has proved one of the most successful to account for various experimental observations. Note that the predicted gap anisotropy in the plane perpendicular to the hexagonal axis (c-axis) is small at low temperatures in the B-phase but becomes significant in the A-phase.

The experimentally measured [38] flux-line lattices grown at different temperatures in a low field are shown in Fig 18.7. When the lattice is formed at low temperature in the B-phase a single hexagonal FLL is observed with

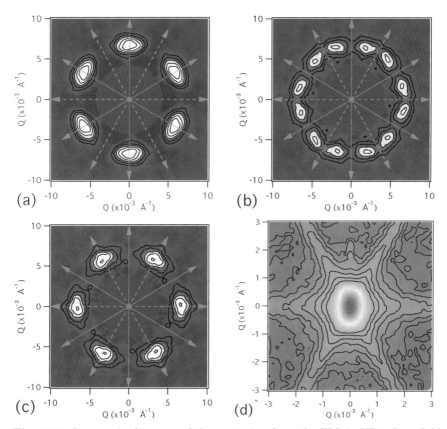

Fig. 18.7. Symmetrized images of the scattering from the FLL in UPt$_3$ for a field of 0.19 Tesla parallel to the c-axis. In all the images an a-axis is vertical and an a^*-axis is horizontal. The images (**a**)–(**c**) represent measurements made at low temperature (100–150 mK) for different field histories. (**a**) corresponds to a FLL formed at low temperature in the B-phase after the sample was cooled in zero field. In (**b**) the field was applied at 475 mK and the sample was subsequently cooled to low temperature in a constant field. If the arrangement of the flux lines is frozen when cooling in a constant field the image represents the FLL arrangement in the A-phase. Two differently orientated hexagonal lattices are seen to exist, aligned at 45° to the crystal axes. (**c**) shows the FLL that results when the crystal was cooled in a constant field from above $T_c(H) \approx 520$ mK. In this case the FLL is nucleated at T_c and appears to be aligned with extended metallurgical defects responsible for the background small-angle scattering shown in (**d**).

nearest neighbours along a^*-directions. However when the FLL is formed in the A-phase the sample is divided equally between two differently oriented approximately hexagonal lattices inclined at ±45° to the crystal axes.

To understand these results it is important to note that, unlike for the 4-fold Fermi surface anisotropies of tetragonal materials, the hexagonal structure of the Fermi surface for UPt$_3$ cannot lead to any deformation from an ideal hexagonal FLL. The Fermi surface can however align the orientation of the FLL [23], and the observed orientation at a low temperature appears to be consistent with that expected from the known Fermi surface morphology.

There must then be a second source of anisotropy at higher temperatures to change the FLL alignment. The formation of two equally populated orthogonal FLLs suggests that the underlying source of this anisotropy has a four-fold symmetry, in analogy with the previously cited results for the borocarbides. The fact that the orientation changes at a temperature close to T_{c-} suggests that the source is related to the anisotropy of the superconducting order that develops as the A-phase is approached. Most of the order parameters allowed by symmetry for the A-phase have either two fold or 6 fold symmetries. They would therefore give rise to a single preferred orientation of the FLL. The experiment therefore strongly supports the identification of the order parameter as belonging to the E_{2u} representation, which naturally gives rise to a near four-fold anisotropy of the order parameter. The lowest order spherical harmonic functions that transform according to this representation are $k_z k_x k_y \hat{z}$ and $k_z(k_x^2 - k_y^2)\hat{z}$. In the A-phase, due to the weak symmetry breaking, only one of these two components survives, at least in zero field. To explain the actual orientation of the FLL realised, and identify which of these two components predominates in the A-phase, requires a slightly more delicate analysis. In zero field in the A-phase the minority component of the order parameter is zero, but it can be induced near to the vortex cores once a field is applied (see the next section). However, for the moment, we neglect this contribution and apply the analysis for a single component order parameter referred to at the beginning of this section. For both the above choices for the order parameter the effective mass tensor m_{ij} is isotropic if the Fermi–velocity is rotationally invariant. An indication that the anisotropy of the Fermi–velocity is very small is provided by measurements of the upper critical field in the basal plane [43]. These measurements show that any anisotropy in the resistive transition as the field is rotated, is smaller than the transition width itself, which defines an experimental upper limit of $\delta H_{c2}/H_{c2} < 10^{-3}$. Furthermore, for the anisotropic London model (Sect. 18.6), when the field is applied along a principal axis, any anisotropy of the effective mass tensor by itself does not favour a specific orientation of the FLL, although it can lead to a deformation from the ideal hexagonal structure. It is therefore still necessary to look at higher order non-local terms to distinguish between competing structures with different orientations. The effective mass will therefore be considered to be isotropic (for a more complete analysis see reference [44]). With T of the order of T_c we then get (see [33]) an equation similar to (18.12),

$$F = \frac{B^2}{8\pi} \sum_{q=G} \frac{g(q)}{1 + \lambda^2 q^2 + d\lambda^2 \xi_0^2 (2q^4 \pm (q^4 - 8q_x^2 q_y^2))} \ . \qquad (18.13)$$

Here d is a positive temperature dependent constant of order 1 and ξ_0 is the microscopic coherence length. The positive sign in the equation applies to a angular dependence for the order parameter perpendicular to c of $\sin(2\phi)$ and the negative sign to a $\cos(2\phi)$ dependence; the term is absent for a rotationally symmetric order parameter (e.g. well inside the B-phase). Allowing the FLL to distort and rotate, the free energy is minimised for a flux line lattice that is only slightly distorted from hexagonal with a unit cell vector inclined at $-45°$ or $+45°$ to the crystal axes, for the first choice of order parameter at the fields and temperatures at which the lattice was formed (notably $H/H_{c2} \approx 1/2$ and $T \approx 0.95 T_{c+}$). As for the borocarbides, there are two equivalent orientations that are inclined at $90°$ to each other. However, the energy difference between these orientations and the two orientations for lattices with a unit cell vector aligned parallel to a crystal axis is very small. Experimentally it would then be interesting to see if different FLL orientations occur at both very low fields, perhaps in decoration experiments, and also in the high field C-phase. The choice of $k_x k_y k_z \hat{z}$ to describe the order parameter in the A-phase is not however unreasonable. For this choice the lobes of the order parameter are almost parallel to directions connecting nearest neighbour uranium sites with identical spin in the anti-ferromagnetic structure, while the nodes lie along directions connecting nearest neighbours with opposite spin. Such a choice might be expected for a spin triplet superconductor in zero field. In fact, symmetry does not require that the order parameter be exactly 4-fold symmetric in the A-phase. A higher order harmonic component can be mixed with the angular dependence considered above [45], which would lead directly to an anisotropy of the effective mass tensor, and could incline the lobes more closely to the equal spin directions.

Finally, we note that the actual order parameter in the sample could be divided into domains corresponding to equivalent choices for the axis of the symmetry-breaking field. However, since these equivalent axes are oriented at $60°$ to each other, all the domains would individually favour the same FLL orientations if the FLL is hexagonal. If, however, the size of the domains is smaller than the typical correlation length of the FLL, the problem is more complicated, since the deformation and orientation of the FLL are intricately coupled.

For YBCO a similar orientation of the FLL with respect to the order parameter to that observed in the A-phase of UPt$_3$ has been predicted from the De Gennes Bugulubiov equations [39]. For YBCO $\Delta = \cos(2\phi)$ and the FLL is aligned parallel to a crystal axis. Experiment agrees with these predictions for fields almost parallel with the c-axis (Fig. 18.4).

18.9 Two Component Superconductors

So far we have ignored the second component of the order parameter, which we now discuss. A two component order parameter which is a function of position r can be written,

$$\Delta = \nu_1(r)\widehat{e}_1(k) + \nu_2(r)\widehat{e}_2(k) . \tag{18.14}$$

This formula applies to both singlet and triplet order parameters. For the triplet case, the basis states ($\widehat{e}_1(k)$ and $\widehat{e}_2(k)$) are vectors which point along the direction of zero spin projection, while for a singlet state they are scalar functions. For unitary order parameters (considered here) the different components of each basis vector have the same phase. For UPt$_3$ an appropriate choice is $\widehat{e}_1 = k_x k_y k_z \widehat{z}$ and $\widehat{e}_2 = (k_x^2 - k_y^2)k_z\widehat{z}$, while for Sr$_2RuO_4$, $\widehat{e}_1 = k_x\widehat{z}$ and $\widehat{e}_2 = k_y\widehat{z}$ are often considered.

The Ginzburg–Landau expansion of the free energy density in the absence of any tetragonal anisotropy of the Fermi surface (as for hexagonal UPt$_3$), writing $\vec{\nu} = (\nu_1, \nu_2)$ is,

$$\begin{aligned}f = &\alpha \vec{\nu}.\vec{\nu}^* + \beta_1(\vec{\nu}.\vec{\nu}^*)^2 + \beta_2(\vec{\nu}.\vec{\nu})(\vec{\nu}.\vec{\nu})^* \\ &+ K_1(D_i\nu_j)(D_i\nu_j)^* + K_2(D_i\nu_i)(D_j\nu_j)^* + K_3(D_i\nu_j)(D_j\nu_i)^* \\ &+ K_4(D_z\nu_j)(D_z\nu_j)^* + \frac{h^2}{8\pi} - \frac{hH}{4\pi} .\end{aligned} \tag{18.15}$$

For UPt$_3$ the coefficients of the 4th order homogeneous terms can be related to the jumps in the specific heat in zero field by $\beta_2/\beta_1 = (\Delta C/T)_{Tc-}/(\Delta C/T)_{Tc+}$. Combined with the general stability requirements ($\beta_1 + \beta_2 > 0$, $\beta_1 > 0$), this forces the solution in zero field to be proportional to either $\vec{\nu}_+$ or $\vec{\nu}_-$ (where $\vec{\nu}_\pm = (1, \pm i)$). The symmetry breaking field gives a small splitting of the α-term which becomes $\alpha_1 \nu_1^2 + \alpha_2 \nu_2^2$ and a similar splitting of the K_1 term. The interesting terms are K_2 and K_3 since these mix the different components of the order parameters in the inhomogenus mixed phase. However, to explain the phase diagram of UPt$_3$ and the observation of a tetracritical point for all field orientations, Sauls [35] has argued that in UPt$_3$, K_2 and K_3 are small, a particularity of his proposition for the order parameter when the basal plane Fermi–velocity anisotropy is small. In general, and for other models of the superconductivity in UPt$_3$, these terms need not be small, and then a wide range of flux line structures, including non-axially-symmetric vortex cores, become possible [40,41]. Although the parameters for UPt$_3$ suggest that the cores are axially symmetric, the existence of any asymmetry would modify the interpretation of the results reported in the previous section. Also, the presence of the second component of the order parameter could be important. For the case of YBCO it has been shown that an induced s-wave component can give rise to terms similar in form to the non-local corrections [42]. Calculations relevant to the A-phase of UPt$_3$, could also shed additional light on the unusual FLL orientations observed.

A more direct consequence of the two component order parameter in UPt$_3$ was examined in low temperature measurements of the FLL as a function of field directed this time perpendicular to the c-axis [46]. The effective mass anisotropy for this field orientation leads to a deformed hexagonal FLL. Experimentally, the distortion was observed to be strongly field dependent. Recall that, in general, the mass anisotropy should include the anisotropy of the order parameter as well as of the Fermi–velocity ($m_\parallel/m_\perp = \langle \Delta^2 v_{f\perp}^2 \rangle / \langle \Delta^2 v_{f\parallel}^2 \rangle$). However, in a one component superconductor, the gap anisotropy would not be expected to evolve strongly with field. In this light, the strong evolution of the distortion of the FLL with field was attributed by Joynt [47] to the change in the proportions of the two components of the order parameter as the C-phase was approached.

Another interesting case concerns the tetragonal material strontium ruthenate, where a square FLL has been observed for a large range of temperatures and fields parallel to the c-axis [48,49]. The superconducting order in Sr$_2$RuO$_4$ is believed [50] to resemble that in the B-phase of UPt$_3$, with an order parameter $\hat{z} e^{\pm /Ik\phi}$, (compared to $\hat{z} \cos(\theta) \sin^2(\theta) e^{\pm ik\phi}$ for UPt$_3$ in the absence of the symmetry-breaking field). Although a square FLL might occur due to the underlying square Fermi surface anisotropy, the symmetry of the superconductivity could also stabilise this choice. For Sr$_2$RuO$_4$, since the crystal symmetry is tetragonal, an additional homogeneous 4th order term, $\eta_1 |\nu_x|^2 |\nu_y|^2$, and a gradient term, $\eta_2(|D_x \nu_x|^2 + |D_y \nu_y|^2)$, must be included in the expression for the free energy [51,52]. These terms induce a marked 4-fold anisotropy of the order parameter around the vortex cores, and favour a square FLL. The orientation of the FLL in this case depends on both the Fermi surface anisotropy, which determines the signs of the η's, and on the orientation of the order parameter. Since the above gradient terms are second order derivatives, whereas the non-local corrections due to the Fermi surface alone involve 4th order derivatives, it has been argued that for a two component order parameter the above terms should be more significant. However, for the specific case of the field parallel to the c-axis in Sr$_2$RuO$_4$, κ is only slightly larger than unity, which means that large non-local corrections should be anticipated. Another test for the marked square symmetry of the vortex cores has recently been made [53] by analysing higher order diffraction peaks that probe the field distribution closer to the vortex core. The field distribution found appears to correlate well with that predicted for the two component description.

18.10 Conclusions

We have illustrated how small-angle neutron scattering can give direct and detailed information about the field distribution in bulk superconductors. We have concentrated the discussion on measurements at the length scale of the flux-line spacing, but it should also be recalled that the SANS technique can probe both the correlation of the FLL at much larger scales also the smaller length scales appropriate to the vortex cores. Fermi surface anisotropy, pinning, an anisotropy of the superconducting gap and the existence of multiple components of the order parameter can all influence the geometry and orientation of the flux-line lattice. The interpretation of a measured FLL geometry then requires a careful analysis to determine the most important factors appropriate on a case by case basis as a function of the field and temperature. However, in some special cases it appears that even a simple observation of the geometry and orientation of the FLL can give valuable information on the symmetry of the underlying superconductivity.

Theoretically a rich variety of predictions exist for novel vortex structures in non-conventional superconductors, such as doubly quantized vortices. SANS offers an important window through which to search for the microscopic realisation of such states. This search is fuelled by the growing number of unusual superconducting compounds that continue to be discovered.

Acknowledgements. It is a pleasure to acknowledge many useful discussions, in particular, with the following people: T. Champel, R. Cubitt, C. D. Dewhurst, N. Van Dijk, E. M. Forgan, S. Levitt, V. P. Mineev, D. McK Paul, and P. Rodière.

References

1. E.M. Forgan, P.G. Kealey, T.M. Riseman, S.L. Lee, D.McK. Paul, C.M. Aegerter, R. Cubitt, P. Schleger, A. Pautrat, Ch. Simon and S.T. Johnson: Physica B **268**, 115–121 (1999)
2. S.L. Lee, P.G. Kealey, E.M. Forgan, S.H. Lloyd, T.M. Riseman, D.McK. Paul, S.T. Johnson, Ch. Simon, C. Goupil, A. Pautrat, R. Cubitt, P. Schleger, C. Dewhurst, C.M. Aegerter and C. Ager: Physica B **276–278**, 752–755 (2000)
3. E.M. Forgan, P.G. Kealey, S.T. Johnson, A. Pautrat, Ch. Simon, S.L. Lee, C.M. Aegerter, R. Cubitt, B. Farago, and P. Schleger: submitted to Phys. Rev. Lett.
4. R. Cubitt, E.M. Forgan, D.McK. Paul, S.L. Lee, J.S. Abell, H. Mook, P.A. Timmins: Physica B **180–181**, 377–379 (1992)
5. P. Harris, B. Lebech and J. S. Pedersen: J. Appl. Crys. **28**, 209–222 (1995)
6. T. Giamarchi and P. Le Doussal: Phys. Rev. Lett. **72**, 1530–1532 (1994)
7. U. Yaron, P.L. Gammel, D.A. Huse, R.N. Kleiman, C.S. Oglesby, E. Bucher, B. Batlogg, D.J. Bishop K. Mortensen, K. Clausen C.A. Bolle, F. De La Cruz: Phys. Rev. Lett. **73**, 2748-51 (1994)
8. A. Huxley, R. Cubitt, D. McPaul, E. Forgan, M. Nutley, H. Mook, M. Yethiraj, P. Lejay, D. Caplan, J.M. Penisson: Physica B **223–224**, 169–171 (1996)
9. F. Pardo, F. de la Cruz, P.L. Gammel, E. Bucher and D.J. Bishop: Nature **396**, 348–350 (1998)
10. E.H. Brandt: Phys. Rev. Lett. **78**, 2208-11 (1997)
11. W. Pesch and L. Kramer: J. of Low Temp. Phys. **15**, 367 (1974)

12. J.R. Clem: J. Low Temp. Phys. **18**, 427–434 (1974)
13. A. Yaouanc, P. Dalams de Rotier and E. H. Brandt: Phys. Rev. B **55**, 11107–11110 (1997)
14. W.H. Kleiner, L.M. Roth and S.H. Autler: Phys. Rev. **133**, A1226-7 (1964)
15. M. Tinkham: 'Introduction to superconductivity' (McGraw-Hill, New York 1986)
16. S.L. Thiemann, Z. Radovic and V.G. Kogan: Phys. Rev. B **39**, 11406–11412 (1989)
17. L.J. Campbell, M. M. Doria and V.G. Kogan: Phys. Rev. B **38** 2439–2443 (1988)
18. S.T. Johnson, E.M. Forgan, S.H. Lloyd, C.M. Aegerter, S.L. Lee, R. Cubitt, P.G. Kealey, C. Ager, S. Tajima, A. Rykov and D.McK. Paul: Phys. Rev. Lett. **82**, 2792–2795 (1999)
19. D.K. Christen, H.R. Kerchner, S.T. Sekula and P. Thorel: Phys. Rev. B **21**, 102–117 (1979)
20. D.K. Christen, H.R. Kerchner, S.T. Sekula and Y.K. Chang: Physica **135B**, 369–373 (1985)
21. D. McK. Paul, C.V. Tomy, C.M. Aegerter, R. Cubitt, S.H. Lloyd, E.M. Forgan, S.L. Lee and M. Yethiraj: Phys. Rev. Lett. **80**, 1517-20 (1998)
22. V. P. Mineev: Ann. Phys. Fr. **19**, 367 (1994)
23. K. Takanaka: Prog. Theor. Phys. **50**, 365-69 (1973)
24. K. Takanaka and T. Nagashima: Physica C **218**, 379–386 (1993)
25. V.G. Kogan, A. Gurevich, J.H. Cho, D.C. Johnston, Ming Xu, J.R. Thompson, A. Martynovich: Phys. Rev. B **54**, 12386-96 (1996)
26. V.G. Kogan, M. Bullock, B. Harmon, P. Miranovic-acute, Lj. Dobrosavljevic-acute-Grujic-acute, P.L. Gammel, D.J. Bishop: Phys. Rev. B **55**, R8693-6 (1997)
27. A. Knigavko, V.G. Kogan, B. Rosenstein and T.-J. Yang: Phys. Rev. B **62**, 111–114 (2000)
28. P.L. Gammel, D.J. Bishop, M.R. Eskildsen, K. Mortensen, N.H. Andersen, I.R. Fisher, K.O. Cheon, P.C. Canfield, V.G. Kogan: Phys. Rev.Lett. **82**, 4082–4085 (1999)
29. M.R. Eskildsen, K. Harada, P.L. Gammel, A.B. Abrahamsen, N.H. Andersen, G. Ernst, A.P. Ramirez, D.J. Bishop, K. Mortensen, D.G. Naugle, K.D.D. Rathnayaka, P.C. Canfield: Nature **393**, 242-5 (1998)
30. G.E. Volovik: J.E.T.P. Letters **58**, 469 (1993)
31. C.J. Pethick and D. Pines: Phys. Rev. Lett. **57**, 118–121 (1986)
32. L. Coffey, T.M. Rice and K. Ueda: J.Phys. C:Solid State Phys. **18**, L813–L816 (1985)
33. M. Franz, I. Affleck and M.H.S. Amin: Phys. Rev. Lett. **79**, 1555–1558 (1997)
34. V.P. Mineev and K.V. Samokhin: 'Introduction to Unconventional Superconductivity.' (Gordon and Breach Science publishers, Amsterdam 1999)
35. J.A. Sauls: Adv. in Phys. **43**, 113–141 (1994)
36. G. Aeppli, E. Bucher, C. Broholm, J.K. Kjems, J. Baumann, J. Hufnagl: Phys. Rev. Lett. **60**, 615-8 (1988)
37. H. Tou, Y. Kitaoka, K. Asayama, N. Kimura, Y. Onuki, E. Yamamoto, and K. Maezawa: Phys. Rev. Lett. **77**, 1374–1377 (1996)
38. A. Huxley, P. Rodiere, D.M. Paul, N. van-Dlik, R. Cuhitt, J. Flouquet: Nature **406**, 160-4 (2000)

39. M. Ichioka, N. Hayashi, N. Enomoto and K. Machida: Phys. Rev. B **53**, 15316–15326 (1996)
40. I.A. Luk'yanchuk and M.E. Zhitomirsky: Superconductivity Review **1**, 207–256 (1995)
41. T.A. Tokuyasu, D.W. Hess and J.A. Sauls: Phys. Rev. B **41**, 8891–8903 (1990)
42. I. Affleck, M. Franz and M.H.S. Amin: Phys. Rev. B **55**, R704–R707 (1997)
43. P. Rodière: PhD thesis, 'Supraconductivité et magntisme dans le composé à électrons fortement corrélés UPt$_3$', université Grenoble I (2001)
44. T. Champel and V.P. Mineev: In preparation. (2001)
45. S. Yip and A. Garg: Phys. Rev. B **48**, 3304-8 (1993)
46. U. Yaron, P.L. Gammel, G.S. Boebinger, G. Aeppli, P. Schiffer, E. Bucher, D.J. Bishop, C. Broholm, K. Mortensen: Phys. Rev. Lett. **78**, 3185-8 (1997)
47. R. Joynt: Phys. Rev. Lett. **78**, 3189-92 (1997)
48. T.M. Riseman, P.G. Kealey, E.M. Forgan, A.P. Mackenzie, L.M. Galvin, A.W. Tyler, S.L. Lee, C. Ager, D.M. Paul, C.M. Aegerter, R. Cubitt, Z.Q. Mao, T. Akima, Y. Maeno: Nature **396**, 242-5 (1998)
49. T.M. Riseman, P.G. Kealey, E.M. Forgan, A.P. Mackenzie, L.M. Galvin, A.W. Tyler, S.L. Lee, C. Ager, D.M. Paul, C.M. Aegerter, R. Cubitt, Z.Q. Mao, T. Akima, Y. Maeno: Nature **404**, 629 (2000)
50. E.M. Forgan, A.P. Mackenzie and Y. Maeno: Journal of Low Temp. Phys. **117**, 1567–1573 (1999)
51. D.F. Agterberg: Phys. Rev. Lett. **80**, 5184–5187 (1998)
52. R. Heeb and D.F. Agterberg: Phys. Rev. B. **59**, 7076–7082 (1999)
53. P.G. Kealey, T.M. Riseman, E.M. Forgan, L.M. Galvin, A.P. Mackenzie, S.L. Lee, D.McK. Paul, R. Cubitt, D.F. Agterberg, R. Heeb, Z.Q. Mao and Y. Maeno: Phys. Rev. Lett. **84**, 6094–6097 (2000)

19 Vortex Dynamics in a Temperature Gradient

Axel Freimuth

Summary. The motion of vortices in a type II superconductor leads to a heat current, which has two main contributions in the hydrodynamic regime: one is due to the heat carried by the vortices themselves and leads to transverse thermomagnetic effects such as the Nernst and Ettingshausen effects. The second contribution is of comparable magnitude and perpendicular to the direction of vortex motion. It arises from what is called the spectral flow effect, which couples the motion of vortices to that of the normal fluid. The existence of this coupling simultaneously explains the small Hall angle and the occurance of longitudinal thermomagnetic effects like the thermopower and the Peltier effect. We briefly summarize the experimental results on thermomagnetic effects in superconductors. We discuss the origin of the two contributions to the heat current and present an analysis of vortex motion in a temperature gradient. We derive in particular an equation of motion for a vortex in a temperature gradient. Our analysis of vortex motion is in excellent agreement with experimental results.

19.1 Introduction

It is well established that the motion of vortices in a type II superconductor gives rise to both an electrical voltage and a heat current. Therefore a variety of thermomagnetic effects occur, such as the Nernst, Ettingshausen, Peltier and Seebeck effects [1–4]. One contribution to the heat current associated with moving vortices stems from the heat carried by the vortices themselves, as is well known already from conventional superconductors [1,2]. This contribution gives rise to large transverse thermomagnetic effects like the Nernst and Ettingshausen effects. A further contribution to the heat current of comparable magnitude arises from the coupling of vortex motion to the normal fluid. It determines the longitudinal thermomagnetic effects such as the thermopower and the Peltier effect. It has recently been pointed out [5] that this latter contribution is related to the spectral flow effect [6–9]. This effect, known from relativistic quantum field theory, occurs in Fermi superfluids and superconductors if quantized vortices are present. It provides a coupling between the vortices and the normal fluid and explains simultaneously the small Hall angle (in the hydrodynamic limit, usually exploited in experiments) and the occurence of large longitudinal thermomagnetic effects.

In this article we briefly summarize the experimental results on thermomagnetic effects in type II superconductors. We discuss the origin of the various contributions to the heat current with emphasis on the spectral flow effect and we present an analysis of vortex motion in a temperature gradient. Our analysis of vortex motion is in excellent agreement with experimental results.

19.2 Definition of Transport Coefficients

We define transport coefficients according to [10]

$$\begin{aligned}\boldsymbol{E} &= \varrho\boldsymbol{j} - \varrho_\mathrm{H}\,\boldsymbol{j}\times\boldsymbol{B} + S\nabla T + Q\,\nabla T\times\boldsymbol{B}\\ \boldsymbol{j}_h &= \Pi\boldsymbol{j} - \varepsilon k\,\boldsymbol{j}\times\boldsymbol{B} - k\nabla T - k_\mathrm{H}\,\nabla T\times\boldsymbol{B}\,.\end{aligned} \qquad(19.1)$$

For simplicity we consider isotropic materials. $\boldsymbol{E}$ and $\boldsymbol{B}$ denote the electric [11] and magnetic field, respectively, and T the temperature. $\boldsymbol{j}$ and $\boldsymbol{j}_h$ are the electrical and heat current densities and ϱ, ϱ_H, k, and k_H are the electrical resistivity, the Hall-resistivity, the thermal conductivity, and the thermal Hall-conductivity, respectively. S is the thermopower (or Seebeck coefficient) and Q is the Nernst coefficient. S and Q are related to the Peltier coefficient Π and the Ettingshausen coefficient ε by Onsager relations:

$$\begin{aligned}\Pi &= ST \qquad \text{(Kelvin-relation)}\\ \varepsilon k &= QT \qquad \text{(Bridgeman-relation)}\,.\end{aligned} \qquad(19.2)$$

As long as these relations hold, it is equivalent to discuss either the electrical voltage resulting from an applied temperature gradient or the heat current resulting from an applied electrical current.

The ratio of the transverse (parallel to $\boldsymbol{j}\times\boldsymbol{B}$ or $\nabla T\times\boldsymbol{B}$) to the longitudinal (parallel to $\boldsymbol{j}$ or ∇T) effects defines the Hall angle. We distinguish an electrical (α_H) and a thermal (α_th) Hall angle according to [12]:

$$\tan\alpha_\mathrm{H} = \frac{\varrho_\mathrm{H}}{\varrho} \quad \text{and} \quad \tan\alpha_\mathrm{th} = \frac{QB}{S}\,. \qquad(19.3)$$

In a superconductor a voltage may be present only in association with a time dependence of the macroscopic wavefunction. Such time dependence is associated with phase slip processes, which in type-II superconductors in a magnetic field are provided by the motion of vortices. For vortices moving with velocity $\boldsymbol{v}_\mathrm{L}$ the Josephson-relation yields an electric field

$$\boldsymbol{E} = \boldsymbol{B}\times\boldsymbol{v}_\mathrm{L}\,. \qquad(19.4)$$

As a result, in the mixed state all transport coefficients involving a voltage (i.e. ϱ, ϱ_H, S, and Q) are finite only if vortices move. Equation (19.2) then shows that $\Pi\neq 0$ and $\varepsilon\neq 0$ requires also that vortices move. Thus, among the transport effects described by (19.1) only the thermal conductivity and the thermal Hall effect do not rely on vortex motion and occur also for pinned vortices.

19.3 Summary of Experimental Results

We briefly present representative examples for the thermomagnetic effects occurring in the mixed state of superconductors. For a more detailed account of experimental data we refer the reader to [1–4]. We focus in the presentation of data on the high-T_c superconductors. A characteristic feature of these systems is a pronounced broadening of the resistive transition in applied magnetic fields [3,14]. This broadening arises from a combination of effects: the quasi-2 dimensionality of the materials, the high temperatures operating in the superconducting state close to T_c, which lead to pronounced thermal fluctuations, and the unusual range of materials parameters as, in particular, the short in-plane coherence length ξ_{ab} of the order of 10 Å. One consequence of these special circumstances is a rich structure of the (B, T) phase diagram with molten and disordered vortex phases [3,13,14]. Another consequence is that there exists a wide range of temperatures and magnetic fields above the so called irreversibility line [3,13,14], in which vortex motion can experimentally be studied without significant complications from pinning effects. All thermomagnetic effects have carefully been studied in this regime of the phase diagram [3,4]. The data presented in the following refer to a configuration where $\boldsymbol{B}$ is applied along the crystallographic c-direction, perpendicular to the CuO_2-planes, and all currents, voltages and the temperature gradient are perpendicular to c [15].

19.3.1 Transverse Effects

Nernst Effect. If a temperature gradient and a magnetic field are present in a sample, a voltage is found along $\nabla T \times \boldsymbol{B}$. This is the Nernst effect. It is measured under the condition $\boldsymbol{j} = 0$. Taking $\nabla T = (\partial_x T, 0, 0)$ and $\boldsymbol{B} = (0, 0, B)$ the Nernst coefficient Q according to (19.1) is obtained as

$$Q = -\frac{E_y}{B \partial_x T} . \tag{19.5}$$

Representative experimental results for the Nernst coefficient of high-T_c superconductors are shown in Fig. 19.1. The normalized Nernst electric field $E_y/\partial T_x = QB$ shows a pronounced maximum in the superconducting state. Remarkably, QB is much larger in the superconducting than in the normal state. Comparison to the electrical resistivity measured on the same sample (see Fig. 19.1) shows clearly that the finite Nernst-voltage below T_c correlates with a finite resistivity. This demonstrates that both effects are due to moving vortices. The vanishing of ϱ and QB at low temperatures is due to the pinning of vortices. Results similar to the ones shown in Fig. 19.1 have been obtained for all major classes of high-T_c materials [3,4,16–23]. The characteristic behaviour of the Nernst effect due to vortex motion occurs also in conventional superconductors and has been studied there in detail [1,2].

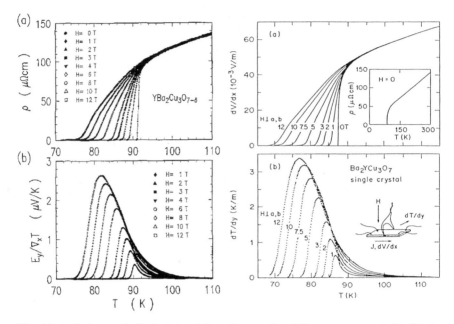

Fig. 19.1. Left panel: Resistivity (a) and normalized Nernst electric field (b) for an epitaxial c-axis oriented $YBa_2Cu_3O_{7-\delta}$ film at different magnetic fields $\boldsymbol{B} \parallel c$. Taken from Ri et al. [16]. Right panel: Longitudinal voltage gradient (a) and transverse temperature gradient $\partial_y T = \varepsilon B j_x$ (b) induced by an electric current of 14 A/cm^2 in a single crystal of $YBa_2Cu_3O_{7-\delta}$ for $\boldsymbol{B} \parallel c$. Insets: The zero-field resistivity up to 300K and the lead configuration. Taken from Palstra et al. [25].

The Nernst effect is a good probe for superconducting fluctuations above T_c, since the superconducting contribution to Q is much larger than the normal state contribution. Note that the data shown in Fig. 19.1 indeed signal a significant increase of QB already 10–20 K above T_c. We mention that an anomalously enhanced Nernst-voltage has been observed recently in the pseudogap regime of underdoped $La_{2-x}Sr_xCuO_4$ and has been interpreted as evidence for vortex-like excitations at temperatures much larger than T_c [24].

Ettingshausen Effect. An electrical current $\boldsymbol{j}$ leads to a heat current in the direction of $\boldsymbol{j} \times \boldsymbol{B}$. This is the Ettingshausen effect. It is usually measured by requiring that the total heat current along $\boldsymbol{j} \times \boldsymbol{B}$ vanishes ("adiabatic conditions"). In this case, a temperature gradient occurs along $\boldsymbol{j} \times \boldsymbol{B}$. Taking $\boldsymbol{j} = (j_x, 0, 0)$, $\boldsymbol{B} = (0, 0, B)$ and $j_{h,y} = 0$ we find from (19.1)

$$\varepsilon = \frac{\partial_y T}{B j_x}. \tag{19.6}$$

We show in Fig. 19.1 results on the Ettingshausen effect for a single crystal of $YBa_2Cu_3O_{7-\delta}$. Obviously, these results are very similar to those found for the Nernst coefficient. The transverse temperature gradient $\partial_y T$ is significantly enhanced in the superconducting state. Comparison with the resistive transition demonstrates that the Ettingshausen effect occurs only in the resistive regime, i.e. that it is intimately related to vortex motion. Similar to the Nernst effect, a large contribution to the Ettingshausen effect is present above T_c, due to superconducting fluctuations.

The data shown in Fig. 19.1 confirm the validity of the Bridgeman-relation (19.2): at $B = 10T$ and $T = 80K$ we find $Q = 0.25 \cdot 10^{-6}$ V/KT and $\varepsilon \simeq 0.2 \cdot 10^{-5}$ Km/TA. The thermal conductivity k measured on the same sample is $k \simeq 10$ W/Km [25]. These values yield $\varepsilon k \simeq QT \simeq 0.2 \cdot 10^{-4}$V/T.

19.3.2 Longitudinal Effects

Thermopower. In a temperature gradient a voltage develops parallel to ∇T. This is the well known Seebeck effect. The measurements are performed under the condition $\boldsymbol{j} = 0$. Then, taking $\nabla T = (\partial_x T, 0, 0)$, (19.1) yields

$$E_x = S\partial_x T . \tag{19.7}$$

We show, in the left panel of Fig. 19.2, the thermopower as a function of temperature in the mixed state of $(Bi,Pb)_2Sr_2Ca_2Cu_3O_{10+\delta}$. To the best of our knowledge these measurements provide the first observation of the thermopower in the mixed state of a type II superconductor. One observes a pronounced broadening of the superconducting transition, similar to what is found for the resistive transition. Similar results have been obtained for all major classes of high-T_c materials for polycrystals, single crystals

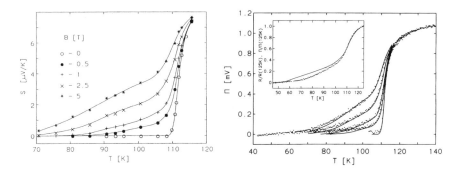

Fig. 19.2. Left panel: Thermopower of a polycrystal of $Bi_{1.76}Pb_{0.24}Sr_2Ca_2Cu_3O_\delta$ measured for $\nabla T \perp \boldsymbol{B}$ at various fixed magnetic fields given in the figure. After Galffy et al. [19]. Right panel: Peltier coefficient Π versus temperature for the same magnetic fields (increasing from the bottom to the top curve) and the same sample. Solid lines: Π as obtained from the thermopower via $\Pi = ST$. Inset: Π/Π_n and ϱ/ϱ_n (solid line) at $B = 5T$. For Π_n and ϱ_n the values measured at 120K have been used. Taken from Galffy et al. [26]

and epitaxial films [3,4,19,27,20–22,16,23]. For conventional superconductors there exist no data on the thermopower to our knowledge. The thermopower and the Nernst voltage are of comparable magnitude in the magnetic field range exploited experimentally. For example, for the sample of Fig. 19.2 the Nernst voltage is $QB \approx 1\mu V/K$ at $T = 90K$ and $B = 5T$, comparable in magnitude to the thermopower [19].

Peltier Effect. An electrical current j leads to a heat current $j_h \parallel j$. This is the Peltier effect. It is usually measured under the condition $j_h = 0$. Then, taking $j = (j_x, 0, 0)$, (19.1) yields for the Peltier coefficient:

$$\Pi = k \frac{\partial_x T}{j_x}. \qquad (19.8)$$

A measurement of the Peltier coefficient of $(Bi,Pb)_2Sr_2Ca_2Cu_3O_{10+\delta}$ is shown in Fig. 19.2. Similar to the thermopower and the resistivity (see inset of Fig. 19.2) the Peltier coefficient shows a strongly broadened superconducting transition in finite magnetic fields. Also shown in the figure is the product ST measured on the same sample. Apparently, $\Pi = ST$ as required by the Onsager–Kelvin relation holds within the experimental accuracy, i.e. the Onsager relations are valid, as expected. In contrast to the thermopower, a Peltier effect has been observed in conventional superconductors [28].

19.3.3 Thermal and Electrical Hall Angle

It is instructive to compare the electrical and the thermal Hall angles [19] defined in (19.3). The Nernst and Seebeck voltages are of comparable magnitude, yielding a thermal Hall angle $\tan\alpha_{th} = QB/S \simeq 1$. The electrical Hall angle $\tan\alpha_H$ of the HTSCs below T_c has also been studied carfully. It shows a complicated temperature and magnetic field dependence, which is not well understood [3]. However, in the same magnetic field range where $\tan\alpha_{th} \simeq 1$, the absolute magnitude of the electrical Hall angle is small, typically $\tan\alpha_H \approx 10^{-3}$, comparable to the value in the normal state. This implies that the Hall voltage is much smaller than the resistive voltage in this low field range. Thus, the experiment shows that

$$\tan\alpha_H \ll \tan\alpha_{th} \quad \text{for} \quad T < T_c. \qquad (19.9)$$

This result holds for the so called hydrodynamic limit of vortex motion (see below), in which the experiments were performed.

19.4 Implications for the Dynamics of Vortices

The experimental results presented in the last section give a clear picture for the dynamics of vortices in a temperature gradient and for the heat current associated with moving vortices.

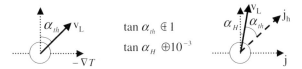

Fig. 19.3. Vortex motion in the hydrodynamic limit ($\omega_0\tau \ll 1$). The magnetic field is in the z direction perpendicular to the plane of the paper. Left panel: Vortex motion in a temperature gradient ∇T. $\boldsymbol{v}_\mathrm{L}$ is at the large angle $\alpha_\mathrm{th} \simeq 45°$ with respect to ∇T. Right panel: Current driven vortex motion. Typically $\tan\alpha_\mathrm{H} \ll 1$, so that $\boldsymbol{v}_\mathrm{L}$ is almost perpendicular to $\boldsymbol{j}$. Note that $\boldsymbol{v}_\mathrm{L}$ and $\boldsymbol{j}_h$ are not parallel. (See text)

The velocity $\boldsymbol{v}_\mathrm{L}$ of vortex motion in a temperature gradient can be obtained from the measured electrical field. Equation (19.1) with $\nabla T \neq 0$ and $\boldsymbol{j} = 0$ combined with (19.4) yields

$$\boldsymbol{E} = S\nabla T + Q\nabla T \times \boldsymbol{B} = \boldsymbol{B} \times \boldsymbol{v}_\mathrm{L} . \tag{19.10}$$

Solving for $\boldsymbol{v}_\mathrm{L}$ we find

$$\boldsymbol{v}_\mathrm{L} = \frac{1}{B}\left(S\nabla T \times \boldsymbol{z} - QB\nabla T\right) . \tag{19.11}$$

Here $\boldsymbol{z}$ is a unit vector in the direction of the magnetic field. Thus, the component of $\boldsymbol{v}_\mathrm{L}$ parallel to $(-\nabla T)$ is determined by the Nernst coefficient and that perpendicular to ∇T by the thermopower. Since the experimental data show that $S \approx QB$ in the low field regime, it follows that the components of $\boldsymbol{v}_\mathrm{L}$ parallel and perpendicular to ∇T are of the same order of magnitude, i.e. vortices move at an angle of the order 45° with respect to the applied temperature gradient. This is shown schematically in Fig. 19.3.

The heat current resulting from an applied electrical current (for $\nabla T = 0$) according to (19.1) and (19.2) is given by

$$\boldsymbol{j}_h = \Pi \boldsymbol{j} - \varepsilon k \boldsymbol{j} \times \boldsymbol{B} = T(S\boldsymbol{j} - QB\boldsymbol{j} \times \boldsymbol{z}) . \tag{19.12}$$

The heat current parallel to $\boldsymbol{j}$ is determined by the thermopower and that perpendicular to $\boldsymbol{j}$ by the Nernst coefficient. Both heat currents are of the same order of magnitude, since $S \approx QB$, so that the heat current is at a large thermal Hall angle with respect to the electrical current, as shown in Fig. 19.3. Also shown in Fig. 19.3 is the velocity $\boldsymbol{v}_\mathrm{L}$ of vortex motion. Since the electrical Hall angle is small, of the order $\tan\alpha_\mathrm{H} \approx 10^{-3}$, vortices move almost perpendicular to $\boldsymbol{j}$. Thus, the heat current is *not* parallel to $\boldsymbol{v}_\mathrm{L}$!

19.5 Origin of the Heat Current

In this section we discuss the origin of the heat current in a current driven resistive state. In a superconductor the electronic heat current is related to the spectrum of quasiparticle (QP) excitations, since the superfluid condensate does not carry heat. Therefore it is necessary to discuss first the excitation spectrum in the presence of vortices. We restrict our discussion to the case of an s-wave superconductor. Vortex motion in d-wave superconductors such as the HTSCs [29], taking into account the spectral flow effect, has been discussed in [30,31].

19.5.1 Excitation Spectrum in the Presence of Vortices

In an s-wave superconductor the order parameter is zero at the center of the vortex and approaches rapidly its equilibrium value Δ_∞ over distances of the order of the coherence length ξ. The region of suppressed order parameter defines the vortex core. The excitation spectrum may be divided into (1) bound excitations with energy $E < \Delta_\infty$ localized at the vortex core and (2) unbound excitations (scattering states), with energy $E > \Delta_\infty$, which spread throughout the bulk of the superconductor. These unbound excitations may be viewed as a gas of "normal excitations" (or a "normal fluid") similar to the case of normal excitations in zero magnetic field.

The spectrum $E_n(k_z, L_z)$ of low lying bound excitations in the vortex core has been discussed in detail in [32,33]. It may be classified with a discrete radial (and spin) quantum number n, an angular momentum L_z and the momentum projection $k_z = k_F \cos\alpha$ on the vortex axis. Here $L_z = m + 1/2$ with m integer. For $\alpha = \pi/2$ (i.e. for a two dimensional vortex) the spectrum is depicted as a function of the angular momentum L_z in Fig. 19.4a. This spectrum has a so called anomalous (chiral) branch, which is given by

$$E_0(k_z, \mu) = L_z \omega_0(k_z). \tag{19.13}$$

Here $\omega_0 \approx \Delta_\infty^2 / E_F$ is the interlevel distance of bound states. (We use units $\hbar = c = 1$.) At $T = 0$ all states with negative energy are occupied and all other states are empty. The level spacing is usually small, since $\Delta_\infty \ll E_F$. In this case the level broadening $\hbar/\tau$ resulting, for example, from the scattering of core excitations by free excitations outside the core may be of comparable magnitude so that the spectrum may be regarded as continuous. In fact, one distinguishes two limits [35], the collisionless and hydrodynamic limits, characterized by $\omega_0 \tau \gg 1$ and $\omega_0 \tau \ll 1$, respectively.

At low temperatures the bound excitations of the vortex core contribute to various physical properties, e.g. to the specific heat, which is expected to have an additional contribution linear in temperature and magnetic field (for an s-wave superconductor). The bound excitations and, in particular, the existence of an anomalous branch of the excitation spectrum have also important consequences for the thermomagnetic effects.

The bound states in the vortex core may be thought of as standing waves set up by Andreev reflected QPs [9,36]. Since this picture is helpful for an understanding of the spectral flow effect (see below), we discuss it in somewhat

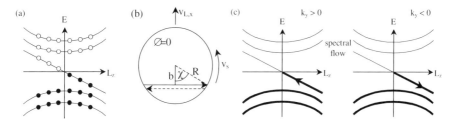

Fig. 19.4. (a) Spectrum of bound QP excitations in the vortex core of an s-wave superconductor. The filled circles show occupied states. In the limit $\omega_0\tau \ll 1$ the angular momentum L_z can be treated as a continuous parameter. (b) Schematic picture of the vortex core with radius R and $\Delta(r) = 0$ for $r < R$. A bound state may be viewed as a QP bounding back and forth with repeated Andreev reflection along the straight full (electron) and dotted (hole) lines shown in the figure. The energy of the bound state varies with the impact parameter b: it is zero for $b = 0$ and of the order Δ_∞ for $b \simeq R$. Motion of the vortex with velocity $v_{L,x}$ in the x-direction leads to a time dependence of the impact parameter and thus to spectral flow along the anomalous branch of the excitation spectrum (c). The rate of change of L_z in the moving vortex core is $\partial L_z/\partial t = v_x k_y$ (see (19.17) with $v_n = 0$). The spectral flow has opposite senses for $k_y > 0$ and $k_y < 0$, so that excitation momentum is created independent of the sign of k_y. (See text.) After Bevan et al. [34] and Stone [9].

more detail following the discussion presented in [9]. We first consider a one-dimensional situation. We use an order parameter $\Delta(x) = 0$ for $0 < x < L$, which defines a one-dimensional "vortex core", and $\Delta(x) = \Delta_\infty$ elsewhere. We take the phase difference of the order parameter to be $\Theta(x) = \Theta_l$ for $x < 0$ and $\Theta(x) = \Theta_r$ for $x > L$. Solving the corresponding one-dimensional Bogoluibov–de Gennes eigenvalue problem in the Andreev approximation, one finds for the energy of states deep in the gap, $E \ll \Delta_\infty$, that

$$E_n = \frac{v_F}{2L}\left(\Delta\Theta + 2\pi\left(n + \frac{1}{2}\right)\right). \tag{19.14}$$

Here $\Delta\Theta = \Theta_r - \Theta_l$. As shown in [9] it is possible to apply this result to a two-dimensional vortex core with radius R and $\Delta = 0$ inside the core. Each QP trajectory in the vortex core then coincides with a chord of a circle of radius $R \simeq \xi$ as shown in Fig. 19.4b. The length of the chord is given by $L = 2R\sin\chi$ and the difference in order parameter phase at the two ends resulting from the flow of supercurrent around the core is $\Delta\Theta = 2\chi$. Using (19.14) one obtains from this the energies $E_n(l)$ of core states with $E \ll \Delta_\infty$. Here $l = bk_F$ is determined by the impact factor b (see Fig. 19.4b). For $n = -1$ and small l this reduces to $E_{-1}(l) = -\omega_0 l$, equivalent to the anomalous branch of the bound state spectrum given by (19.13). Note that the spectrum is now classified by the impact parameter b. The zero energy bound state occurs at $b = 0$.

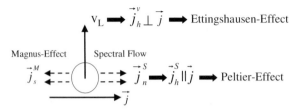

Fig. 19.5. Current driven vortex motion in the hydrodynamic regime. We neglect the finite but small Hall angle for simplicity, so that $\boldsymbol{v}_\mathrm{L} \perp \boldsymbol{j}$. The magnetic field is in the z direction. (See text).

We note that Andreev reflection of QPs in the vortex core is not perfectly retroreflective due to the presence of the finite supercurrent outside the core. Therefore, the orientation of a chord will not be definite, but it precesses with angular frequency ω_0 in the opposite sense to the superflow around the vortex. A detailed discussion of the bound state spectrum in a vortex core using the picture of Andreev reflection has been given in [9].

19.5.2 Heat Current

We discuss now the heat current in the case of moving vortices. We consider a current driven resistive state, as shown in Fig. 19.5. In the hydrodynamic limit, vortices move with velocity $\boldsymbol{v}_\mathrm{L}$ almost perpendicular to the applied electrical current $\boldsymbol{j}$. We neglect the small, but finite electrical Hall angle at this point. The heat current then has two contributions: one is due to the motion of vortices itself and the other arises from the spectral flow effect, which couples the motion of vortices to that of the normal fluid.

Vortex Heat Current

It is well known that vortices carry an entropy s_v (per unit length). This leads to a finite heat current $\boldsymbol{j}_h^v$ parallel to $\boldsymbol{v}_\mathrm{L}$ and therefore perpendicular to the electrical current. This heat current is unique to vortex motion and determines the transverse Ettingshausen and Nernst effects.

The transport entropy can be obtained experimentally from measurements of the Ettingshausen or Nernst coefficients and of the resistivity. This follows directly from (19.1) and (19.4). We take $\boldsymbol{j} = (j_x, 0, 0)$ and $\boldsymbol{B} = (0, 0, B)$. A vortex moving with velocity $v_{\mathrm{L},y}$ in the y-direction carries heat $s_v T$. The total heat current density in the y-direction is then given by

$$j_{h,y}^v = -s_v T n_v v_{L,y} = -s_v T \frac{B}{\Phi_0} \frac{\varrho}{B} j_x \,. \tag{19.15}$$

Here $n_v = B/\Phi_0$ is the vortex (areal) density and we have used that $E_x = v_{\text{L},y} B_z = \varrho j_x$, which results from (19.4) and the definition of the resistivity. Comparison of (19.15) with the transport equation (19.1) yields immediately

$$\frac{s_v}{\Phi_0} = \frac{\varepsilon k B}{T \varrho} = \frac{QB}{\varrho}. \tag{19.16}$$

In the last step (19.2) was used.

Experimentally it is found that s_v vanishes at $T = T_c$ (apart from a contribution due to superconducting fluctuations) [2–4]. It also must vanish for $T \to 0$. For the HTSCs typical values are in the range of $10^{-14} - 10^{-15}$ J/Km at $T = 0.95 T_c$ and $B = 1$ Tesla.

The existence of a finite transport entropy of vortices – resulting e.g. from the presence of low lying QP excitations in the vortex cores – is intuitively clear. Obviously s_v should be positive, since the vortex should have higher entropy than the superconducting surrounding. Theoretical results for the transport entropy have been derived on the basis of time dependent Ginzburg Landau theory. In these approaches it is usually found that s_v is related to the spatially averaged magnetization. However, it has been pointed out that the various theoretical approaches do not give a consistent and complete picture. For a detailed discussion the reader is referred to [37].

Spectral Flow

The second contribution to the heat current arises from the coupling of the normal fluid to the motion of vortices. This contribution is less obvious and its existence has been realized only recently [5]. It is related to the so called spectral flow effect [6–9,34]. The physical picture underlying this effect is the following: Let the vortex shown in Fig. 19.4b move with velocity $(\boldsymbol{v}_\text{L} - \boldsymbol{v}_n)$ with respect to the normal fluid. Here $\boldsymbol{v}_n$ is the velocity of the normal fluid. Then the impact parameter b and thus the angular momentum L_z of the QP excitations in the core change with time according to [8]

$$L_z(t) = L_z(0) + t\left[(\boldsymbol{v}_\text{L} - \boldsymbol{v}_n) \times \boldsymbol{k}\right] \cdot \boldsymbol{z}. \tag{19.17}$$

This leads to a flow of QPs from negative to positive levels of the spectrum $E_0(k_z, L_z)$ (see Fig. 19.4c). The occupied levels cross the zero of energy at a rate $\partial L_z/\partial t = (\boldsymbol{v}_\text{L} - \boldsymbol{v}_n) \cdot (\boldsymbol{k} \times \boldsymbol{z})$. When the energy of the QPs reaches Δ_∞ the QPs leave the vortex core in a direction perpendicular to $(\boldsymbol{v}_\text{L} - \boldsymbol{v}_n)$. Since each level carries momentum $\boldsymbol{k}$, this leads to a transfer of momentum from the moving vortex to the heat bath and thus to a force between the vortex and the normal fluid, given by

$$\boldsymbol{F}_S = -\pi C (\boldsymbol{v}_\text{n} - \boldsymbol{v}_\text{L}) \times \boldsymbol{z}. \tag{19.18}$$

The coefficient C depends on the regime of vortex dynamics and is given by [8]:

$$\frac{C}{n} = 1 - \frac{\omega_0^2 \tau^2}{1 + \omega_0^2 \tau^2} \tanh\left(\frac{\Delta(T)}{2 k_\text{B} T}\right). \tag{19.19}$$

In the hydrodynamic limit ($\omega_0\tau \ll 1$) spectral flow is fully active and one obtains $C/n \approx 1$ and $(n-C)/n \simeq (\Delta_\infty/E_\mathrm{F})^2 \ll 1$ [8,38]. In the collisionless limit characterized by $\omega_0\tau \gg 1$ (19.19) yields $C/n \to 0$ for $T \to 0$ and $C/n \approx 1$ for $T \to T_\mathrm{c}$. Spectral flow is suppressed in this limit with decreasing temperature and vanishes for $T \to 0$. Note that the suppression of the spectral flow in the collisionless limit may be attributed to the rotation of the QP bound states with angular velocity ω_0 (see above), which for $\omega_0\tau \gg 1$ restores the rotational symmetry, i.e. the QPs leave the vortex core in arbitrary directions so that the net momentum transferred to the normal fluid vanishes. The spectral flow force has recently been observed experimentally in ^{3}He [34,39].

According to these considerations a vortex moving with velocity $\boldsymbol{v}_\mathrm{L}$ generates a stream of QPs leaving the vortex core in a direction perpendicular to $\boldsymbol{v}_\mathrm{L}$. On the other hand, it is well known that a moving vortex generates also a phase slippage perpendicular to $\boldsymbol{v}_\mathrm{L}$ which accelerates the superfluid and therefore generates a supercurrent perpendicular to $\boldsymbol{v}_\mathrm{L}$. This is the Magnus effect (see e.g. [40]). Remarkably, this supercurrent induced by the Magnus effect is (almost) of equal magnitude but opposite to the normal current generated by the spectral flow. Therefore, if spectral flow is active, the total force on a vortex perpendicular to $\boldsymbol{v}_\mathrm{L}$ is almost zero and the electrical Hall angle is small. On the other hand, as has been pointed out in [5], the normal current generated by the spectral flow carries heat, whereas the supercurrent generated by the Magnus effect does not. Thus, there is a net heat current perpendicular to direction of vortex motion. This is illustrated in Fig. 19.5. The heat current is parallel to the electrical current, which drives the vortex motion, so that it gives rise to a longitudinal Peltier effect.

19.6 Analysis of Vortex Motion

After the discussion of the basic physical mechanisms leading to thermomagnetic effects in the mixed state of a superconductor we turn now to a quantitative discussion of vortex motion in a temperature gradient. We follow the discussion given in [5].

19.6.1 Forces on Vortices

We write the equation of motion for a vortex as:

$$0 = \boldsymbol{F}_\mathrm{M} + \boldsymbol{F}_\mathrm{I} + \boldsymbol{F}_\mathrm{S} + \boldsymbol{F}_\mathrm{th} + \boldsymbol{F}_d . \tag{19.20}$$

We discuss the forces in this equation briefly.

(i) $\boldsymbol{F}_\mathrm{M}$ and $\boldsymbol{F}_\mathrm{I}$ are the Magnus and Iordanskii forces, respectively, given by

$$\boldsymbol{F}_\mathrm{M} + \boldsymbol{F}_\mathrm{I} = \pi n_\mathrm{s}(\boldsymbol{v}_\mathrm{s} - \boldsymbol{v}_\mathrm{L}) \times \boldsymbol{z} + \pi n_\mathrm{n}(\boldsymbol{v}_\mathrm{n} - \boldsymbol{v}_\mathrm{L}) \times \boldsymbol{z} . \tag{19.21}$$

Here $\boldsymbol{v}_\mathrm{s}$ and $\boldsymbol{v}_\mathrm{n}$ are the velocities of the superfluid and normal components of the liquid, respectively, and n_s and n_n are the corresponding densities. The total density is given by

$$n = n_\mathrm{s} + n_\mathrm{n}. \qquad (19.22)$$

$\boldsymbol{F}_\mathrm{M}$ and $\boldsymbol{F}_\mathrm{I}$ describe the coupling of vortex motion to the superfluid and normal components of the liquid. They are well known for superconductors and superfluid He. A detailed discussion can be found in [40].

(ii) $\boldsymbol{F}_\mathrm{S}$ is the spectral flow force discussed in the preceding section [6–9]. It provides an additional coupling between the motion of vortices and that of the normal fluid.

(iii) $\boldsymbol{F}_\mathrm{th}$ in (19.20) describes the force on a vortex by a temperature gradient [41]. It is given by

$$\boldsymbol{F}_\mathrm{th} = -s_\mathrm{v}\nabla T. \qquad (19.23)$$

The existence of this force is well known from experiment [1–4]. It is due to the finite transport entropy s_v: If a moving vortex transports entropy it experiences, vice versa, a force in a temperature gradient.

(iv) $\boldsymbol{F}_d$ is a dissipative friction force, given by

$$\boldsymbol{F}_d = D(\boldsymbol{v}_\mathrm{n} - \boldsymbol{v}_\mathrm{L}). \qquad (19.24)$$

The friction coefficient D is of the order [42–46]

$$\frac{D}{n} \simeq \frac{\omega_0 \tau}{1 + \omega_0^2 \tau^2}. \qquad (19.25)$$

Therefore $D/n \ll 1$ in both, the hydrodynamic and the collisionless limits.

We assume that vortices move freely in response to the driving forces. The influence of pinning will be discussed below.

Summing all contributions, the equation of motion may be written as

$$\pi n_\mathrm{s} \boldsymbol{v}_\mathrm{s} \times \boldsymbol{z} + (\gamma - \pi n_\mathrm{s})\boldsymbol{v}_\mathrm{n} \times \boldsymbol{z} + D\boldsymbol{v}_\mathrm{n} - s_\mathrm{v}\nabla T = D\boldsymbol{v}_\mathrm{L} + \gamma \boldsymbol{v}_\mathrm{L} \times \boldsymbol{z}, \qquad (19.26)$$

where $\gamma = \pi(n - C)$. Note that $\gamma \ll 1$ in the hydrodynamic limit and $\gamma \approx \pi n$ in the collisionless limit at low temperatures, where spectral flow is suppressed.

19.6.2 Current Driven Vortex Motion

Expressions for the resistivity and the Hall angle are obtained from (19.26) by requiring $\nabla T = 0$ and $\boldsymbol{v}_\mathrm{n} \approx 0$ as usual for current driven vortex motion. Then, solving (19.26) for $\boldsymbol{v}_\mathrm{L}$ and using $\boldsymbol{E} = \boldsymbol{B} \times \boldsymbol{v}_\mathrm{L} = \varrho \boldsymbol{j}_\mathrm{s} - \varrho_\mathrm{H} \boldsymbol{j}_\mathrm{s} \times \boldsymbol{z}$ with $\boldsymbol{j}_\mathrm{s} = n_\mathrm{s} e \boldsymbol{v}_\mathrm{s}$ we find:

$$\varrho = \frac{B\Phi_0 D}{D^2 + \gamma^2} \quad \text{and} \quad \varrho_\mathrm{H} = \frac{B\Phi_0 \gamma}{D^2 + \gamma^2}. \qquad (19.27)$$

The Hall angle α_H is given by $\tan\alpha_H = \gamma/D$. In the hydrodynamic limit $\tan\alpha_H$ is small. As discussed above the spectral flow force almost "cancels" the $\boldsymbol{v}_L$-dependent part of the Magnus (and Iordanskii) force, and vortices move approximately at right angles with respect to the supercurrent. In contrast, in the collisionless limit at low temperatures spectral flow is suppressed yielding $\gamma \approx \pi n \gg D$ and we obtain $\varrho \to 0$ and $\varrho_H \simeq B\Phi_0/\pi n$. In this limit vortices move with the superfluid, i.e. $\boldsymbol{v}_L \simeq \boldsymbol{v}_s$.

19.6.3 Vortex Motion in a Temperature Gradient

Experiments on vortex dynamics in a temperature gradient are usually performed such that $\nabla T \neq 0$ and $\boldsymbol{j} = \boldsymbol{j}_s + \boldsymbol{j}_n = 0$. It is, however, important to realize that although the total current $\boldsymbol{j}$ vanishes the components $\boldsymbol{j}_s$ and $\boldsymbol{j}_n$ are finite in general if $\nabla T \neq 0$. The reason is that a temperature gradient sets up a normal QP current [10,47]

$$\boldsymbol{j}_n = en_n\boldsymbol{v}_n = -\mathcal{L}_{QP}\nabla T, \tag{19.28}$$

which is compensated by a supercurrent $\boldsymbol{j}_s = -\boldsymbol{j}_n$. Here $\mathcal{L}_{QP}$ is the thermoelectric coefficient of the QPs, related to the resistivity (ϱ_{QP}) and the thermopower (S_{QP}) of the QPs by $\mathcal{L}_{QP} = S_{QP}/\varrho_{QP}$. The compensation of $\boldsymbol{j}_s$ and $\boldsymbol{j}_n$ is well established for a superconductor in a temperature gradient *without* an applied magnetic field. However, in the mixed state the counterflow model must be extended, taking into account the inhomogeneous situation [48]. The coefficient $\mathcal{L}_{QP}$ in a superconductor for elastic scattering is given by [10,47]

$$\mathcal{L}_{QP} = \mathcal{L}_n G\left(\frac{\Delta}{k_B T}\right) \simeq \mathcal{L}_n \frac{n_n(T)}{n}, \tag{19.29}$$

where $\mathcal{L}_n = S_n/\varrho_n$. Here S_n and ϱ_n are the normal state thermopower and resistivity, respectively. The function $G(\Delta/k_B T)$ is the same function which governs the behaviour of the electronic thermal conductivity in the superconducting state. Its main temperature variation is due to the decrease of the numbers of QPs with decreasing temperature (at least in the case of elastic impurity scattering) [10]. From (19.28) and (19.29) and using that $\boldsymbol{j} = 0$ requires $n_n\boldsymbol{v}_n = -n_s\boldsymbol{v}_s$ we obtain

$$\boldsymbol{v}_n = -\frac{\mathcal{L}_n}{en}\nabla T = -\frac{n_s}{n_n}\boldsymbol{v}_s. \tag{19.30}$$

Inserting this into (19.26) we obtain the equation of motion of a vortex in a temperature gradient:

$$\Phi_0\mathcal{L}_n\left(1 - \frac{\gamma}{\pi n}\right)\nabla T \times \boldsymbol{z} - \left(s_v + \frac{D\mathcal{L}_n}{en}\right)\nabla T = D\boldsymbol{v}_L + \gamma\boldsymbol{v}_L \times \boldsymbol{z}. \tag{19.31}$$

Using $\gamma = \pi(n-C)$ one obtaines $\Phi_0\mathcal{L}_n(C/n)\nabla T \times \boldsymbol{z}$ for the force perpendicular to ∇T. Apparently, the spectral flow plays the central role for this force.

In particular, the force is absent in the collisionless limit at low temperatures, where $C \to 0$. Note, however, that in this limit the Hall-term $\propto \gamma \boldsymbol{v}_\mathrm{L} \times \boldsymbol{z}$ is large, since the Magnus effect is not compensated any more by the spectral flow.

Solving (19.31) for $\boldsymbol{v}_\mathrm{L}$ and using (19.10) we obtain for the thermopower

$$S = \varrho \frac{S_\mathrm{n}}{\varrho_\mathrm{n}} + \tan \alpha_\mathrm{H} \varrho \frac{s_\mathrm{v}}{\varPhi_0}, \qquad (19.32)$$

where (19.27) has been used. The interpretation is straightforward: the first contribution to S results from the motion of the normal fluid along the temperature gradient. The second contribution is due to the heat carried by the vortices which has a component perpendicular to ∇T because of the finite Hall angle α_H. For the Nernst coefficient we find

$$QB = \varrho \left(\frac{s_\mathrm{v}}{\varPhi_0} + \frac{D}{\pi n} \frac{S_\mathrm{n}}{\varrho_\mathrm{n}} - \tan \alpha_\mathrm{H} \frac{C}{n} \frac{S_\mathrm{n}}{\varrho_\mathrm{n}} \right). \qquad (19.33)$$

The three contributions to Q have the following origin: The first term arises from the motion of the vortex along the temperature gradient $\propto s_\mathrm{v} \nabla T$. The second term arises since the vortices have a component of motion perpendicular to the temperature gradient due to the thermopower: the coupling of this motion to the normal fluid via $D\boldsymbol{v}_\mathrm{n}$ drags normal fluid in this direction and therefore contributes to the transverse voltage. The third term is the Nernst effect of the normal fluid, as is apparent from comparison with the result for the thermopower.

In order to derive results appropriate for the hydrodynamic and collisionless limits we estimate the magnitude of the various contributions to S and QB. Using $s_\mathrm{v} \approx 10^{-14} - 10^{-13}$ J/Km [2–4]. We find $s_\mathrm{v}/\varPhi_0 \approx 10 - 100$ A/Km. Using $S_\mathrm{n} \approx 1\mu$V/K and $\varrho_\mathrm{n} \approx 10^{-2} \mu\Omega$m we obtain $S_\mathrm{n}/\varrho_\mathrm{n} \approx 100$ A/Km, comparable in magnitude to $s_\mathrm{v}/\varPhi_0$. D/n is small in both the hydrodynamic and the collisionless limits (see (19.25)).

Hydrodynamic Limit. In this limit $\tan \alpha_\mathrm{H} \ll 1$ and we find

$$\frac{S}{S_\mathrm{n}} \simeq \frac{\varrho}{\varrho_\mathrm{n}} \qquad (19.34)$$

$$QB \simeq \varrho \frac{s_\mathrm{v}}{\varPhi_0}. \qquad (19.35)$$

Equation (19.34) agrees very well with experimental results on high-T_c superconductors [3,4,19,27,20,22,16,23,26,48]. (see Fig. 19.6). Minor deviations are related to the finite Hall angle [49]. Equation (19.35) is the usual expression relating the Nernst coefficient, the flux flow resistivity and the transport entropy (see (19.16)).

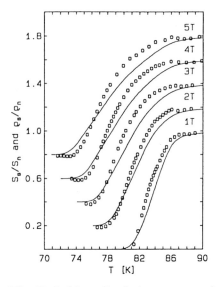

Fig. 19.6. Normalized thermopower S/S_n (open circles) and normalized resistivity ϱ/ϱ_n (solid lines) versus temperature for a c-axis oriented epitaxial film of YBa$_2$Cu$_3$O$_{7-\delta}$. $S_n(T < T_c)$ and $\varrho_n(T < T_c)$ were extrapolated from their normal state behaviour. The curves for different magnetic fields are shifted by 0.2 along the y-axis. Taken from Hohn et al. [20].

Collisionless limit. New results are obtained in this limit at low temperatures, where the spectral flow is suppressed so that $\gamma \approx \pi n$. This yields $\tan \alpha_H \gg 1$ and we obtain

$$S \simeq \tan \alpha_H\, QB \simeq \tan \alpha_H\, \varrho \frac{s_v}{\Phi_0} = \frac{B s_v}{\pi n} \qquad (19.36)$$

Apparently, the absence of spectral flow at low temperatures leaves the heat carried by the moving vortices as the only source of thermomagnetic effects. Note, however, that, since $\tan \alpha_H \gg 1$, vortices move almost perpendicular to ∇T and the thermopower is larger than the Nernst coefficient.

It is instructive to consider also the heat flow in this limit in the case of current driven vortex motion. The heat current associated with the moving vortices is given by $\boldsymbol{j}_h^v = n_v s_v T \boldsymbol{v}_L$ (compare to (19.15)). At low temperatures we have $n_s \approx n$ and $\boldsymbol{v}_L \approx \boldsymbol{v}_s = \boldsymbol{j}_s/ne$. This yields

$$\boldsymbol{j}_h^v \approx \frac{B s_v T}{\pi n} \boldsymbol{j}_s = \Pi \boldsymbol{j}_s\,, \qquad (19.37)$$

where Π is the Peltier coefficient. Comparison to (19.36) shows that $\Pi = ST$, as is required by the Onsager relations [10].

19.6.4 Pinning Effects

In HTSCs pinning effects are weak above the so called irreversibility line and vortex motion induced by a temperature gradient occurs in a broad range of temperatures and magnetic fields. In contrast, in conventional superconductors pinning effects are usually much more pronounced so that thermomagnetic effects have been observed only in a narrow temperature range (see e.g. [1,2]). We expect that the main effect of pinning is a reduction of the flux flow resistivity, which leads to a corresponding decrease of the Nernst and Seebeck voltages according to (19.34) and (19.35). On the other hand, the transport entropy, the thermal Hall angle as well as the ratio of S and ϱ should be unaffected by pinning. This is consistent with recent experimental results [50].

19.7 Summary

In summary, it is experimentally established that the motion of vortices in type II superconductors leads to a variety of thermomagnetic effects. The transverse Nernst and Ettingshausen effects are characteristic features of vortex motion and are determined by the entropy carried by moving vortices. The longitudinal Peltier effect and the thermopower arise from the spectral flow associated with moving vortices, which couples the motion of vortices to that of the normal fluid and gives rise to a heat current perpendicular to the direction of vortex motion. An analysis of vortex motion in a temperature gradient on this basis yields excellent agreement with experimental results.

We thank R.P. Huebener, A.P. Kampf, P. Kes, R. Kümmel, A. von Otterlo, D. Rainer, M. Stone, P. Wölfle, M. Zirnbauer, and M. Zittartz for useful discussions. We also thank H. Micklitz and M. Zirnbauer for a critical reading of the manuscript and J. Baier for help in preparing the figures. This work was supported by the Deutsche Forschungsgemeinschaft through SFB 341.

References

1. Y.B. Kim, M.J. Stephen: 'Flux Flow and Irreversible Effects'. In *Superconductivity, Vol.2*, ed. by R.D. Parks, (Marcel Dekker, New York 1969), pp. 1107–1165
2. R.P. Huebener: *Magnetic Flux Structures in Superconductors*, (Springer Verlag, Berlin, 1979)
3. A. Freimuth: 'Transport Properties in the Mixed State of High Temperature Superconductors'. In: *Superconductivity, Frontiers in Solid State Sciences Vol. 1*, ed. by L.C. Gupta and M.S. Multani, (World–Scientific, Singapore 1992), pp. 393–449
4. R.P. Huebener: Supercond. Sci. Technol. **8**, 189 (1995)
5. A. Freimuth, M. Zittartz: Phys. Rev. Lett. **84**, 4978 (2000)
6. G.E. Volovik: JETP **77**, 435 (1993)

7. G.E. Volovik: JETP Lett. **57**, 244 (1993)
8. N.B. Kopnin, G.E. Volovik, Ü. Parts: Europhy. Lett. **32**, 651 (1995)
9. M. Stone: Phys. Rev. B **54**, 13222 (1996)
10. A.A. Abrikosov: *Fundamentals of the Theory of Metals* (North-Holland, 1988)
11. More precisely, $\boldsymbol{E}$ denotes the gradient of the electrochemical potential.
12. A third Hall angle α_R (also named "thermal Hall angle") is defined using the thermal conductivity and the thermal Hall effect as $\tan \alpha_R = k_H/k$.
13. G. Blatter, M.V. Feigelman, V.B. Geshkenbein, A.I. Larkin, V.M. Vinokur: Rev. Mod. Phys. **66**, 1125 (1994)
14. M. Tinkham: *Introduction to Superconductivity*, 2nd edn (McGraw-Hill, 1996)
15. Due to the strong anisotropy of the HTSCs results on polycrystals are also representative for this configuration.
16. H.-C. Ri, R. Gross, F. Gollnik, A. Beck, R.P. Huebener: Phys. Rev. B **50**, 3312 (1994)
17. M. Zeh, H.-C. Ri, F. Kober, R.P. Huebener, A.V. Ustinov, J. Mannhart, R. Gross, A. Gupta: Phys. Rev. Lett. **64**, 3195 (1990)
18. S.J. Hagen, C.J. Lobb, R.L. Greene, M.G. Forrester, J. Talvacchio: Phys. Rev. B **42**, 6777 (1990)
19. M. Galffy, A. Freimuth, U. Murek: Phys. Rev. B **41**, 11029 (1990)
20. C. Hohn, M. Galffy, A. Dascoulidou, A. Freimuth, H. Soltner, U. Poppe: Z. Phys. B **85**, 161 (1991)
21. N.V. Zavaritsky, A.V. Samoilov, A.A. Yurgens: Physica C **180**, 417 (1991)
22. A. Dascoulidou, M. Galffy, C. Hohn, N. Knauf, A. Freimuth: Physica C **201**, 202 (1992)
23. C. Hohn, M. Galffy, A. Freimuth: Phys. Rev. B **50**, 15875 (1994)
24. Z.A. Xu, N.P. Ong, Y. Wang, T. Kakeshita, S. Uchida: Nature **406**, 486 (2000)
25. T.T.M. Palstra, B. Batlogg, L.F. Schneemeyer, J.V. Waszczak: Phys. Rev. Lett. **64**, 3090 (1990)
26. M. Galffy, Ch. Hohn, A. Freimuth: Ann. Phys. **3**, 215 (1994)
27. H.-C. Ri, F. Kober, R. Gross, R.P. Huebener, A. Gupta: Phys. Rev. B **43**, 13739 (1991)
28. A.T. Fiory, B. Serin: Phys. Rev. Lett. **19**, 227 (1966)
29. For a recent review see e.g. C.C. Tsuei, J.R. Kirtley: Rev. Mod. Phys. **72**, 969 (2000)
30. N.B. Kopnin, G.E. Volovik: Phys. Rev. Lett. **79**, 1377 (1997)
31. Yu.G. Makhlin: Phys. Rev. B **56**, 11872 (1997)
32. C. Caroli, P.G. de Gennes, J. Matricon: Phys. Lett. **9**, 307 (1964)
33. C. Caroli, J. Matricon: Phys. kondens. Mat. **3**, 380 (1965)
34. T.D.C. Bevan, A.J. Manninen, J.B. Cook, J.R. Hook, H.E. Hall., T. Bachaspate, G.E. Volovik: Nature **386**, 689 (1997)
35. In superconductors actually 3 limits are distinguished characerized by ω_0 and ω_c, where ω_c is the cyclotron frequency. Note that $\omega_c \ll \omega_0$ since $B \ll H_{c2}$.
36. D. Rainer, J.A. Sauls, D. Waxman: Phys. Rev. B **54**, 10094 (1996) and references therein
37. N.B. Kopnin: J. Low Temp. Phys. **93**, 117 (1993) and references therein
38. An additional contribution to $n-C$ of the same order of magnitude exists due to charging effects in the vortex core as discussed by D.I. Khomskii and A. Freimuth, Phys. Rev. Lett. **75**, 1384 (1995) and M.V. Feigelman et al.,Physica C **235–240**, 3127 (1994)

39. T.D.C. Bevan, A.J. Manninen, J.B. Cook, H. Alles, J.R. Hook, H.E. Hall: J. Low Temp. Phys. **109**, 423 (1997)
40. E.B. Sonin: Phys. Rev. B **55**, 485 (1997)
41. M.J. Stephen: Phys. Rev. Lett. **16**, 801 (1966)
42. N.B. Kopnin, V.E. Kravtsov: Pis'ma Zh. Eksp. Toer. Fiz. **23**, 631 (1976) (JETP Lett. **23**, 578 (1976))
43. N.B. Kopnin, M.M. Salomaa: Phys. Rev. B. **44**, 9667 (1991)
44. N.B. Kopnin: Phys. Rev. B. **47**, 14354 (1993)
45. N.B. Kopnin: Physica B **210**, 267 (1995)
46. N.B. Kopnin, A.V. Lopatin: Phys. Rev. B **51**, 15291 (1995)
47. V.L. Ginzburg, G.F. Zharkov: Sov. Phys. Usp. **21**, 381 (1978)
48. R.P. Huebener, A.V. Ustinov, V.K. Kaplunenko: Phys. Rev. B **42**, 4831 (1990)
49. H.-C. Ri, F. Kober, A. Beck, L. Alff, R. Gross, R.P. Huebener: Phys. Rev. B **47**, 12312 (1993)
50. T.W. Clinton, Wu Liu, X. Jiang, A.W. Smith, M. Rajeswari, R.L. Greene, C.J. Lobb: Phys. Rev. B **54**, R9670 (1996)

20 Electric Field Dependent Flux–Flow Resistance in the Cuprate Superconductor $Nd_{2-x}Ce_xCuO_y$

R.P. Huebener

20.1 Introduction

The flux–flow resistance generated by current-induced vortex motion only results in linear resistance phenomena if the vortices do not change because of their motion. In this case the flux–flow resistance is independent of the vortex velocity, i.e., independent of the electric field. However, nonlinear resistive phenomena, such as hysteretic resistance steps, negative differential resistance and spontaneous resistance oscillations, can arise, if the vortices change due to their motion. An important mechanism leading to such nonlinearities is the shift of the quasiparticle energy distribution to higher energy in the presence of the electric field generated by vortex motion. Based on this mechanism there are different possible scenarios depending on the electronic vortex structure and on some details of the quasiparticle scattering. These scenarios are associated with specific temperature regimes.

At temperatures near the critical temperature T_c the quasiparticle energy smearing $\delta\varepsilon = \hbar/\tau$ is of the order of or larger than the superconducting energy gap Δ (dirty limit). Here $\hbar$ is Planck's constant devided by 2π, and τ is the quasiparticle scattering time. In this case features of the electronic vortex structure on an energy scale less than Δ are irrelevant. The shift of the quasiparticle energy distribution due to the electric field only involves the fact that the quasiparticles leave the vortex core region when they reach the energy Δ and that the vortex core shrinks. This shrinking of the vortex core due to the vortex motion has been treated theoretically by Larkin and Ovchinnikov (LO) [1]. They predicted flux–flow instabilities which have been experimentally observed both in classical and in high-temperature superconductors. References can be found in [2,3].

In the low-temperature limit, $T \ll T_c$, features of the electronic vortex structure on an energy scale less than the gap energy Δ can become important for the flux–flow resistance. Such features can develop in the following way. We note that the electronic structure of the core of an isolated vortex consists of the discrete energy levels ε_n of the quasiparticles given by

$$\varepsilon_n = \left(n + \frac{1}{2}\right)\frac{\Delta^2}{\varepsilon_F} = \left(n + \frac{1}{2}\right)\frac{2\hbar^2}{m\xi^2} \tag{20.1}$$

where n is an integer, ε_F the Fermi energy, and m the quasiparticle mass. ξ is the superconducting coherence length. The result (20.1) was first obtained by Caroli, De Gennes, and Matricon [4] by solving the corresponding Bogoliubov–De Gennes equations. If the quasiparticle energy smearing is less than Δ but still larger than the level distance $\varepsilon_{n+1} - \varepsilon_n = \Delta^2/\varepsilon_F$, we are in the quasiclassical limit. This limit is generally valid for $k_F \xi \gg 1$, which is also referred to as the "geometric optics limit" (k_F = Fermi wave vector). In this limit the flux–flow resistance was theoretically analysed also by LO [5]. They predicted a logarithmic singularity in the flux–flow resistance and the proportionality between the flux–flow voltage V and the current I

$$V \sim I/\ln\left(\frac{I^*}{I}\right). \tag{20.2}$$

The singularity appears for $I \to I^*$. The instability current I^* depends on the quasiparticle energy relaxation rate τ_ε^{-1} with the proportionality $I^* \sim \tau_\varepsilon^{-1/2}$ (τ_ε is the quasiparticle energy relaxation time). The physics behind this instability lies in the fact that the bound-state wave functions of the quasiparticles extend to larger distances from the vortex center with increasing energy ε_n. Since the higher energy levels become thermally more populated with increasing temperature, this leads to an effective vortex core radius which increases proportional to T. This temperature dependence of the vortex core radius has been predicted by Kramer and Pesch from solving the appropriate Eilenberger equations (Kramer–Pesch effect) [6,7]. According to LO [5], in the electric field generated by vortex motion at sufficiently high currents, the quasiparticle distribution function is not given any more by the bath temperature T, but instead by an effective temperature $T^\star > T$. The condition $I = I^\star$ for the appearance of the singularity in the flux–flow resistance is identical to the condition $k_B T^\star = \Delta$, where k_B is Boltzmann's constant. A more detailed discussion is given in [8].

Continuing our discussion of the low-temperature limit, an interesting case arises if the quasiparticle density of states (DOS) shows a step structure as a function of energy between the Fermi energy and the gap energy. Since the DOS provides the available phase space for quasiparticle scattering, the electric field induced shift of the quasiparticle distribution to higher energies can result in a strongly electric field dependent resistance. For such an effect to become important, the quasiparticle energy smearing $\delta\varepsilon = \hbar/\tau$ must remain smaller than the energy scale of the quasiparticle DOS step structure. Another interesting possibility for a strongly field dependent resistance is the development of subbands between the Fermi energy and the gap energy. We note that this can be looked at as a special case of the more general case of an energy-dependent quasiparticle DOS. However, in the second case we deal with the quantum limit ($k_F \cdot \xi \approx 1$).

In this chapter we discuss recent flux–flow resistance measurements performed in the low-temperature limit with the electron-doped cuprate superconductor $Nd_{2-x}Ce_xCuO_y$ (NCCO). At $T \ll T_c$ the flux–flow resistance in NCCO shows an intrinsic electric field dependence resulting in voltage steps or negative differential resistance and spontaneous resistance oscillations, depending on the bias conditions. Noting that the quasiparticle energy destribution is shifted upwards in the electric field generated by vortex motion, we conclude that the quasiparticle density of states, providing the phase space for scattering, must be strongly energy dependent. We discuss the possibility of subbands between the Fermi energy and the gap energy, originating from the Andreev bound states in the core of an isolated vortex due to the interaction between vortices. We also discuss the more general possibility of a characteristic step structure in the energy dependence of the quasiparticle density of states for explaining our experimental results. The symmetry of the pair wave function in NCCO is expected to be crucial for the development of a satisfactory model.

This chapter is organized as follows: In Sect. 20.2 we summarize the theoretical concepts dealing with the damping of the vortex motion, and we emphasize specifically the role played by the quasiparticle scattering. In Sect. 20.3 we briefly present the results of the flux–flow resistance measurements performed recently with NCCO at temperatures $T \ll T_c$, demonstrating the highly nonlinear behaviour. Experimental results on the nucleation and growth of a high-electric-field domain are discussed in Sect. 20.4. Section 20.5 contains a general model discussion of the shifted nonequilibrium quasiparticle distribution and of the expected behaviour if this shift leads to an electric field dependence of the resistivity. In the following section we discuss our model based on the appearance of subbands between the Fermi energy and the gap energy. This model looks attractive if s-wave symmetry of the pair wave function prevails in NCCO. In Sect. 20.7 we investigate another more general model in which a pronounced (step-like) structure in the energy dependence of the quasiparticle DOS between the Fermi energy and the gap energy is assumed. This second model looks more attractive in the case of d-wave symmetry of the pair wave function in NCCO. Section 20.8 contains the summary and conclusions.

20.2 Damping of the Vortex Motion

The vortex motion effected by the electric current density j is governed by the phenomenological force equation

$$j \times \varphi_0 - \eta v_\varphi - \alpha(v_\varphi \times n) = 0 . \tag{20.3}$$

From left to right the three terms describe the Lorentz force, the damping force, and the Hall force, respectively. Here, the forces are given per unit length of vortex line. For simplicity we have neglected flux pinning. φ_0 is the magnetic flux quantum, v_φ the vortex velocity, and n a unit vector in the direction of the magnetic field B. η and α are damping coefficients. We assume that j is applied in x-direction and B in z-direction. Equation (20.3) has been proposed many years ago by Vinen and Warren [9]. A microscopic derivation has been presented by Kopnin and coworkers [10–12]. A simple heuristic argument for (20.3) has been given by Blatter and Ivlev [13]. For η and α one obtains the expressions [10–13]

$$\eta = \frac{\pi \hbar n^\star \omega_0 \tau}{1 + (\omega_0 \tau)^2} \tag{20.4}$$

and

$$\alpha = -\frac{\pi \hbar n^\star (\omega_0 \tau)^2}{1 + (\omega_0 \tau)^2} \ . \tag{20.5}$$

Here $n^\star$ is the charge carrier concentration. $\hbar \omega_0$ is the level spacing in the core according to (20.1): $\hbar \omega_0 = \Delta^2 / \varepsilon_F$.

The electric field $\boldsymbol{F}$ generated by vortex motion is given by

$$\boldsymbol{F} = -\boldsymbol{v}_\varphi \times \boldsymbol{B} \ . \tag{20.6}$$

The x-component of the force equation (20.3) yields the Hall angle θ

$$tg\,\theta = \frac{F_y}{F_x} = \frac{v_{\varphi x}}{v_{\varphi y}} = -\frac{\alpha}{\eta} = \omega_0 \cdot \tau \ . \tag{20.7}$$

From the y-component we obtain the flux-flow resistivity

$$\rho_f = \frac{F_x}{j_x} = \frac{\varphi_0 B}{\eta \left(1 + \frac{\alpha^2}{\eta^2}\right)} \ . \tag{20.8}$$

For strong electron scattering ($\omega_0 \tau \ll 1$), the Hall angle is small, and the vortices move nearly perpendicular to the current. In the opposite limit ($\omega_0 \tau \gg 1$) the Hall angle approaches $\pi/2$, the vortices move nearly parallel to the current, and the dissipation is strongly reduced. In this case an effective damping coefficient $\eta_{\text{eff}} = \eta \left(1 + \alpha^2/\eta^2\right)$ appears in (20.8), which is much larger than the value of η valid in the limit $\omega_0 \tau \ll 1$. We emphasize that, according to the force equation (20.3), large values of the damping coefficients η and/or α always lead to small values of the electric field because of (20.6).

The expressions for the damping coefficients η and α in (20.4) and (20.5), respectively, show the role of the quasiparticle scattering time τ. Since in the cuprate superconductors this scattering time is dominated by the electron-electron interaction [14–16], the quasiparticle DOS in the mixed state providing the available phase space for scattering will be crucial. In Sects. 20.6 and 20.7 we will discuss how, in particular, the energy dependence of the quasiparticle DOS can be invoked for explaining the observed electric field dependence of the flux-flow resistance in NCCO.

20.3 Flux–Flow Resistance Measurements in $Nd_{2-x}Ce_xCuO_y$

Recently, an unexpected intrinsic electric field dependence of the flux–flow resistance in thin films of the electron-doped cuprate superconductor NCCO has been discovered [17]. Under current bias the voltage-current characteristics (VIC's) display hysteretic voltage steps. Depending upon the magnetic field, one or two steps are observed. On the other hand, under quasi voltage bias negative differential resistance (NDR) and spontaneous resistance oscillations appear [18–20]. The experiments were performed with epitaxial c-axis oriented NCCO films (close to optimal doping) with 90–100 nm thickness. The magnetic field was oriented along the c-axis. During most of the measurements the samples were immersed in liquid helium (below the temperature $T_\lambda = 2.17$ K in superfluid helium) in order to minimize effects from Joule heating. The critical temperature of the samples investigated were in the range $T_c = 21.3$–24.0 K. Further experimental details are given in [17,18].

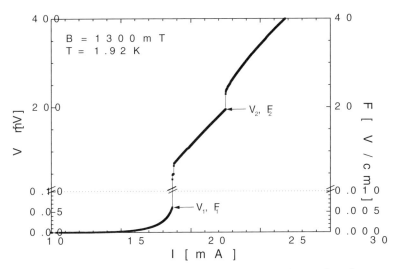

Fig. 20.1. Typical example of the two voltage steps observed at the onset voltages V_1 and V_2, respectively, for current bias with increasing current. The corresponding electric field is indicated on the right vertical axis. Note the change of scale on the voltage axis at $V = 0.10$ mV. $T = 1.92$ K, $B = 1300$ mT.

In Fig. 20.1 we show a typical VIC measured for *current bias* with increasing current at $T = 1.92$ K and $B = 1300$ mT. Two steps can be seen in the VIC at the onset voltages V_1 and V_2, respectively. At 1.92 K, depending on the magnetic field, the voltages V_1 were in the range of 10–400 µV, corresponding to the electric field range $F_1 = 1$–40 mV/cm for the typical sample length $L = 100$ µm. The voltages V_2 were about thousand times higher, $V_2 =$

50–100 mV, corresponding to electric fields $F_2 = 5$–10 V/cm. Between both voltage steps a nearly straight section of the VIC can be seen. The slope of this straight section is denoted by R_{SS}. The magnetic field and temperature dependence of the threshold voltages V_1 and V_2 observed under current bias with increasing current can be summarized as follows [17,18, 21–23]:

(a) V_1 increases nearly exponentially with B: $V_1 \sim \exp(B/B_0)$,
(b) V_2 increases only weakly with increasing B,
(c) At $T = 1.92$ K the appearance of two steps is restricted to the magnetic-field range of about 950–1850 mT. At lower (higher) fields only the lower (higher) step is observed,
(d) V_1 shows thermally activated behaviour with $V_1 \sim [\exp(\varepsilon_A/k_BT) + 1]^{-1}$ and $\varepsilon_A = 0.42$–0.37 meV for magnetic fields in the range $B = 300$–1300 mT and temperatures below T_λ (where heating effects are strongly suppressed),
(e) V_2 shows the proportionality $V_2 \sim [1 - \exp(-\varepsilon_\Delta/k_BT)]$ with $\varepsilon_\Delta = 2.0$–2.2 meV for magnetic fields ranging between 1000 and 2600 mT and temperatures below T_λ,
(f) the slope R_{SS} is about 1.6 times the normal state resistance and nearly independent of B,
(g) below T_λ the slope R_{SS} is independent of T.

Because of the appearance of the voltage steps in the VIC's under current bias, under *quasi voltage bias* we expect NDR and spontaneous resistance oscillations. The latter effects have also been studied [18–20]. A typical series of VIC's measured at 1.92 K and magnetic fields ranging between 1000 and 3000 mT under current bias for *increasing current* and under quasi voltage bias is shown in Fig. 20.2. We see that the voltage steps for the current bias originate exactly at the *lower* end of the NDR regions observed for quasi voltage bias. In the same way, for *decreasing current* the voltage steps for current bias coincide exactly with the *upper* end of the NDR region. For clarity, the curves for decreasing current are not included in Fig. 20.2. However, the hysteretic behaviour is indicated schematically in the inset. We emphasize that perfect voltage bias (with a horizontal load line and zero load resistance in the plots of Fig. 20.3) is impossible because of the finite load resistance. Because of this limitation, the NDR branch associated with the lower voltage step could not be observed separately.

20.4 Nucleation and Growth of High-Electric-Field Domains

The hysteretic step structure of the VIC's observed under current control and the NDR branch found under quasi voltage bias clearly indicate the appearance of a different state of the sample with a higher electric resistivity

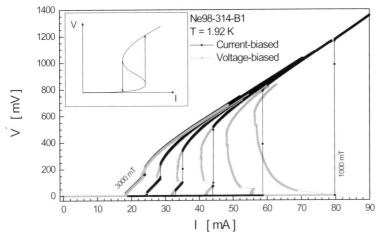

Fig. 20.2. Voltage plotted versus current under current bias for increasing current (*black color*) and under quasi voltage bias for increasing voltage (*grey color*). $T = 1.92$ K. From right to left the magnetic flux density is as follows: 1.00, 1.40, 1.80, 2.20, 2.60, 3.00 T. The hysteresis is indicated schematically in the inset.

than the previous state. The underlying mechanism of this transition is expected to be associated with the electric-field dependence of the resistivity, as we will discuss in Sect. 20.5. First, we turn to some spatially resolved resistance measurements [20]. The spatial resolution was achieved by fabricating a sample geometry with four voltage leads placed along the bridge. The inner distance between two neighboring leads was 100 μm. The total sample length between the two outer leads was 360 μm. The width and thickness of the sample was 40 μm and 90–100 nm, respectively. The geometry of this sample is shown schematically in the inset of Fig. 20.3. A typical result is presented in the main part of Fig. 20.3 for $T = 1.92$ K and $B = 800$ mT oriented in c-direction. The quasi voltage control was applied to the total sum of the voltages of the three sample sections in series. The nucleation of a high-electric-field domain is seen to start in section 12. In this section the VIC first displays the negative slope of the load line and above about 50 mV turns upwards with a steeper negative slope. Up to the value $V_{12} \approx 150$ mV the voltages in the other two sample sections remain undetectable on this plot. (We will discuss the results obtained at much higher voltage resolution further below.) At $I \approx 54$ mA section 23 shows the first appearance of resistance, leading to a kink in the $V_{12}(I)$ curve due to the overall voltage bias. In the current range below $I \approx 53.5$ mA the curve $V_{23}(I)$ develops a steep negative slope simultaneously with the appearance of a minimum of $V_{12}(I)$. Section 34 becomes resistive below about $I = 52$ mA. Simultaneously with the steep negative slope of $V_{34}(I)$ a second minimum in $V_{12}(I)$ and a kink in $V_{23}(I)$ appears. The detailed pattern of all three curves in Fig. 20.3 can be

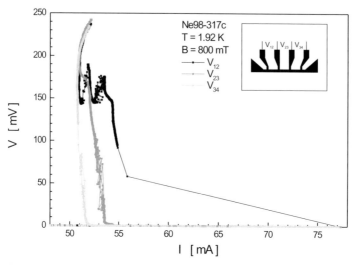

Fig. 20.3. Spatially resolved voltage measurement: Voltages V_{12} (*black color*), V_{23} (*dark grey color*), and V_{34} (*light grey color*) of the three sample sections as shown in the inset plotted versus current for quasi voltage bias applied to the total sample length (voltage increasing). $T = 1.92$ K, $B = 800$ mT.

well understood from the overall voltage bias of the sum of the three voltages V_{12}, V_{23}, and V_{34}. The other important message from the results shown in Fig. 20.3 lies in the fact that a *single high-electric-field domain* is nucleated initially in section 12 and subsequently grows to fill the next and then the second next section. Reversing the current kept the three voltages in Fig. 20.3 exactly the same. This apparently indicates that the domain nucleation site represents a sample inhomogeneity and that the growth process of the domain is not affected by the current direction. We note that, in principle, the nearly simultaneous nucleation of several or many high-field domains in the three sample sections also could have been possible. However, such a scenario is clearly eliminated by our experiments.

Turning next to the results obtained at much *higher voltage resolution*, in Fig. 20.4 we show the three voltages V_{12}, V_{23}, and V_{34} versus current at $T = 1.92$ K and $B = 1000$ mT for the same sample as in Fig. 20.3. The voltage resolution is now increased by a factor larger than 10^4 and, again, quasi voltage bias is applied. With increasing current the three sample sections show the onset of a resistive voltage at about the same current, and initially the three VIC's of the three sections are quite similar. This behaviour agrees exactly with our previous results obtained under current control [17]. At $V \approx 0.005$ mV in section 12 a high-electric-field domain nucleates and the voltage $V_{12}(I)$ increases rapidly with a steep negative slope. Due to the quasi voltage bias the current decreases, and the sections 23 and 34 return to the zero-voltage state. Only at a much higher voltage V_{12}, when the high-

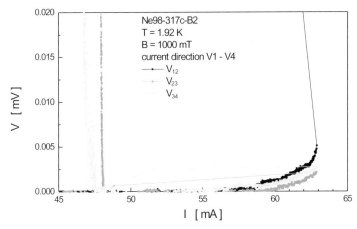

Fig. 20.4. Voltages V_{12} (*black color*), V_{23} (*dark grey color*), and V_{34} (*light grey color*) of the three sample sections shown in the inset of Fig. 20.3 plotted versus current for quasi voltage bias applied to the total sample length (voltage increasing). $T = 1.92$ K, $B = 1000$ mT.

field domain enters section 23, the voltage V_{23} increases rapidly. The same occurs again for section 34 and the voltage V_{34}, when the high-field domain enters section 34, and we are back to our discussion of Fig. 20.3. Again, upon reversing the current direction the data of Fig. 20.4 reproduced nearly exactly.

At $T = 1.92$ K we have performed quasi-voltage-biased measurements for two samples with the geometry shown in the inset of Fig. 20.3 at different magnetic fields in the range $B = 600\text{--}1200$ mT. The results were qualitatively similar to those of Fig. 20.3 and Fig. 20.4.

20.5 Electric Field Dependent Flux–Flow Resistance

Before turning to more specific models for explaining our results obtained with the NCCO films, we discuss some general features involved in nonlinear VIC's. We emphasize that here we are not concerned with effects due to Joule heating of the sample, and that we only concentrate on the *electronic nonequilibrium*. A key element in our discussion is the energy shift of the quasiparticles away from the Fermi energy ε_F in the presence of the electric field $\boldsymbol{F}$, yielding the quasiparticle energy

$$\varepsilon = \varepsilon_F \pm e\boldsymbol{F}\boldsymbol{v}_F \tau \tag{20.9}$$

(v_F = Fermi velocity). As we see from (20.9), the energy shift is proportional to the quasiparticle scattering time τ. If the time τ is energy-dependent, the electric resistance is dependent on the electric field, leading to a nonlinear

VIC. Since electron-electron scattering is dominant in the cuprate superconductors [14–16], the energy dependence of the quasiparticle DOS providing the available phase space for scattering is expected to be crucial. In Sects. 20.6 and 20.7 we discuss two possible scenarios for a strongly energy-dependent quasiparticle scattering rate.

A general discussion of the nonlinear VIC resulting from an electric-field dependent resistivity $\rho(F)$ has been presented in [24]. As an example we consider a single stepwise increase of $\rho(F)$ as shown schematically in Fig. 20.5a (solid line).

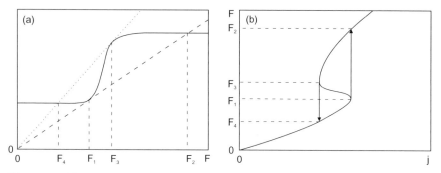

Fig. 20.5. Electric resistivity ρ showing a single stepwise increase as a function of the electric field F (**a**) and the resulting field plotted versus the electric current density j (**b**). The field values F_1–F_4 are discussed in the text.

Since ρ increases stronger than linearly with F, NDR and hysteresis develop. This is shown in Fig. 20.5b, where the field F is plotted versus the current density j. For *current bias*, with increasing current F jumps from F_1 to F_2 at the field F_1. The field F_1 is given by the point on the $\rho(F)$ curve touched from below by the dashed straight line passing through the origin in Fig. 20.5a. The field F_2 represents the crossing point of this straight line with the S-shaped $\rho(F)$ curve. At F_3 the field F jumps from F_3 to F_4 for decreasing j. F_3 is given by the point where the dotted straight line passing through the origin in Fig. 20.5a touches the $\rho(F)$ curve from above. F_4 represents the crossing point of this straight line with the $\rho(F)$ curve. All fields F_1–F_4 are marked in Fig. 20.5. If the resistivity would increase only proportional to F, $\rho = \alpha \cdot F$, between F_1 and F_3, the current density would just remain constant, $j = F/(\alpha \cdot F) = 1/\alpha$, and the hysteresis would disappear. For perfect voltage bias, the whole nonlinear S-shaped $F(j)$ curve would be followed in both directions.

This discussion can be extended easily to the case where two or more steps occur in the function $\rho(F)$ [24].

20.6 Subbands, Bloch Oscillations, and Zener Breakdown

As the first possible scenario for explaining the observed nonlinear VIC's in the NCCO films, we discuss the quasiparticle dynamics in an electronic vortex structure with subbands appearing between the Fermi energy and the gap energy. Such subbands develop in a perfect vortex lattice from the Andreev bound states in the core of an isolated vortex, if the bound state wave functions of two neighboring vortices overlap. Theoretical discussions of this subject have been carried out for some time [25-28]. As shown by Pöttinger and Klein [28], the development of these subbands must be taken into account if the intervortex distance becomes smaller than about 7ξ. As an important prerequisite for these concepts, the quasiparticles must propagate coherently through many unit cells of the vortex lattice, and the energy smearing $\delta\varepsilon = \hbar/\tau$ must remain smaller than the characteristic energy scale of the underlying electronic structure. Taking for this characteristic energy scale the level distance $\varepsilon_n - \varepsilon_{n-1} = \Delta^2/\varepsilon_F$ in the core of an isolated vortex according to (20.1), and using the values $\Delta = 4$ meV [29] and $\varepsilon_F = 30$ meV [30] for NCCO, the condition $\hbar/\tau < \Delta^2/\varepsilon_F$ yields for the scattering time $\tau > 1.2 \cdot 10^{-12}$ s. With $v_F = 10^7$ cm/s and $\tau = 1.2 \cdot 10^{-12}$ s the mean free path is $v_F \cdot \tau = 1.2 \cdot 10^{-5}$ cm. On the other hand, at $B = 1T$ (the typical magnetic field of our experiments) the intervortex distance a is $a \approx 4.5 \cdot 10^{-6}$ cm, which is only about three times smaller than this value of $v_F \cdot \tau$. Hence, for coherent propagation through many unit cells of the vortex lattice, the quasiparticle life time τ must be clearly larger than the value $\tau = 1.2 \cdot 10^{-12}$ s. We will return to this point further below.

Experimentally, the establishment of a perfect vortex lattice in NCCO is difficult because of flux pinning. However, in the flux–flow regime, due to an electric transport current of sufficient magnitude, the quality of the vortex crystal can improve considerably, such that the ansatz with Bloch wave functions may be justified. This phenomenon is referred to as *dynamic correlation*, and more details can be found in [31].

The arguments for the development of subbands between the Fermi energy and the gap energy appear straightforward and convincing in the case of *s*-wave symmetry of the pair wave function. In the case of *d*-wave symmetry the development of such subbands appears less likely. Whereas the experimental evidence has been in favor of *s*-wave symmetry in NCCO for some time [29, 32-36], recently some evidence for *d*-wave symmetry was reported [37,38]. Although the subject of the symmetry still appears controversial, in the following we summarize the main features of a model based on subbands, Bloch oscillations, and Zener breakdown [18, 21–23]. In the next section we then discuss a more general model where the energy dependence of the quasiparticle DOS between the Fermi energy and the gap energy represents the key feature without the explicit appearance of subbands.

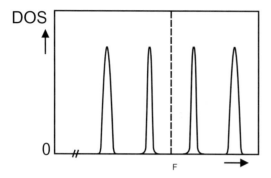

Fig. 20.6. Quasiparticle density of states versus energy near the Fermi energy ε_F for the case of two subbands above and below ε_F, respectively.

As an example we assume two subbands above and below the Fermi energy, respectively, located near ε_F within an energy distance less than the gap energy Δ (see Fig. 20.6). In the limit $T = 0$ all subbands below (above) the Fermi energy are occupied (unoccupied). In this limit the quasiparticle scattering rate τ^{-1} is near zero because of the minigap between the fully occupied and unoccupied states. Then the vortex motion due to an applied electric current is parallel to the current direction, and according to (20.7) the Hall angle θ is $\theta \approx \pi/2$. At $T > 0$ the subbands above (below) the Fermi energy become thermally populated (depopulated) and the scattering rate τ^{-1} is finite. The current-induced vortex motion has a component perpendicular to the current direction, the Hall angle is $\theta < \pi/2$, and a resistive electric field $\boldsymbol{F}$ appears. Due to the force $e \cdot \boldsymbol{F}$ acting on the quasiparticles, the equilibrium Fermi distribution function f_0 is replaced by the nonequilibrium distribution $f = f_0 + \delta f$, where the deviation δf is given by

$$\delta f = \frac{\partial f}{\partial \varepsilon} \cdot \frac{\partial \varepsilon}{\partial t} \cdot \tau = \frac{\partial f}{\partial \varepsilon} e \boldsymbol{F} \boldsymbol{v}_F \tau . \tag{20.10}$$

The quasiparticle current density is

$$\boldsymbol{j} = \int e \cdot \boldsymbol{v}_k \cdot \delta f \cdot d\boldsymbol{k} . \tag{20.11}$$

If the scattering rate τ^{-1} is sufficiently small (super/clean limit), the temporal momentum change of the quasiparticles $\hbar \dot{\boldsymbol{k}} = e \boldsymbol{F}$ can last long enough, such that the wave vector is shifted up to the Brillouin zone boundary $k = \frac{\pi}{a}$ of the vortex lattice. Then the quasiparticles experience Bragg reflection and become localized due to Bloch oscillations with the frequency $\omega_B = eFa/\hbar$. This is similar to the situation in semiconducting superlattices [39,40].

At this stage it is important to note that the excitation of an electron into a subband above ε_F is always accompanied by the creation of a hole in the subband below ε_F, from which the electron originated. In the field $\boldsymbol{F}$ the holes move in the direction opposite to that of the electrons. This can

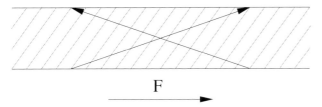

Fig. 20.7. Propagation of electrons (*arrow pointing to the left*) and holes (*arrow pointing to the right*) in a subband and in the electric field **F**.

be visualized schematically as shown in Fig. 20.7, where the (negative) hole energy below ε_F is turned over to the positive side above ε_F. The fact that electrons and holes always appear together suggests that in this discussion we must replace e by $2e$, yielding for the Bloch frequency

$$\omega_B = 2eFa/\hbar \tag{20.12}$$

Because of the localization of the quasiparticles due to Bloch oscillations, the resistivity becomes electric field dependent, following the relation [40]

$$\rho = \rho_0 \left(1 + \omega_B^2 \tau^2\right) = \rho_0 \left(1 + \frac{F^2}{F^{*2}}\right) . \tag{20.13}$$

Here ρ_0 is the resistivity in the limit $\omega_B \to 0$, $F \to 0$, and F^* is given by

$$F^* = \hbar/2ea\tau . \tag{20.14}$$

From (20.13) we have

$$j = \frac{F}{\rho} = \frac{F}{\rho_0 \left(1 + \frac{F^2}{F^{*2}}\right)} . \tag{20.15}$$

According to (20.15) the current density j passes through a maximum at $F = F^*$, NDR sets in or a hysteretic voltage step appears, depending on the bias conditions.

Transitions from the lowest subband to a higher energy range with nonzero DOS become possible via Zener breakdown. Denoting the gap between the two subsequent regions where the DOS is finite by Δ_{dos}, the transition rate P is given by [41]

$$P = \exp\left(-\frac{\pi^2}{4} \frac{\Delta_{dos}^2}{\varepsilon_0 eFa}\right) . \tag{20.16}$$

Zener breakdown occurs for

$$eFa = \Delta_{dos}^2/\varepsilon_0 \tag{20.17}$$

where $\varepsilon_0 = \hbar^2 \pi^2/(2ma^2)$.

Having discussed the *quasiparticle dynamics*, next we turn to the *dynamics of the pair condensate* in the electric field generated by vortex motion.

The oscillation of the pair wave function is given by the Josephson voltage-frequency relation applied to a single unit cell of the moving vortex lattice and yielding the Josephson frequency

$$\omega_J = \frac{2eFa}{\hbar} \qquad (20.18)$$

(Taking $\omega_J = 2\pi v_\varphi/a$ together with the Josephson relation $F \cdot a = \hbar \omega_J/2e$ we recover (20.6) for the flux–flow electric field.) From (20.12) and (20.18) we see that $\omega_B = \omega_J$: there exists a close connection between the Bloch dynamics of the quasiparticles and the Josephson dynamics of the pair condensate. Going one step further, the phase coherence of the pair wave function in a perfect vortex lattice leads to an itinerant pair wave function with an energy band in the spirit of the tight binding approximation. The Josephson oscillation can then be interpreted as the Bloch oscillation of the pair condensate within this band.

As we have discussed in detail in [18, 21-23, 42], our results on the nonlinear VIC's observed in the NCCO films can be well explained by a model assuming the existence of two subbands between the Fermi energy and the gap energy and the appearance of Bloch oscillations in both subbands, respectively. In particular, the condition for the appearance of subbands, $a \leq 7\xi$, according to Pöttinger and Klein [28], is well fulfilled. With $\xi_{ab} = 8$ nm in NCCO [29] $a \leq 7\xi$ corresponds to $B \geq 0.58$ T, which is the magnetic field range of our experiments. In our model the onset voltages V_1 and V_2 are interpreted in terms of the fields F_1^* and F_2^* for the lower and the upper subband, respectively, according to (20.13) and (20.14). The width of the lower subband is expected to strongly increase with decreasing intervortex distance (increasing B). This is accompanied by a corresponding increase of the phase space for scattering and, hence, of F_1^* and explains observation (a) of Sect. 20.3. The increase of the widths of the subbands with increasing magnetic field can lead to the merging of the two subbands (Mott transistion) [42]. This can explain observation (c) for the high-magnetic-field side. Since τ^{-1} increases proportional to the number of quasiparticles thermally excited into the lower subband, we also understand observation (d). Above the first voltage step the upper subband can become populated by means of Zener breakdown. The upper subband is expected to be much wider than the lower subband, resulting also in a much larger scattering rate τ^{-1}. This can explain the factor of about 10^3 between V_2 and V_1. An electronic band structure with two subbands based on the experimental data at $B = 1.3$ T and $T = 1.9$ K, has been proposed in [23].

Calculating the scattering times τ from the instability fields F_1^* and F_2^* according to (20.14), one finds the following [18]: Depending on the magnetic field, F_1^* yields for the lower subband values in the range $\tau = (2.8 - 52) \times 10^{-9}$ s. The values obtained from F_2^* for the upper subband are in the range $\tau = (2.2 - 3.8) \times 10^{-11}$ s. From these values of τ we conclude that the quasiparticles propagate coherently over many unit cells of the vortex lattice (which has been an underlying assumption in our discussion).

Finally, we emphasize that at the onset of NDR the transition to the state with higher resistivity takes place via the nucleation and growth of a single high-electric-field domain, as has been demonstrated by our spatially resolved measurements discussed in Sect. 20.4.

20.7 Step Structure in the Quasiparticle DOS

The subband model we have discussed in the last section appears favorable only in the case of s-wave symmetry of the pair wave function. In view of the recent evidence for d-wave symmetry in the electron doped superconductor NCCO [37,38] a more general model may be necessary. We have discussed such a model in [17,24]. It is based again on the quasiparticle nonequilibrium distribution in the electric field expressed in (20.10), and on the energy dependent scattering rate $\tau(\varepsilon)^{-1}$. In this model the latter energy dependence is assumed to arise from the quasiparticle DOS providing the phase space available for scattering.

In the mixed state of a superconductor with d-wave symmetry of the pair wave function, a strong increase of the quasiparticle DOS with energy is expected near the energy $\varepsilon = \varepsilon_F \pm \Delta_0$, similar to the case of s-wave symmetry. Here Δ_0 is the maximum gap value for the gap modulus $\Delta = \Delta_0 \sin(2\alpha)$, where the angle α indicates the position of the gap nodes in momentum space.

For the s-wave case, calculations of the DOS in the mixed state based on the quasiclassical approximation have been reported about 35 years ago [43-45]. On the other hand, in the d-wave case for $B = 0$ the node lines of the energy gap in momentum space result in a linear increase of the quasiparticle DOS with energy near the Fermi energy [46]. However, in the mixed state, in addition to the strong energy dependent behaviour near $\varepsilon = \varepsilon_F \pm \Delta_0$, another sharp increase of the DOS with energy is expected at an energy much closer to ε_F than Δ_0 [47]. Here, two energy scales play an important role. As discussed by Kopnin and Volovik [48,49], there exists an *average minigap* of the same order $\Delta_0^2/\varepsilon_F \equiv E_1$ as in the s-wave case. Furthermore, there is a characteristic resonant energy $E_2 = (E_1 \hbar \omega_c)^{1/2}$, where ω_c is the cyclotron frequency [49]. From these results we conclude that the phase space for quasiparticle scattering is likely to show two steps as a function of energy: a first step at the energy value $\varepsilon \approx \varepsilon_F \pm E_1$ or $\varepsilon \approx \varepsilon_F \pm E_2$, and a second step at $\varepsilon \approx \varepsilon_F \pm \Delta_0$. This expected energy dependence of the quasiparticle DOS is shown schematically in Fig. 20.8 for $\varepsilon > \varepsilon_F$.

As an important result we find that the quasiparticle energy shift of (20.9) causes two steps in the electric field dependent resistivity, namely at the field values satisfying the equations

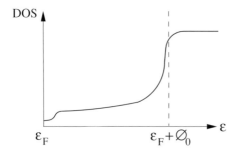

Fig. 20.8. Schematics of the quasiparticle DOS plotted versus energy and showing two steps near the Fermi energy and the maximum gap energy, respectively.

$$eFv_F\tau = E_1 \text{ or } E_2 \tag{20.19}$$

and

$$eFv_F\tau = \Delta_0 , \tag{20.20}$$

respectively. Denoting the field values satisfying (20.19) and (20.20) by F_1 and F_2, respectively, we identify these values with the electric fields of the two observed steps in the VIC's under current bias (see Fig. 20.1).

Using the values $\Delta_0 = 4$ meV and $\varepsilon_F = 30$ meV for NCCO, we obtain $E_1 = 0.53$ meV. At the typical magnetic field of our experiments, $B = 1$ T, we have $\hbar\omega_c = 1.16 \cdot 10^{-4}$ eV and $E_2 = 0.25$ meV. We see that E_2 is not much different from E_1. Taking a value of 0.5 meV at the rhs of (20.19), and using the values $F_1 = 1$–40 mV/cm observed at 1.92 K in the magnetic field range $B = 0.5$–1.8 T [21], with $v_F = 10^7$ cm/s we obtain $\tau = (1.25 - 50) \times 10^{-9}$ s from (20.19). From the values $F_2 = 5$–10 V/cm measured at 1.92 K in the range $B = 1$–4 T [21] we find $\tau = (4.0 - 8.0) \times 10^{-11}$ s from (20.20). In both given ranges of the scattering time, τ decreases with increasing B. These values of τ calculated from (20.19) and (20.20) appear reasonable. Further, they are not much different from the values we have calculated at the end of Sect. 20.6.

The energy values we have used on the rhs of (20.19) and (20.20) are based on theoretical estimates valid for static vortices. In our discussion we have tacitly assumed that the result for the static case is also valid in good approximation for the moving vortices we deal with in our experiments. A detailed quantum mechanical treatment of the electronic structure of moving vortices based on the time-dependent Bogoliubov-De Gennes equations has recently been performed by Hofmann and Kümmel [50]. Their results indicate that the static electronic structure is still approximately valid for moving vortices, as long as the Doppler energy shift $\hbar k_F v_s$ is small compared to the minigap E_1. Here v_s is the pair velocity due to the applied supercurrent density j. As we have discussed in [24], this condition is well satisfied in our experiments.

Most of the experimental observations discussed in Sect. 20.3 can be qualitatively explained by our model assuming there are two steps in the energy dependence of the quasiparticle DOS, as indicated schematically in Fig. 20.8. However, at this stage further theoretical work is clearly necessary for a quantitative treatment and for understanding such features as, for example, the expontential dependence of V_1 upon B.

20.8 Summary and Conclusions

In epitaxial c-axis oriented films of the electron doped cuprate superconductor NCCO an intrinsic hysteretic step structure of the flux–flow resistance has been observed under current bias. As expected from these results, under quasi voltage bias NDR and spontaneous resistance, oscillations have been found. Spatially resolved measurements have shown that in the NDR regime a single high-electric-field domain is nucleated, growing subsequently until it fills the total length of the sample. During the experiments the samples were imbedded in superfluid helium with the reduced temperature $\frac{T}{T_c}$ kept near $\frac{T}{T_c} = 0.087$, in this way promoting the detection of possible quantum effects influencing the electronic structure of the moving vortex system. This electronic quasiparticle structure is distinctly different for s-wave and d-wave symmetry of the pair wave function. Since the symmetry of the pair wave function in NCCO is still controversial, we discuss two models for explaining the observed electric-field dependence of the flux–flow resistance. In the first model applicable to the s-wave case we assume the existence of two subbands between the Fermi energy and the gap energy, such that the quasiparticles experience Bloch oscillations in the electric field generated by vortex motion. The second (more general) model applicable to the d-wave case starts from the assumption that the energy dependence of the quasiparticle DOS shows two steps near the Fermi energy and the maximum gap energy, respectively. The nonequilibrium quasiparticle energy shift in the flux–flow electric field then leads to two steps in the flux–flow resistance because of the abrupt increments in the phase space available for scattering.

Acknowledgements. The author benefitted from discussions with A.A. Abrikosov, T. Dahm, U.R. Fischer, T. Kita, R. Kümmel, Yu.N. Ovchinnikov, N. Schopohl, O.M. Stoll, Z. Tesanovic, C.C. Tsuei, and A. Wacker.

References

1. A.I. Larkin, Yu.N. Ovchinnikov: Zh. Eksp. Teor. Fiz **68**, 1915 (1975) [Sov. Phys. JETP **41**, 960 (1976)]
2. S.G. Doettinger, R.P. Huebener, R. Gerdemann, A. Kühle, S. Anders, T.G. Träuble, J.C. Villegier: Phys. Rev. Lett. **73**, 1691 (1994)

3. S.G. Doettinger, S. Kittelberger, R.P. Huebener, C.C. Tsuei: Phys. Rev. B **56**, 14157 (1997)
4. C. Caroli, P.G. De Gennes, J. Matricon: Phys. Lett. **9**, 307 (1964)
5. A.I. Larkin, Yu.N. Ovchinnikov: Zh. Eksp. Teor. Fiz. **73**, 299 (1977) [Sov. Phys. JETP **46**, 155 (1977)]
6. L. Kramer, W. Pesch: Z. Phys. **269**, 59 (1974)
7. W. Pesch, L. Kramer: J. Low Temp. Phys. **15**, 367 (1973)
8. S.G. Doettinger, R.P. Huebener, S. Kittelberger: Phys. Rev. B **55**, 6044 (1997)
9. W.F. Vinen, A.C. Warren: Proc. Phys. Soc. London **91**, 409 (1967)
10. N.B. Kopnin, V.E. Kravtsov: Zh. Eksp. Teor. Fiz. **71**, 1644 (1976) [Sov. Phys. JETP **44**, 861 (1976)]
11. N.B. Kopnin, V.E. Kratsov: Zh. Eksp. Teor. Fiz. Pis' ma Red. **23**, 631 (1976) [JETP Lett. **23**, 578 (1976)]
12. N.B. Kopnin, M.M. Salomaa: Phys. Rev. B **44**, 9667 (1991)
13. G. Blatter, B.I. Ivlev: Phys. Rev. B **50**, 10272 (1994)
14. A. Virosztek, J. Ruvalds: Phys. Rev. B **42**, 4064 (1990); **45**, 347 (1992)
15. C.C. Tsuei, A. Gupta, G. Koren: Physica C **161**, 415 (1989)
16. C.T. Rieck, W.A. Little, J. Ruvalds, A. Virosztek: Phys. Rev. B **51**, 3772 (1995)
17. O.M. Stoll, S. Kaiser, R.P. Huebener, M. Naito: Phys. Rev. Lett. **81**, 2994 (1998)
18. O.M. Stoll, R.P. Huebener, S. Kaiser, M. Naito: J. Low Temp. Phys. **118**, 59 (2000)
19. O.M. Stoll, A. Wehner, R.P. Huebener, M. Naito: Physica B **284-288**, 827 (2000)
20. A. Wehner, O.M. Stoll, R.P. Huebener, M. Naito: to be published
21. O.M. Stoll, R.P. Huebener, S. Kaiser, M. Naito: Phys. Rev. B **60**, 12424 (1999)
22. R.P. Huebener, O.M. Stoll, M. Naito: Physica B **280**, 237 (2000)
23. R.P. Huebener, O.M. Stoll, A. Wehner, M. Naito: Physica C **332**, 187 (2000)
24. R.P. Huebener, S. Kaiser, O.M. Stoll: Europhys. Lett. **44**, 772 (1998)
25. E. Canel: Phys. Lett. **16**, 101 (1965)
26. A.P. Van Gelder: Phys. Rev. **181**, 787 (1969)
27. R. Kümmel: Phys. Rev. B **3**, 3787 (1971)
28. P. Pöttinger, U. Klein: Phys. Rev. Lett. **70**, 2806 (1993)
29. S.M. Anlage, D.H. Wu, J. Mao, X.X. Xi, T. Venkatesan, J.L. Peng, R.L. Greene: Phys. Rev. B **50**, 523 (1994)
30. C.P. Poole, H.A. Parach, R.J. Creswick: *Superconductivity* (Academic, San Diego 1995)
31. R.P. Huebener: *Magnetic Flux Structures in Superconductors*, 2nd ed. (Springer, Berlin 2001)
32. D.H. Wu, J. Mao, S.N. Mao, J.L. Peng, X.X. Xi, T. Venkatesan, R.L. Greene, S.M. Anlage: Phys. Rev. Lett. **70**, 85 (1993)
33. C.W. Schneider, Z.H. Barber, J.E. Evetts, S.N. Mao, X.X. Xi, T. Venkatesan: Physica C **233**, 77 (1994)
34. L. Alff, A. Beck, R. Gross, A. Marx, S. Kleefisch, Th. Bauch, H. Sato, M. Naito, G. Koren: Phys. Rev. B **58**, 11197 (1998)
35. S. Kashiwaya, T. Ito, K. Oka, S. Ueno, H. Takashima, M. Koyanagi, Y. Tanaka, K. Kajimura: Phys. Rev. B **57**, 8680 (1998)
36. L. Alff, S. Meyer, S. Kleefisch, U. Schoop, A. Marx, H. Sato, M. Naito, R. Gross: Phys. Rev. Lett. **83**, 2644 (1999)

37. C.C. Tsuei, J.R. Kirtley: Phys. Rev. Lett. **85**, 182 (2000)
38. J.D. Kokales, P. Fournier, L.V. Mercaldo, V.V. Talanov, R.L. Greene, S.M. Anlage: Phys. Rev. Lett. **85**, 3696 (2000)
39. L. Esaki, R. Tsu: IBM J. Res. Div. **14**, 61 (1970)
40. M. Helm: Semicond. Sci. Technol. **10**, 557 (1995)
41. J.M. Ziman: *Principles of the Theory of Solids* (Cambridge University Press, 1972)
42. R.P. Huebener, O.M. Stoll, A. Wehner, M. Naito: Physica C **332**, 187 (2000)
43. M. Cyrot: Phys. Kond. Matter **3**, 374 (1965)
44. U. Brandt, W. Pesch, L. Tewordt: Z. Phys. **201**, 209 (1967)
45. K. Maki: Phys. Rev. **156**, 437 (1967)
46. D. Xu, S.K. Yip, J.A. Sauls: Phys. Rev. B **51**, 16233 (1995)
47. N. Schopohl, K. Maki: Phys. Rev. B **52**, 490 (1995)
48. N.B. Kopnin, G.E. Volovik: Phys. Rev. Lett. **79**, 1377 (1997)
49. N.B. Kopnin: Phys. Rev. B **57**, 11775 (1998)
50. S. Hofmann, R. Kümmel: Phys. Rev. B **57**, 7904 (1998)

Springer Series in Solid-State Sciences

Editors: M. Cardona P. Fulde K. von Klitzing H.-J. Queisser

1. **Principles of Magnetic Resonance**
   3rd Edition By C. P. Slichter
2. **Introduction to Solid-State Theory**
   By O. Madelung
3. **Dynamical Scattering of X-Rays in Crystals** By Z. G. Pinsker
4. **Inelastic Electron Tunneling Spectroscopy**
   Editor: T. Wolfram
5. **Fundamentals of Crystal Growth I**
   Macroscopic Equilibrium and Transport Concepts
   By F. E. Rosenberger
6. **Magnetic Flux Structures in Superconductors**
   2nd Edition By R. P. Huebener
7. **Green's Functions in Quantum Physics**
   2nd Edition By E. N. Economou
8. **Solitons and Condensed Matter Physics**
   Editors: A. R. Bishop and T. Schneider
9. **Photoferroelectrics** By V. M. Fridkin
10. **Phonon Dispersion Relations in Insulators** By H. Bilz and W. Kress
11. **Electron Transport in Compound Semiconductors** By B. R. Nag
12. **The Physics of Elementary Excitations**
    By S. Nakajima, Y. Toyozawa, and R. Abe
13. **The Physics of Selenium and Tellurium**
    Editors: E. Gerlach and P. Grosse
14. **Magnetic Bubble Technology** 2nd Edition
    By A. H. Eschenfelder
15. **Modern Crystallography I**
    Fundamentals of Crystals
    Symmetry, and Methods of Structural Crystallography
    2nd Edition
    By B. K. Vainshtein
16. **Organic Molecular Crystals**
    Their Electronic States By E. A. Silinsh
17. **The Theory of Magnetism I**
    Statics and Dynamics
    By D. C. Mattis
18. **Relaxation of Elementary Excitations**
    Editors: R. Kubo and E. Hanamura
19. **Solitons** Mathematical Methods for Physicists
    By. G. Eilenberger
20. **Theory of Nonlinear Lattices**
    2nd Edition By M. Toda
21. **Modern Crystallography II**
    Structure of Crystals 2nd Edition
    By B. K. Vainshtein, V. L. Indenbom, and V. M. Fridkin
22. **Point Defects in Semiconductors I**
    Theoretical Aspects
    By M. Lannoo and J. Bourgoin
23. **Physics in One Dimension**
    Editors: J. Bernasconi and T. Schneider
24. **Physics in High Magnetics Fields**
    Editors: S. Chikazumi and N. Miura
25. **Fundamental Physics of Amorphous Semiconductors** Editor: F. Yonezawa
26. **Elastic Media with Microstructure I**
    One-Dimensional Models By I. A. Kunin
27. **Superconductivity of Transition Metals**
    Their Alloys and Compounds
    By S. V. Vonsovsky, Yu. A. Izyumov, and E. Z. Kurmaev
28. **The Structure and Properties of Matter**
    Editor: T. Matsubara
29. **Electron Correlation and Magnetism in Narrow-Band Systems** Editor: T. Moriya
30. **Statistical Physics I** Equilibrium
    Statistical Mechanics 2nd Edition
    By M. Toda, R. Kubo, N. Saito
31. **Statistical Physics II** Nonequilibrium
    Statistical Mechanics 2nd Edition
    By R. Kubo, M. Toda, N. Hashitsume
32. **Quantum Theory of Magnetism**
    2nd Edition By R. M. White
33. **Mixed Crystals** By A. I. Kitaigorodsky
34. **Phonons: Theory and Experiments I**
    Lattice Dynamics and Models of Interatomic Forces By P. Brüesch
35. **Point Defects in Semiconductors II**
    Experimental Aspects
    By J. Bourgoin and M. Lannoo
36. **Modern Crystallography III**
    Crystal Growth
    By A. A. Chernov
37. **Modern Chrystallography IV**
    Physical Properties of Crystals
    Editor: L. A. Shuvalov
38. **Physics of Intercalation Compounds**
    Editors: L. Pietronero and E. Tosatti
39. **Anderson Localization**
    Editors: Y. Nagaoka and H. Fukuyama
40. **Semiconductor Physics** An Introduction
    6th Edition By K. Seeger
41. **The LMTO Method**
    Muffin-Tin Orbitals and Electronic Structure
    By H. L. Skriver
42. **Crystal Optics with Spatial Dispersion, and Excitons** 2nd Edition
    By V. M. Agranovich and V. L. Ginzburg
43. **Structure Analysis of Point Defects in Solids**
    An Introduction to Multiple Magnetic Resonance Spectroscopy
    By J.-M. Spaeth, J. R. Niklas, and R. H. Bartram
44. **Elastic Media with Microstructure II**
    Three-Dimensional Models By I. A. Kunin
45. **Electronic Properties of Doped Semiconductors**
    By B. I. Shklovskii and A. L. Efros
46. **Topological Disorder in Condensed Matter**
    Editors: F. Yonezawa and T. Ninomiya

Springer Series in Solid-State Sciences

Editors: M. Cardona   P. Fulde   K. von Klitzing   H.-J. Queisser

47 **Statics and Dynamics of Nonlinear Systems**
Editors: G. Benedek, H. Bilz, and R. Zeyher

48 **Magnetic Phase Transitions**
Editors: M. Ausloos and R. J. Elliott

49 **Organic Molecular Aggregates**
Electronic Excitation and Interaction Processes
Editors: P. Reineker, H. Haken, and H. C. Wolf

50 **Multiple Diffraction of X-Rays in Crystals**
By Shih-Lin Chang

51 **Phonon Scattering in Condensed Matter**
Editors: W. Eisenmenger, K. Laßmann,
and S. Döttinger

52 **Superconductivity in Magnetic and Exotic Materials**   Editors: T. Matsubara and A. Kotani

53 **Two-Dimensional Systems, Heterostructures, and Superlattices**
Editors: G. Bauer, F. Kuchar, and H. Heinrich

54 **Magnetic Excitations and Fluctuations**
Editors: S. W. Lovesey, U. Balucani, F. Borsa,
and V. Tognetti

55 **The Theory of Magnetism II**   Thermodynamics and Statistical Mechanics   By D. C. Mattis

56 **Spin Fluctuations in Itinerant Electron Magnetism**   By T. Moriya

57 **Polycrystalline Semiconductors**
Physical Properties and Applications
Editor: G. Harbeke

58 **The Recursion Method and Its Applications**
Editors: D. G. Pettifor and D. L. Weaire

59 **Dynamical Processes and Ordering on Solid Surfaces**   Editors: A. Yoshimori and
M. Tsukada

60 **Excitonic Processes in Solids**
By M. Ueta, H. Kanzaki, K. Kobayashi,
Y. Toyozawa, and E. Hanamura

61 **Localization, Interaction, and Transport Phenomena**   Editors: B. Kramer, G. Bergmann, and Y. Bruynseraede

62 **Theory of Heavy Fermions and Valence Fluctuations**   Editors: T. Kasuya and T. Saso

63 **Electronic Properties of Polymers and Related Compounds**
Editors: H. Kuzmany, M. Mehring, and S. Roth

64 **Symmetries in Physics**   Group Theory
Applied to Physical Problems   2nd Edition
By W. Ludwig and C. Falter

65 **Phonons: Theory and Experiments II**
Experiments and Interpretation of
Experimental Results   By P. Brüesch

66 **Phonons: Theory and Experiments III**
Phenomena Related to Phonons
By P. Brüesch

67 **Two-Dimensional Systems: Physics and New Devices**
Editors: G. Bauer, F. Kuchar, and H. Heinrich

68 **Phonon Scattering in Condensed Matter V**
Editors: A. C. Anderson and J. P. Wolfe

69 **Nonlinearity in Condensed Matter**
Editors: A. R. Bishop, D. K. Campbell,
P. Kumar, and S. E. Trullinger

70 **From Hamiltonians to Phase Diagrams**
The Electronic and Statistical-Mechanical Theory
of sp-Bonded Metals and Alloys   By J. Hafner

71 **High Magnetic Fields in Semiconductor Physics**
Editor: G. Landwehr

72 **One-Dimensional Conductors**
By S. Kagoshima, H. Nagasawa, and T. Sambongi

73 **Quantum Solid-State Physics**
Editors: S. V. Vonsovsky and M. I. Katsnelson

74 **Quantum Monte Carlo Methods in Equilibrium and Nonequilibrium Systems**   Editor: M. Suzuki

75 **Electronic Structure and Optical Properties of Semiconductors**   2nd Edition
By M. L. Cohen and J. R. Chelikowsky

76 **Electronic Properties of Conjugated Polymers**
Editors: H. Kuzmany, M. Mehring, and S. Roth

77 **Fermi Surface Effects**
Editors: J. Kondo and A. Yoshimori

78 **Group Theory and Its Applications in Physics**
2nd Edition
By T. Inui, Y. Tanabe, and Y. Onodera

79 **Elementary Excitations in Quantum Fluids**
Editors: K. Ohbayashi and M. Watabe

80 **Monte Carlo Simulation in Statistical Physics**
An Introduction   3rd Edition
By K. Binder and D. W. Heermann

81 **Core-Level Spectroscopy in Condensed Systems**
Editors: J. Kanamori and A. Kotani

82 **Photoelectron Spectroscopy**
Principle and Applications   2nd Edition
By S. Hüfner

83 **Physics and Technology of Submicron Structures**
Editors: H. Heinrich, G. Bauer, and F. Kuchar

84 **Beyond the Crystalline State**   An Emerging Perspective   By G. Venkataraman, D. Sahoo,
and V. Balakrishnan

85 **The Quantum Hall Effects**
Fractional and Integral   2nd Edition
By T. Chakraborty and P. Pietiläinen

86 **The Quantum Statistics of Dynamic Processes**
By E. Fick and G. Sauermann

87 **High Magnetic Fields in Semiconductor Physics II**
Transport and Optics   Editor: G. Landwehr

88 **Organic Superconductors**   2nd Edition
By T. Ishiguro, K. Yamaji, and G. Saito

89 **Strong Correlation and Superconductivity**
Editors: H. Fukuyama, S. Maekawa,
and A. P. Malozemoff

Springer Series in Solid-State Sciences

Editors: M. Cardona   P. Fulde   K. von Klitzing   H.-J. Queisser

Managing Editor: H. K. V. Lotsch

90 **Earlier and Recent Aspects of Superconductivity**
    Editors: J. G. Bednorz and K. A. Müller

91 **Electronic Properties of Conjugated Polymers III** Basic Models and Applications
    Editors: H. Kuzmany, M. Mehring, and S. Roth

92 **Physics and Engineering Applications of Magnetism** Editors: Y. Ishikawa and N. Miura

93 **Quasicrystals** Editors: T. Fujiwara and T. Ogawa

94 **Electronic Conduction in Oxides** 2nd Edition
    By N. Tsuda, K. Nasu, F. Atsushi, and K. Siratori

95 **Electronic Materials**
    A New Era in Materials Science
    Editors: J. R. Chelikowsky and A. Franciosi

96 **Electron Liquids** 2nd Edition By A. Isihara

97 **Localization and Confinement of Electrons in Semiconductors**
    Editors: F. Kuchar, H. Heinrich, and G. Bauer

98 **Magnetism and the Electronic Structure of Crystals** By V. A. Gubanov, A. I. Liechtenstein, and A. V. Postnikov

99 **Electronic Properties of High-T_c Superconductors and Related Compounds**
    Editors: H. Kuzmany, M. Mehring, and J. Fink

100 **Electron Correlations in Molecules and Solids** 3rd Edition By P. Fulde

101 **High Magnetic Fields in Semiconductor Physics III** Quantum Hall Effect, Transport and Optics By G. Landwehr

102 **Conjugated Conducting Polymers**
    Editor: H. Kiess

103 **Molecular Dynamics Simulations**
    Editor: F. Yonezawa

104 **Products of Random Matrices**
    in Statistical Physics By A. Crisanti, G. Paladin, and A. Vulpiani

105 **Self-Trapped Excitons**
    2nd Edition By K. S. Song and R. T. Williams

106 **Physics of High-Temperature Superconductors**
    Editors: S. Maekawa and M. Sato

107 **Electronic Properties of Polymers**
    Orientation and Dimensionality
    of Conjugated Systems Editors: H. Kuzmany, M. Mehring, and S. Roth

108 **Site Symmetry in Crystals**
    Theory and Applications 2nd Edition
    By R. A. Evarestov and V. P. Smirnov

109 **Transport Phenomena in Mesoscopic Systems** Editors: H. Fukuyama and T. Ando

110 **Superlattices and Other Heterostructures**
    Symmetry and Optical Phenomena 2nd Edition
    By E. L. Ivchenko and G. E. Pikus

111 **Low-Dimensional Electronic Systems**
    New Concepts
    Editors: G. Bauer, F. Kuchar, and H. Heinrich

112 **Phonon Scattering in Condensed Matter VII**
    Editors: M. Meissner and R. O. Pohl

113 **Electronic Properties of High-T_c Superconductors**
    Editors: H. Kuzmany, M. Mehring, and J. Fink

114 **Interatomic Potential and Structural Stability**
    Editors: K. Terakura and H. Akai

115 **Ultrafast Spectroscopy of Semiconductors and Semiconductor Nanostructures**
    2nd Edition By J. Shah

116 **Electron Spectrum of Gapless Semiconductors**
    By J. M. Tsidilkovski

117 **Electronic Properties of Fullerenes**
    Editors: H. Kuzmany, J. Fink, M. Mehring, and S. Roth

118 **Correlation Effects in Low-Dimensional Electron Systems**
    Editors: A. Okiji and N. Kawakami

119 **Spectroscopy of Mott Insulators and Correlated Metals**
    Editors: A. Fujimori and Y. Tokura

120 **Optical Properties of III–V Semiconductors**
    The Influence of Multi-Valley Band Structures
    By H. Kalt

121 **Elementary Processes in Excitations and Reactions on Solid Surfaces**
    Editors: A. Okiji, H. Kasai, and K. Makoshi

122 **Theory of Magnetism**
    By K. Yosida

123 **Quantum Kinetics in Transport and Optics of Semiconductors**
    By H. Haug and A.-P. Jauho

124 **Relaxations of Excited States and Photo-Induced Structural Phase Transitions**
    Editor: K. Nasu

125 **Physics and Chemistry of Transition-Metal Oxides**
    Editors: H. Fukuyama and N. Nagaosa

Volumes 1–89 are listed at the end of the book

You are one **click** *away from a* **world of physics** *information!*

Come and visit Springer's

Physics Online Library

Books

- Search the Springer website catalogue
- Subscribe to our free alerting service for new books
- Look through the book series profiles

You want to order?   Email to: orders@springer.de

Journals

- Get abstracts, ToC´s free of charge to everyone
- Use our powerful search engine LINK Search
- Subscribe to our free alerting service LINK *Alert*
- Read full-text articles (available only to subscribers of the paper version of a journal)

You want to subscribe?   Email to: subscriptions@springer.de

Electronic Media

- Get more information on our software and CD-ROMs

You have a question on
an electronic product?   Email to: helpdesk-em@springer.de

Bookmark now:

www.springer.de/phys/

Springer

Springer · Customer Service
Haberstr. 7 · 69126 Heidelberg, Germany
Tel: +49 (0) 6221 - 345 - 217/8
Fax: +49 (0) 6221 - 345 - 229 · e-mail: orders@springer.de

d&p · 6437.MNT/SFb

Printing (Computer to Film): Saladruck Berlin
Binding: Stürtz AG, Würzburg